FUNCTIONAL FOODS

The Connection Between Nutrition,
Health, and Food Science

FUNCTIONAL FOODS

The Connection Between Nutrition, Health, and Food Science

Edited by
Leah Coles, PhD

Apple Academic Press

TORONTO   NEW JERSEY

Apple Academic Press Inc. | Apple Academic Press Inc.
3333 Mistwell Crescent | 9 Spinnaker Way
Oakville, ON L6L 0A2 | Waretown, NJ 08758
Canada | USA

©2014 by Apple Academic Press, Inc.

First issued in paperback 2021

Exclusive worldwide distribution by CRC Press, a member of Taylor & Francis Group
No claim to original U.S. Government works

ISBN 13: 978-1-77463-316-8 (pbk)
ISBN 13: 978-1-926895-94-9 (hbk)

Library of Congress Control Number: 2013949390

Library and Archives Canada Cataloguing in Publication

Functional foods: the connection between nutrition, health, and food science/edited by Leah Coles, PhD.

Includes bibliographical references and index.
ISBN 978-1-926895-94-9
1. Functional foods. I. Coles, Leah, writer of introduction, editor of compilation

QP144.F85F85 2013   613.2   C2013-906175-4

Apple Academic Press also publishes its books in a variety of electronic formats. Some content that appears in print may not be available in electronic format. For information about Apple Academic Press products, visit our website at **www.appleacademicpress.com** and the CRC Press website at **www.crcpress.com**

ABOUT THE EDITOR

LEAH COLES, PhD

Leah Coles, PhD, completed her PhD in Human Nutrition at the Riddet Institute, Massey University, New Zealand. She is presently a Research Fellow in the Nutritional Interventions Lab at Baker IDI Heart and Diabetes Institute, Melbourne, Australia. Her current research involves clinical trials focused on functional foods and weight loss, particularly in relation to diabetes and cardiovascular disease. She has also published several peer-reviewed articles in the area of *in vitro* and *in vivo* (animal and human) digestibility studies and linked these with mathematical models to predict the available energy (ATP) content of foods.

CONTENTS

Part II: Functional Foods: The Connection Between Health and Food Science

ACKNOWLEDGMENT AND HOW TO CITE

The chapters in this book were previously published in various places and in various formats. By bringing them together here in one place, we offer the reader a comprehensive perspective on recent investigations of functional foods. Each chapter is added to and enriched by being placed within the context of the larger investigative landscape.

We wish to thank the authors who made their research available for this book, whether by granting their permission individually or by releasing their research as Open Source articles. When citing information contained within this book, please do the authors the courtesy of attributing them by name, referring back to their original articles, using the credits provided at the end of each chapter.

LIST OF CONTRIBUTORS

Wataru Aoi
Research Center for Sports Medicine, Doshisha University, Kyoto 602-8580, Japan and Department of Inflammation and Immunology, Graduate School of Medical Science, Kyoto Prefectural University of Medicine, Kyoto 602-8566, Japan

Alan Albert Aragon
California State University, Northridge, CA, USA

Pankaj K. Bagul
Division of Pharmacology and Chemical Biology, Indian Institute of Chemical Technology (IICT), Hyderabad-500607, India

Sanjay K. Banerjee
Division of Pharmacology and Chemical Biology, Indian Institute of Chemical Technology (IICT), Hyderabad-500607, India

Inger M.E. Björck
Department of Food Technology, Engineering and Nutrition, Lund University P.O. Box 124, SE-221 00 Lund, Sweden

Eliete Bouskela
Physiological Sciences and Clinical Medicine Departments; Clinical and Experimental Research Laboratory in Vascular Biology - BioVasc; Rua São Francisco Xavier, 524 Rio de Janeiro, CEP:20550-013 -Brazil

Urs Brügger
Institute of Health Economics, Zurich University of Applied Sciences, St. Georgenstrasse, 70 P.O. Box, Winterthur 8401, CH, Switzerland

Susanne Bügel
Department of Human Nutrition, Faculty of Life Sciences, University of Copenhagen, Rolighedsvej 30, DK-1958 Frederiksberg C, Denmark

Cristina M. Caperchione
School of Health and Exercise Sciences, The University of British Columbia, 3333 University Way, Kelowna, British Columbia, V1V 1V7, Canada

Catherine L. Carpenter
Center for Human Nutrition, David Geffen School of Medicine at UCLA, Los Angeles, CA 90095, USA

Wiyada Charoensiriwatana
Department of Medical Sciences, Ministry of Public Health, Nonthaburi, Thailand

Steve Chen
Center for Human Nutrition, David Geffen School of Medicine at UCLA, Los Angeles, CA 90095, USA

Cecilia L. de Oliveira
Nutrition Applied Department; Nutrition Institute - Rua São Francisco Xavier, 524, Rio de Janeiro, CEP: 20550-013 - Brazil

Suparna Dutta
Griffin Hospital, 130 Division Street, Derby, CT 06418, USA

Klaus Eichler
Institute of Health Economics, Zurich University of Applied Sciences, St. Georgenstrasse, 70 P.O. Box, Winterthur 8401, CH, Switzerland

Mark W. Empie
Compliance, Archer Daniels Midland Company, 1001 North Brush College Road, Decatur, IL 62521, USA

Zubaida Faridi
Yale-Griffin Prevention Research Center, 130 Division Street, Derby, CT 06418, USA

Emma S. George
School of Science and Health, University of Western Sydney, Locked Bag 1797, Penrith, New South Wales 2751, Australia

Iosif Gergen
Faculty of Food Technologies, Banat's University of Agricultural Science and Veterinary Medicine from Timisoara, Calea Aradului 119, Timisoara, RO 300645, Romania

Anjelica L. Gonzalez-Simon
Department of Biomedical Engineering, Yale University, 55 Prospect St, New Haven, CT 06511-6816, USA

Yvonne E. Granfeldt
Department of Food Technology, Engineering and Nutrition, Lund University P.O. Box 124, SE-221 00 Lund, Sweden

Machteld Huber
Louis Bolk Institute, Hoofdstraat 24, NL-3972 LA Driebergen,The Netherlands

Johannes Kahl
Department of Organic Food Quality and Food Culture, University of Kassel, Nordbahnhofstr. 1a,
D-37213 Witzenhausen, Germany

Mohanraj Karunanithi
The Australian e-Health Research Centre, ICT Centre, CSIRO, Level 5, UQ Health Sciences Building
901/16, Royal Brisbane and Women's Hopsital, Herston, Queensland 4029, Australia

David L. Katz
Yale-Griffin Prevention Research Center, 130 Division Street, Derby, CT 06418, USA and Department
of Epidemiology and Public Health, Yale University School of Medicine, 60 College St, New Haven,
CT 06510-3210, USA

Tarak N. Khatua
Division of Pharmacology and Chemical Biology, Indian Institute of Chemical Technology (IICT),
Hyderabad-500607, India

Gregory S. Kolt
School of Science and Health, University of Western Sydney, Locked Bag 1797, Penrith, New South
Wales 2751, Australia

Josely C. Koury
Study Center for Nutrition and Oxidative Stress; Nutrition Institute; Rua São Francisco Xavier, 524,
Rio de Janeiro, CEP: 20550-013 – Brazil

Luiz G. Kraemer-Aguiar
Endocrinology, Department of Internal Medicine; Clinical and Experimental Research Laboratory in
Vascular Biology - BioVasc; Rua São Francisco Xavier, 524, Rio de Janeiro, CEP:20550-013 - Brazil

Madhusudana Kuncha
Division of Pharmacology and Chemical Biology, Indian Institute of Chemical Technology (IICT),
Hyderabad-500607, India

Maria C.C. Kuschnir
Study Center for Adolescent Health - NESA Av 28 de setembro,. 87, CEP: 20551-030, Rio de Janeiro
- Brazil

Shaheen E. Lakhan
Global Neuroscience Initiative Foundation, Los Angeles, CA, USA

Zhaoping Li
Center for Human Nutrition, David Geffen School of Medicine at UCLA, Los Angeles, CA 90095, USA and Department of Medicine, Greater Los Angeles VA Health Care System, Los Angeles, CA 90073, USA

Priscila A. Maranhão
Clinical and Experimental Research Laboratory in Vascular Biology - BioVasc; Rua São Francisco Xavier, 524, Rio de Janeiro, CEP:20550-013 - Brazil

W. Kerry Mummery
Faculty of Physical Education and Recreation, University of Alberta, Canada W1-34 van Vliet Centre, Edmonton, Alberta, T6G 2H9, Canada

K. Akhilender Naidu
Department of Biochemistry and Nutrition, Central Food Technological Research Institute, Mysore 570 013, India

Yuji Naito
Department of Medical Proteomics, Graduate School of Medical Science, Kyoto Prefectural University of Medicine, Kyoto 602-8566, Japan

Valentine Njike
Yale-Griffin Prevention Research Center, 130 Division Street, Derby, CT 06418, USA

Manny J. Noakes
CSIRO Food and Nutritional Sciences, P.O. Box 10041, Adelaide BC, South Australia 5000, Australia

Raju Padiya
Division of Pharmacology and Chemical Biology, Indian Institute of Chemical Technology (IICT), Hyderabad-500607, India

Catherine M. Phillips
HRB Centre for Diet and Health Research, Department of Epidemiology and Public Health, University College Cork, College Road, Cork, Ireland

Angelika Ploeger
Department of Organic Food Quality and Food Culture, University of Kassel, Nordbahnhofstr. 1a, D-37213 Witzenhausen, Germany

D. Hansi Priscilla
Structural Biology Lab, Centre for Biomedical Research, SBST, VIT University, Vellore, India

Isabelle Rüthemann
Institute of Health Economics, Zurich University of Applied Sciences, St. Georgenstrasse, 70 P.O. Box, Winterthur 8401, CH, Switzerland

Brad Jon Schoenfeld
Department of Health Science, Lehman College, Bronx, NY, USA

H. Mohamed Sham Shihabudeen
Structural Biology Lab, Centre for Biomedical Research, SBST, VIT University, Vellore, India

Maria G.C. Souza
Physiological Sciences Department; Clinical and Experimental Research Laboratory in Vascular Biology - BioVasc; Rua São Francisco Xavier, 524 , Rio de Janeiro CEP:20550-013 - Brazil

Pongsant Srijantr
Department of Soil Science, Kasetsart University, Nakhon Pathom, Thailand

Ducu Sandu Stef
Faculty of Food Technologies, Banat's University of Agricultural Science and Veterinary Medicine from Timisoara, Calea Aradului 119, Timisoara, RO 300645, Romania

Sam Z. Sun
Compliance, Archer Daniels Midland Company, 1001 North Brush College Road, Decatur, IL 62521, USA

Pennie J. Taylor
Clinical Research Unit, CSIRO Food and Nutritional Sciences, P.O. Box 10041, Adelaide BC, South Australia 5000, Australia

Punthip Teeyapant
Department of Medical Sciences, Ministry of Public Health, Nonthaburi, Thailand

Gail Thames
Center for Human Nutrition, David Geffen School of Medicine at UCLA, Los Angeles, CA 90095, USA

Leo Treyzon
Center for Human Nutrition, David Geffen School of Medicine at UCLA, Los Angeles, CA 90095, USA

Kavitha Thirumurugan
Structural Biology Lab, Centre for Biomedical Research, SBST, VIT University, Vellore, India

Corneel Vandelanotte
Centre for Physical Activity Studies, Institute for Health and Social Science Research, Building 18 CQUniversity, Rockhampton, Queensland 4701, Australia

Karen F. Vieira
Global Neuroscience Initiative Foundation, Los Angeles, CA, USA

Yasmine R. Vieira
Clinical and Experimental Research Laboratory in Vascular Biology - BioVasc; Rua São Francisco Xavier, 524, Rio de Janeiro, CEP:20550-013 - Brazil

Simon Wieser
Institute of Health Economics, Zurich University of Applied Sciences, St. Georgenstrasse, 70 P.O. Box, Winterthur 8401, CH, Switzerland

Jintana Wongvilairattana
Department of Medical Sciences, Ministry of Public Health, Nonthaburi, Thailand

Eric Yan
Center for Human Nutrition, David Geffen School of Medicine at UCLA, Los Angeles, CA 90095, USA

Toshikazu Yoshikawa
Department of Inflammation and Immunology, Graduate School of Medical Science, Kyoto Prefectural University of Medicine, Kyoto 602-8566, Japan and Department of Medical Proteomics, Graduate School of Medical Science, Kyoto Prefectural University of Medicine, Kyoto 602-8566, Japan

Aneta Załęcka
Division of Organic Food, Warsaw University of Life Sciences (SGGW), Nowoursynowska 159c , PL-02-787 Warszawa, Poland

INTRODUCTION

"Functional foods" are foods that are given some other role or function; most often this is related to disease prevention and/or health. This book seeks to highlight the role of these foods. The first section, titled, "The Connection Between Nutrition and Health", shows the varying and complicated relationships between nutrition, physical and mental health, and disease. The second section, on the connection between health science and food, presents a number of case studies on the possible uses of effectivity of functional foods. Both sections as a whole seek to discuss methods for nutritional interventions in relation to diseases like obesity and other prominant health concerns in today's modern society.

Chapter 1, by Catherine Phillips, discusses the link between functional foods and obesity. Obesity, particularly central adiposity, is the primary causal factor in the development of insulin resistance, the hallmark of the metabolic syndrome (MetS), a common condition characterized by dyslipidaemia and hypertension, which is associated with increased risk of cardiovascular disease (CVD) and type 2 diabetes (T2DM). Interactions between genetic and environmental factors such as diet and lifestyle, particularly over-nutrition and sedentary behavior, promote the progression and pathogenesis of these polygenic diet-related diseases. Their current prevalence is increasing dramatically to epidemic proportions. Nutrition is probably the most important environmental factor that modulates expression of genes involved in metabolic pathways and the variety of phenotypes associated with obesity, the MetS and T2DM. Furthermore, the health effects of nutrients may be modulated by genetic variants. Nutrigenomics and nutrigenetics require an understanding of nutrition, genetics, biochemistry and a range of "omic" technologies to investigate the complex interaction between genetic and environmental factors relevant to metabolic health and disease. These rapidly developing fields of nutritional science hold much promise in improving nutrition for optimal personal and public health. This review presents the current state of the art in nutrigenetic research illustrating the significance of gene-nutrient interactions in the context of metabolic disease.

In Chapter 2, Taylor and colleagues discuss how energy excess, low fruit and vegetable intake and other suboptimal dietary habits contribute to an increased poor health and the burden of disease in males. However the best way to engage males into nutrition programs remains unclear. This review provides a critical evaluation of the nature and effectiveness of nutrition interventions that target the adult male population. They conducted a search for full-text publications, using The Cochrane Library; Web of Science; SCOPUS; MEDLINE and CINAHL. Studies were included if 1) published from January 1990 to August 2011 and 2) male only studies ($\geq$18 years) or 3) where males contributed to >90% of the active cohort. A study must have described, (i) a significant change (p<0.05) over time in an objective measure of body weight, expressed in kilograms (kg) OR Body Mass Index (BMI) OR (ii) at least one significant change (p<0.05) in a dietary intake measure to qualify as effective. To identify emerging patterns within the research a descriptive process was used. Nine studies were included. Sample sizes ranged from 53 to 5042 male participants, with study durations ranging from 12 weeks to 24 months. Overlap was seen with eight of the nine studies including a weight management component whilst six studies focused on achieving changes in dietary intake patterns relating to modifications of fruit, vegetable, dairy and total fat intakes and three studies primarily focused on achieving weight loss through caloric restriction. Intervention effectiveness was identified for seven of the nine studies. Five studies reported significant positive changes in weight (kg) and/or BMI (kg/m2) changes (p$\leq$0.05). Four studies had effective interventions (p<0.05) targeting determinants of dietary intake and dietary behaviours and/or nutritional intake. Intervention features, which appeared to be associated with better outcomes, include the delivery of quantitative information on diet and the use of self-monitoring and tailored feedback. They found that uncertainty remains as to the features of successful nutrition interventions for males due to limited details provided for nutrition intervention protocols, variability in mode of delivery and comparisons between delivery modes as well as content of information provided to participants between studies. This review offers knowledge to guide researchers in making informed decisions on how to best utilise resources in interventions to engage adult males while highlighting the need for improved reporting of intervention protocols.

Nutrient timing is a popular nutritional strategy that involves the consumption of combinations of nutrients--primarily protein and carbohydrate--in and around an exercise session. Chapter 3 explores how some have claimed that this approach can produce dramatic improvements in body composition. It has even been postulated that the timing of nutritional consumption may be more important than the absolute daily intake of nutrients. The post-exercise period is widely considered the most critical part of nutrient timing. Theoretically, consuming the proper ratio of nutrients during this time not only initiates the rebuilding of damaged muscle tissue and restoration of energy reserves, but it does so in a supercompensated fashion that enhances both body composition and exercise performance. Several researchers have made reference to an anabolic "window of opportunity" whereby a limited time exists after training to optimize training-related muscular adaptations. However, Aragon and Schoenfeld argue that the importance - and even the existence - of a post-exercise 'window' can vary according to a number of factors. Not only is nutrient timing research open to question in terms of applicability, but recent evidence has directly challenged the classical view of the relevance of post-exercise nutritional intake with respect to anabolism. Therefore, the purpose of this paper will be twofold: 1) to review the existing literature on the effects of nutrient timing with respect to post-exercise muscular adaptations, and; 2) to draw relevant conclusions that allow practical, evidence-based nutritional recommendations to be made for maximizing the anabolic response to exercise.

In Chapter 4, Lakhan and Veira argue that according to the Diagnostic and Statistical Manual of Mental Disorders, 4 out of the 10 leading causes of disability in the US and other developed countries are mental disorders. Major depression, bipolar disorder, schizophrenia, and obsessive compulsive disorder (OCD) are among the most common mental disorders that currently plague numerous countries and have varying incidence rates from 26 percent in America to 4 percent in China. Though some of this difference may be attributable to the manner in which individual healthcare providers diagnose mental disorders, this noticeable distribution can be also explained by studies which show that a lack of certain dietary nutrients contribute to the development of mental disorders. Notably, essential vitamins, minerals, and omega-3 fatty acids are often deficient in the general population in America and other developed countries; and are exceptionally deficient in patients suffering from mental disorders. Studies have shown that daily supplements of vital nutrients often effectively reduce patients' symptoms. Supplements that contain amino acids also reduce symptoms, because they are converted to neurotransmitters that alleviate depression and other mental disorders. Based on emerging scientific evidence, this form of nutritional supplement treatment may be appropriate for controlling major depression, bipolar disorder, schizophrenia and anxiety disorders, eating disorders, attention deficit disorder/attention deficit hyperactivity disorder (ADD/ADHD), addiction, and autism. The aim of this manuscript is to emphasize which dietary supplements can aid the treatment of the four most common mental disorders currently affecting America and other developed countries: major depression, bipolar disorder, schizophrenia, and obsessive compulsive disorder (OCD). Most antidepressants and other prescription drugs cause severe side effects, which usually discourage patients from taking their medications. Such noncompliant patients who have mental disorders are at a higher risk for committing suicide or being institutionalized. One way for psychiatrists to overcome this noncompliance is to educate themselves about alternative or complementary nutritional treatments. Although in the cases of certain nutrients, further research needs to be done to determine the best recommended doses of most nutritional supplements, psychiatrists can recommend doses of dietary supplements based on previous and current efficacious studies and then adjust the doses based on the results obtained.

Sun and Empie explore fructose consumption and its implications on public health in Chapter 5. Their work reviewed the metabolic fate of dietary fructose based on isotope tracer studies in humans. The mean oxidation rate of dietary fructose was $45.0\% \pm 10.7$ (mean $\pm$ SD) in non-exercising subjects within 3–6 hours and $45.8\% \pm 7.3$ in exercising subjects within 2–3 hours. When fructose was ingested together with glucose, the mean oxidation rate of the mixed sugars increased to $66.0\% \pm 8.2$ in exercising subjects. The mean conversion rate from fructose to glucose was $41\% \pm 10.5$ (mean $\pm$ SD) in 3–6 hours after ingestion. The conversion amount from fructose to glycogen remains to be further clarified. A small percentage of ingested fructose (<1%) appears to be directly converted to plasma TG. However, hyperlipidemic effects of larger amounts of fructose consumption are observed in studies using infused labeled acetate to quantify longer term de novo lipogenesis. While the mechanisms for the hyperlipidemic effect remain controversial, energy source shifting and lipid sparing may play a role in the effect, in addition to de novo lipogenesis. Finally, approximately a quarter of ingested fructose can be converted into lactate within a few of hours. The reviewed data provides a profile of how dietary fructose is utilized in humans.

In Chapter 6, Naidu researches the relationship between human health and Vitamin C. Ascorbic acid is one of the important water soluble vitamins. It is essential for collagen, carnitine and neurotransmitters biosynthesis. Most plants and animals synthesize ascorbic acid for their own requirement. However, apes and humans can not synthesize ascorbic acid due to lack of an enzyme gulonolactone oxidase. Hence, ascorbic acid has to be supplemented mainly through fruits, vegetables and tablets. The current US recommended daily allowance (RDA) for ascorbic acid ranges between 100–120 mg/per day for adults. Many health benefits have been attributed to ascorbic acid such as antioxidant, anti-atherogenic, anti-carcinogenic, immunomodulator and prevents cold etc. However, lately the health benefits of ascorbic acid has been the subject of debate and controversies viz., Danger of mega doses of ascorbic acid? Does ascorbic acid act as a antioxidant or pro-oxidant? Does ascorbic acid cause cancer or may interfere with cancer therapy? However, the Panel on dietary antioxidants and related compounds stated that the in vivodata do not clearly show a relationship between excess ascorbic acid intake and kidney stone formation, pro-oxidant effects, excess iron absorption. A number of clinical and epidemiological studies on anti-carcinogenic effects of ascorbic acid in humans did not show any conclusive beneficial effects on various types of cancer except gastric cancer. Recently, a few derivatives of ascorbic acid were tested on cancer cells, among them ascorbic acid esters showed promising anticancer activity compared to ascorbic acid. Ascorbyl stearate was found to inhibit proliferation of human cancer cells by interfering with cell cycle progression, induced apoptosis by modulation of signal transduction pathways. However, more mechanistic and human in vivo studies are needed to understand and elucidate the molecular mechanism underlying the anti-carcinogenic property of ascorbic acid. Thus, though ascorbic acid was discovered in 17th century, the exact role of this vitamin/nutraceutical in human biology and health is still a mystery in view of many beneficial claims and controversies.

Nijke and colleagues explore the effect of egg consumption in Chapter 7. Limiting consumption of eggs, which are high in cholesterol, is generally recommended to reduce risk of cardiovascular disease. However, recent evidence suggests that dietary cholesterol has limited influence on serum cholesterol or cardiac risk. Their objective is to assess the effects of egg consumption on endothelial function and serum lipids in hyperlipidemic adults. Randomized, placebo-controlled crossover trial of 40 hyperlipidemic adults (24 women, 16 men; average age = 59.9 ± 9.6 years; weight = 76.3 ± 21.8 kilograms; total cholesterol = 244 ± 24 mg/dL). In the acute phase, participants were randomly assigned to one of the two sequences of a single dose of three medium hardboiled eggs and a sausage/cheese breakfast sandwich. In the sustained phase, participants were then randomly assigned to one of the two sequences of two medium hardboiled eggs and 1/2 cup of egg substitute daily for six weeks. Each treatment assignment was separated by a four-week washout period. Outcome measures of interest were endothelial function measured as flow mediated dilatation (FMD) and lipid panel. They found that single dose egg consumption had no effects on endothelial function as compared to sausage/cheese (0.4 ± 1.9 vs. $0.4 \pm 2.4\%$; $p = 0.99$). Daily consumption of egg substitute for 6 weeks significantly improved endothelial function as compared to egg ($1.0 \pm 1.2\%$ vs. $-0.1 \pm 1.5\%$; $p < 0.01$) and lowered serum total cholesterol (-18 ± 18 vs. -5 ± 21 mg/dL; $p < 0.01$) and LDL (-14 ± 20 vs. -2 ± 19 mg/dL; $p = 0.01$). Study results (positive or negative) are expressed in terms of change relative to baseline. Egg consumption was found to be non-detrimental to endothelial function and serum lipids in hyperlipidemic adults, while egg substitute consumption was beneficial

Chapter 8 discusses another kind of functional food: this time protein-enriched meal replacements. There is concern that recommending protein-enriched meal replacements as part of a weight management program could lead to changes in biomarkers of liver or renal function and reductions in bone density. Li and colleagues designed the experiment as a placebo-controlled clinical trial utilizing two isocaloric meal plans utilizing either a high protein-enriched (HP) or a standard protein (SP) meal replacement in an outpatient weight loss program. One hundred obese men and women over 30 years of age with a body mass index (BMI) between 27 to 40 kg/m2 were randomized to one of two isocaloric weight loss meal plans 1). HP group: providing 2.2 g protein/kg of lean body mass (LBM)/day or 2). SP group: providing 1.1 g protein/kg LBM/day. Meal replacement (MR) was used twice daily (one meal, one snack) for 3 months and then once a day for 9 months. Body weight, lipid profiles, liver function, renal function and bone density were measured at baseline and 12 months Seventy subjects completed the study. Both groups lost weight (HP -4.29 ± 5.90 kg vs. SP -4.66 ± 6.91 kg, p < 0.01) and there was no difference in weight loss observed between the groups at one year. There was no significant change noted in liver function, bilirubin, alkaline phosphatase, renal function [serum creatinine, urea nitrogen, 24 hour urine creatinine clearance, and calcium excretion] or in bone mineral density by DEXA in either group over one year. These studies demonstrate that protein-enriched meals replacements as compared to standard meal replacements recommended for weight management do not have adverse effects on routine measures of liver function, renal function or bone density at one year.

In Chapter 9, Kahl and colleagues conducted a literature review pertaining to organic and functional food was conducted according its conceptual background. Functional and organic food both belong to fast growing segments of the European food market. Both are food according to the European food regulations, but organic food is further regulated by the European regulation for organic agriculture and food production. This regulation restricts the number of food additives and limits substantial changes in the food. This may cause problems in changing the food based on single constituents or attributes when applying the concept of functional food to organic food production. Claims of the influence of the food positively on health can only be accepted as true when the claims have been tested and then validated by the EU-Commission. Whereas functional food focuses on product comparison based on specific constituents or attributes, organic food as a whole has no placebo for comparison and effects on environment and society are not part of the health claim regulation. Therefore it seems rather difficult to establish the health claims of organic foods. Consumers buy organic food out of an emotional attitude and associate the food with naturalness. In contrast, the decision for buying functional food is related to rationality and consumers associate functional food with a more techno- logical approach. For this reason, the authors conclude that the concept of functional food seems not to support organic food production in Europe.

Chapter 10 begins the second section of the book, which contains case studies on the use of functional foods. Aoi and colleagues conclude that appropriate nutrition is an essential prerequisite for effective improve- ment of athletic performance, conditioning, recovery from fatigue after exercise, and avoidance of injury. Nutritional supplements containing carbohydrates, proteins, vitamins, and minerals have been widely used in various sporting fields to provide a boost to the recommended daily al- lowance. In addition, several natural food components have been found to show physiological effects, and some of them are considered to be useful for promoting exercise performance or for prevention of injury. However, these foods should only be used when there is clear scientific evidence and with understanding of the physiological changes caused by exercise. This article describes various "functional foods" that have been reported to be effective for improving exercise performance or health promotion, along with the relevant physiological changes that occur during exercise.

Eichler and colleagues examine the effects of micronutrient fortified milk and cereal in Chapter 11. Micronutrient deficiency is a common pub- lic health problem in developing countries, especially for infants and chil- dren in the first two years of life. As this is an important time window for child development, micronutrient fortified complementary feeding after 6 months of age, for example with milk or cereals products, in combi- nation with continued breastfeeding, is recommended. The overall effect of this approach is unclear. The authors performed a Systematic Review and Meta-analysis to assess the impact of micronutrient fortified milk and cereal food on the health of infants and little children (aged 6 months to 5 years) compared to non-fortified food. We reviewed randomized con- trolled trials using electronic databases (MEDLINE and Cochrane library searches through FEB 2011), reference list screening and hand searches. Three reviewers assessed 1153 studies for eligibility and extracted data. One reviewer assessed risk of bias using predefined forms. They included 18 trials in our analysis (n = 5'468 children; range of mean hemoglobin values: 9.0 to 12.6 g/dl). Iron plus multi micronutrient fortification is more effective than single iron fortification for hematologic outcomes. Compared to non-fortified food, iron multi micronutrient fortification increases hemoglobin levels by 0.87 g/dl (95%-CI: 0.57 to 1.16; 8 studies) and reduces risk of anemia by 57% (relative risk 0.43; 95%-CI 0.26 to 0.71; absolute risk reduction 22%; number needed to treat 5 [95%-CI: 4 to 6]; 6 Studies). Compared to non-fortified food, fortification increases serum levels of vitamin A but not of zinc. Information about functional health outcomes (e.g. weight gain) and morbidity was scarce and evidence is inconclusive. Risk of bias is unclear due to underreporting, but high quality studies lead to similar results in a sensitivity analysis. They found that multi micronutrient fortified milk and cereal products can be an effective option to reduce anemia of children up to three years of age in developing countries. On the basis of their data, the evidence for functional health outcomes is still inconclusive.

Chapter 12 attempted to evaluate the effects of different medicinal herbs rich in polyphenol (Lemon balm, Sage, St. John's wort and Small-flowered Willowherb) used as dietary supplements on bioaccumulation of some essential metals (Fe, Mn, Zn and Cu) in different chicken meats (liver, legs and breast). Stef and Gergen found that in different type of chicken meats (liver, legs and breast) from chickens fed with diets enriched in minerals and medicinal herbs, beneficial metals (Fe, Mn, Zn and Cu) were analysed by flame atomic absorption spectrometry. Fe is the predominant metal in liver and Zn is the predominant metal in legs and breast chicken meats. The addition of metal salts in the feed influences the accumulations of all metals in the liver, legs and breast chicken meat with specific difference to the type of metal and meat. The greatest influences were observed in legs meat for Fe and Mn. Under the influence of polyphenol-rich medicinal herbs, accumulation of metals in the liver, legs and breast chicken meat presents specific differences for each medicinal herb, to the control group that received a diet supplemented with metal salts only. Great influence on all metal accumulation factors was observed in diet enriched with sage, which had significantly positive effect for all type of chicken meats. Under the influence of medicinal herbs rich in different type of polyphenol, accumulation of metals in the liver, legs and breast chicken meat presents significant differences from the group that received a diet supplemented only with metal salts. Each medicinal herb from diet had a specific influence on the accumulation of metals and generally moderate or poor correlations were observed between total phenols and accumulation of metals. This may be due to antagonism between metal ions and presence of other chelating agents (amino acids and protein) from feeding diets which can act as competitor for complexation of metals and influence accumulation of metals in chicken meat.

Charoensiriwatana and colleagues found that evidence showed that the occurrence of iodine deficiency endemic areas has been found in every provinces of Thailand. Chapter 13 describes a new pilot programme for elimination of iodine deficiency endemic areas at the community level, designed in 2008 by integrating the concept of Sufficient Economic life style with the iodine biofortification of nutrients for community consumption. A model of community hen egg farm was selected at an iodine deficiency endemic area in North Eastern part of Thailand. The process for the preparation of high content iodine enriched hen food was demonstrated to the farm owner with technical transfer in order to ensure the sustainability in the long term for the community. The iodine content of the produced iodine enriched hen eggs were determined and the iodine status of volunteers who consumed the iodine enriched hen eggs were monitored by using urine iodine excretion before and after the implement of iodine enrichment in the model farm. The content of iodine in eggs from the model farm were 93.57 µg per egg for the weight of 55 - 60 g egg and 97.76 µg for the weight of 60 - 65 g egg. The biological active iodo-organic compounds in eggs were tested by determination of the base-line urine iodine of the volunteer villagers before and after consuming a hard boiled iodine enriched egg per volunteer at breakfast for five days continuous period in 59 volunteers of Ban Kew village, and 65 volunteers of Ban Nong Nok Kean village. The median base-line urine iodine level of the volunteers in these two villages before consuming eggs were 7.00 and 7.04 µg/dL respectively. After consuming iodine enriched eggs, the median urine iodine were raised to the optimal level at 20.76 µg/dL for Ban Kew and 13.95 µg/dL for Ban Nong Nok Kean.They found that the strategic programme for iodine enrichment in the food chain with biological iodo-organic compound from animal origins can be an alternative method to fortify iodine in the diet for Iodine Deficiency Endemic Areas at the community level in Thailand.

Chapter 14 looks at the effects of cinnamon when administered to diabetic rats. Shihabudeen and colleagues find that α-glucosidase inhibitors regulate postprandial hyperglycemia (PPHG) by impeding the rate of carbohydrate digestion in the small intestine and thereby hampering the diet associated acute glucose excursion. PPHG is a major risk factor for diabetic vascular complications leading to disabilities and mortality in diabetics. Cinnamomum zeylanicum, a spice, has been used in traditional medicine for treating diabetes. This study evaluated the α-glucosidase inhibitory potential of cinnamon extract to control postprandial blood glucose level in maltose, sucrose loaded STZ induced diabetic rats. The methanol extract of cinnamon bark was prepared by Soxhlet extraction. Phytochemical analysis was performed to find the major class of compounds present in the extract. The inhibitory effect of cinnamon extract on yeast α-glucosidase and rat-intestinal α-glucosidase was determined in vitro and the kinetics of enzyme inhibition was studied. Dialysis experiment was performed to find the nature of the inhibition. Normal male Albino wistar rats and STZ induced diabetic rats were treated with cinnamon extract to find the effect of cinnamon on postprandial hyperglycemia after carbohydrate loading. Phytochemical analysis of the methanol extract displayed the presence of tannins, flavonoids, glycosides, terpenoids, coumarins and anthraquinones. In vitro studies had indicated dose-dependent inhibitory activity of cinnamon extract against yeast α-glucosidase with the IC 50 value of 5.83 μg/ml and mammalian α-glucosidase with IC 50 value of 670 μg/ml. Enzyme kinetics data fit to LB plot pointed out competitive mode of inhibition and the membrane dialysis experiment revealed reversible nature of inhibition. In vivo animal experiments are indicative of ameliorated postprandial hyperglycemia as the oral intake of the cinnamon extract (300 mg/kg body wt.) significantly dampened the postprandial hyperglycemia by 78.2% and 52.0% in maltose and sucrose loaded STZ induced diabetic rats respectively, compared to the control. On the other hand, in rats that received glucose and cinnamon extract, postprandial hyperglycemia was not effectively suppressed, which indicates that the observed postprandial glycemic amelioration is majorly due to α-glucosidase inhibition. The current study demonstrates one of the mechanisms in which cinnamon bark extract effectively inhibits α-glucosidase leading to suppression of postprandial hyperglycemia in STZ induced diabetic rats loaded with maltose, sucrose. This bark extract shows competitive, reversible inhibition on α-glucosidase enzyme. Cinnamon extract could be used as a potential nutraceutical agent for treating postprandial hyperglycemia. In future, specific inhibitor has to be isolated from the crude extract, characterized and therapeutically exploited.

In Chapter 15, a similar study is done, this time with garlic instead of cinnamon. Padiya and colleagues explore type 2 diabetes mellitus, characterized by peripheral insulin resistance, is a major lifestyle disorder of the 21st Century. Raw garlic homogenate has been reported to reduce plasma glucose levels in animal models of type 1 diabetes mellitus. However, no specific studies have been conducted to evaluate the effect of raw garlic on insulin resistance or type 2 diabetes mellitus. This study was designed to investigate the effect of raw garlic on fructose induced insulin resistance, associated metabolic syndrome and oxidative stress in diabetic rats. Male Sprague Dawley rats weighing 200-250 gm body weight were divided into 3 groups (n = 7 per group) and fed diet containing 65% cornstarch (Control group) and 65% fructose (Diabetic group) for 8 weeks. The third group (Dia+Garl group) was fed both 65% fructose and raw garlic homogenate (250 mg/kg/day) for 8 weeks. Whole garlic cloves were homogenized with water to make a fresh paste each day. At the end of 8 weeks, serum glucose, insulin, triglyceride and uric acid levels, as well as insulin resistance, as measured by glucose tolerance test, were significantly ($p < 0.01$) increased in fructose fed rats (Diabetic group) when compared to the cornstarch fed (Control) rats. Administration of raw garlic to fructose fed rats (Dia+Garl group) significantly ($p < 0.05$) reduced serum glucose, insulin, triglyceride and uric acid levels, as well as insulin resistance when compared with fructose fed rats. Garlic also normalised the increased serum levels of nitric oxide (NO) and decreased levels of hydrogen sulphide (H2S) after fructose feeding. Although body weight gain and serum glycated haemoglobin levels of fructose fed rats (Diabetic group) were not significantly different from control rats, significant ($p < 0.05$) reduction of these parameters was observed in fructose fed rats after garlic administration (Dia+Garl group). Significant ($p < 0.05$) increase in TBARS and decrease in GSH was observed in diabetic liver. Catalase was not significantly affected in any of the groups. Administration of raw garlic homogenate normalised both hepatic TBARS and GSH levels. The study demonstrates that raw garlic homogenate is effective in improving insulin sensitivity while attenuating metabolic syndrome and oxidative stress in fructose-fed rats.

In traditional medicine, blueberries have been used to facilitate blood glucose regulation in type 2 diabetes. In Chapter 16, Granfeldt and Björck find that recent studies in diabetic mice have indicated facilitated glycaemic regulation following dietary supplementation with extracts from European blueberries, also called bilberries, (Vaccinium myrtillus). The purpose of the present study was to investigate the impact of fermented oat meal drinks containing bilberries or rosehip (Rosa canina) on glycaemic and insulinaemic responses. Glycaemic and insulinaemic responses in young healthy adults were measured in two series. In series 1, two drinks based on oat meal (5%), fermented using Lactobacillus plantarum 299v, and added with fruit (10%); bilberries (BFOMD) or rose hip (RFOMD) respectively, were studied. In series 2, BFOMD was repeated, additionally, a drink enriched with bilberries (47%) was tested (BBFOMD). As control a fermented oat meal drink (FOMD) was served. In series 1 the bilberry- and rosehip drinks, gave high glucose responses similar to that after the reference bread. However, the insulin index (II) after the BFOMD was significantly lower (II = 65) (P < 0.05). In series 2 a favourably low insulin demand to BFOMD was confirmed. FOMD gave high glucose response (GI = 95) but, significantly lower insulin response (II = 76). BBFOMD gave remarkably low insulin response II = 49, and tended to lower glycaemia (GI = 79) (P = 0.0684). They concluded that a fermented oat meal drink added with bilberries induced a lower insulin response than expected from the glycaemic response. The mechanism for the lowered acute insulin demand is still unclear, but may be related to some bio-active component present in the bilberries, or to the fermented oat meal base.

The final chapter looks at how Brazil nuts can influence obesity. Obesity is a chronic disease associated to an inflammatory process resulting in oxidative stress that leads to morpho-functional microvascular damage that could be improved by some dietary interventions. In this study, Maranhão and colleagues investigate the intake of Brazil nuts (Bertholletia excelsa), composed of bioactive substances like selenium, α- e γ- tocopherol, folate and polyunsaturated fatty acids, on antioxidant capacity, lipid and metabolic profiles and nutritive skin microcirculation in obese adolescents.

Obese female adolescents (n = 17), 15.4 ± 2.0 years and BMI of 35.6 ± 3.3 kg/m², were randomized 1:1 in two groups with the diet supplemented either with Brazil nuts [BNG, n = 08, 15-25 g/day (equivalent to 3 to 5 units/day)] or placebo [PG (lactose), n = 09, one capsule/day] and followed for 16 weeks. Anthropometry, metabolic-lipid profiles, oxidative stress and morphological (capillary diameters) and functional [functional capillary density, red blood cell velocity (RBCV) at baseline and peak (RBCVmax) and time (TRBCVmax) to reach it during post-occlusive reactive hyperemia, after 1 min arterial occlusion] microvascular variables were assessed by nailfold videocapillaroscopy at baseline (T0) and after intervention (T1). T0 characteristics were similar between groups. At T1, BNG (intra-group variation) had increased selenium levels (p = 0.02), RBCV (p = 0.03) and RBCVmax (p = 0.03) and reduced total (TC) (p = 0.02) and LDL-cholesterol (p = 0.02). Compared to PG, Brazil nuts intake reduced TC (p = 0.003), triglycerides (p = 0.05) and LDL-ox (p = 0.02) and increased RBCV (p = 0.03). Brazil nuts intake improved the lipid profile and microvascular function in obese adolescents, possibly due to its high level of unsaturated fatty acids and bioactive substances.

— **Leah Coles, PhD**

PART I

THE CONNECTION BETWEEN NUTRITION AND HEALTH

CHAPTER 1

NUTRIGENETICS AND METABOLIC DISEASE: CURRENT STATUS AND IMPLICATIONS FOR PERSONALISED NUTRITION

CATHERINE M. PHILLIPS

1.1 INTRODUCTION

The MetS represents a constellation of metabolic perturbations including central obesity, insulin resistance, dyslipidaemia characterized by raised triglyceride and reduced high density lipoprotein concentrations and hypertension. The MetS and these interrelated risk factors are associated with increased risk of type 2 diabetes (T2DM) and cardiovascular disease (CVD) [1]. Numerous definitions of the MetS have been proposed; initially by the WHO in 1998 [2] and subsequently by the European Group for the Study of Insulin Resistance in 1999 [3], the National Cholesterol Education Program's Adult Treatment Panel III report (NCEP ATP III) in 2001 [4] and in 2005 by the International Diabetes Federation (IDF) [5] and the IDF in conjunction with the American Heart Association and the National Heart, Lung and Blood Institute [6]. While details differ between definitions, all agree on the essential components (central obesity, insulin resistance/glucose intolerance, dyslipidaemia and hypertension). Notwithstanding the varying definitions it is clear that the incidence of the MetS is increasing among men and women of all ages and ethnicities [7]. Recent estimates from the US show that the prevalence of the MetS among adults ranges from 34.3% to 38.5% depending on the criteria used to define abdominal obesity [7]. Thus it is conceivable that 77 to 86 million adults in the US meet current MetS criteria. Individuals with the MetS have a five-fold increased risk of developing T2DM. Coupled with this is a two-fold risk of developing CVD over the next 5 to 10 years compared to individuals without the syndrome. Lifetime risk is even higher. The prevalence of obesity is also increasing worldwide, with the condition predicted to affect more than one billion people by the year 2020 [8]. Obesity and weight-gain are directly related to T2DM risk [9]. Excess adiposity, particularly central adiposity, is a key causal factor in the development of insulin resistance, the hallmark of the MetS. The increasing global prevalence of T2DM in children and adults, and its medical and socio-economic consequences represent a major public health concern. Recent estimates predict in excess of 400 million individuals with T2DM worldwide by the year 2030 [10].

Complex gene-environment interactions are certainly contributing to the current diabesity epidemic. Family and twin studies indicate that up to 80% of the variance in body mass index (BMI) is attributable to genetic factors. Genetic factors also contribute approximately 50% towards T2DM risk. Heritability rates of 10%–30% for the MetS have been estimated [11,12], indicating that these conditions are partly heritable. Nutrition and physical activity are key environmental factors, which potentially interact with genetic predisposition, to promote the progression and pathogenesis of these combined environmental and polygenic, diet-related diseases. Excessive caloric intake and sedentary lifestyle promote the obese phenotype. More than half of adults in Europe and the US are overweight or obese, leading to the MetS, which in turn greatly enhances subsequent risk of cardiometabolic disease [13]. There is no doubt that a genetic component can also impact on the risk of insulin resistance, the sensitivity to which may be further amplified by poor diet. If we have a greater understanding of potential gene-nutrient interactions, then it may be possible to manipulate diet in such a way to minimize the metabolic risk of obesity, to attenuate insulin resistance and development of cardiometabolic disease. At a public health level, more attention must be given to modification of lifestyles of the general public to reduce risk of obesity and T2DM and to increase physical activity. At a clinical level, individual patients with increased metabolic risk need to be identified so that their multiple risk factors can be reduced. Early identification of "at risk" individuals is of paramount importance. Considering the long asymptomatic period often preceding the manifestation of T2DM and CVD, early diagnosis could enable earlier targeted interventions such as implementation of healthy lifestyle changes in nutritional behavior and exercise or pharmacotherapy, thus reducing disease development. From the nutritional/dietary advice perspective, this review will focus on dietary fatty acids, as these are the most energy dense nutrients. A deeper understanding of the underlying gene-nutrient interactions, to aid an understanding of the link between nutrition and health, will provide the evidence base to define whether more targeted nutritional advice is an appropriate public health approach. Nutrigenetics hold much promise in terms of both public health nutrition and for individuals and genetic subgroups. In this review we will present the current state of the art, illustrating the significance of gene-nutrient interactions in the context of diet-related metabolic disease.

1.2 DIETARY FAT AND METABOLIC HEALTH

Dietary fat is an important environmental factor, wherein excessive exposure plays a key role in the development of the MetS [14–18]. High-fat diets, in particular high saturated fatty acid (SFA) diets, have been shown to exert detrimental effects on adiposity, inflammation and insulin sensitivity, promoting the development of insulin resistance, the MetS and T2DM [17,19–21]. Diets rich in monounsaturated fatty acids (MUFA) have been associated with improvements in insulin sensitivity in healthy subjects [22]. In the LIPGENE study, a large pan-European isocaloric dietary intervention study of MetS subjects, substitution of SFA with either MUFA or low-fat, high-complex carbohydrate to improve insulin sensitivity was only effective in individuals whose habitual pre-intervention dietary fat intake was below the median (<36% energy from fat) [23]. Evidence from the MUFA in Obesity (MUFObes) study suggest that MUFA rich diets had beneficial effects on insulin and glucose concentrations and were associated with reduced body fat regain [24,25]. The KANWU study showed that a high-SFA diet reduced insulin sensitivity in overweight subjects, and post-hoc analysis indicated that isoenergetic substitution of SFA for MUFA improved insulin sensitivity but only in subjects whose habitual pre-intervention dietary fat intake was below the median [26]. While cell and animal studies have demonstrated beneficial effects of long chain n-3 polyunsaturated fatty acids (LC n-3 PUFA) on inflammation and insulin sensitivity [27–29] translation of these potentially anti-diabetic effects in humans has proven more difficult with conflicting epidemiological data in relation to their effect on insulin resistance in humans [23]. A limited number of large human dietary intervention studies have been performed to determine the effects of dietary quantity and quality on risk factors associated with metabolic health. Considering that up to 30% of subjects are not responsive to interventions, variable intervention outcomes are perhaps not surprising. Moreover it seems that pre-intervention dietary intake may affect outcome. A range of methods to assess dietary intake including food-frequency questionnaires and weighed or estimated dietary records are currently used. Each has advantages, practical limitations and associated errors. In the context of dietary fatty acids the use of biomarkers of habitual dietary fat intake, such as plasma fatty acids, offers some advantages over self-reported food-frequency questionnaires whereby they are not subject to misclassification of exposure, due to deficiencies in nutrient databases, accuracy of memories or willingness to divulge these details [30]. In contrast to dietary fat measurement, plasma fatty acid composition reflects the combination of dietary fat consumption and endogenous de novo fatty acid biosynthesis and metabolism, thus making direct comparisons between some plasma and dietary fatty acid measurements difficult.

The failure of current dietary guidelines in combating the obesity epidemic provides further evidence that the optimal dietary fat composition (amount and type of fatty acid) for optimal metabolic health is still unknown and that the traditional one size fits all approach does not work in the context of obesity and metabolic health. Indeed such inter-individual differences in response to dietary factors and interventions highlight the role of genetics and the potential of a nutrigenetics approach based on identification of nutrient sensitive or responsive genotypes, whereby nutrient intake is manipulated or optimized based on an individuals' genetic profile to reduce disease risk or improve effectiveness of dietary recommendations. Current evidence to support the nutrigenetics concept with respect to obesity, the MetS and T2DM is largely based on data relating to dietary fat [15,31–33] and is discussed in more detail later.

1.3 GENETIC DETERMINANTS OF METABOLIC DISEASE

It is commonly accepted that the current global diabesity pandemic is being driven by an obesogenic environment which promotes consumption of energy-dense foods and discourages energy expenditure, inevitably leading to an energy imbalance favoring energy storage and weight gain. A number of factors including industrialization and modernization of the built environment (reliance on car use, reduced manual labor, lack of safe pathways or cycle lanes, proximity to fast food and convenience outlets, increased commuting time, etc.) as well as the social environment (socioeconomic status, advertising, consumer pressure, etc.) have led to more sedentary lifestyle behaviors and freely available caloric abundance. However, the familial clustering associated with obesity is not just due to common environmental factors. Studies of twins, adoptees and families indicate that up to 80% of the variance in BMI is attributable to genetic factors. High relative risk ratios [34] and concordance rates for monozygotic twins compared to dizygotic twins have also been estimated for obesity [35]. Interestingly adoption studies revealed that adoptees' weight is more similar to that of the biological parents than the adoptive parents [36]. Heritability rates of 25%–40% for BMI and body fat [36–38] and 10%–30% for the MetS have been estimated [11,12], indicating that these conditions are partly heritable. Genetic factors are also estimated to contribute approximately 50% towards T2DM risk. Family studies have demonstrated that first degree relatives of T2DM individuals are approximately 3 times more likely to develop the disease than individuals without a positive family history [39,40]. Furthermore, twin studies have shown that concordance rates for monozygotic twins, ranging from 60% to 90%, are significantly higher than those for dizygotic twins. Thus, it is clear that genetic differences between individuals also play a role in the risk of becoming obese and developing the MetS and T2DM.

1.3.1 IDENTIFICATION OF GENES ASSOCIATED WITH METABOLIC DISEASE

Monogenic disorders account for up to 5% of all cases of obesity and also diabetes. Over the past 15–20 years, mutations in a number of genes including leptin (LEP), leptin receptor (LEPR), pro-opiomelanocortin (POMC) and melanocortin-4 receptor (MC4R) have been associated with monogenic obesity [41]. Six genes account for the majority of the monogenic forms of diabetes: hepatic nuclear factor 4α (HNF-4α), glucokinase (GCK), hepatic nuclear factor 1α (HNF-1α), insulin promoter factor-1 (IPF-1), hepatic nuclear factor 1β (HNF-1β) and neuro D1 transcription factor (NEUROD1) [40]. However, for most individuals genetic predisposition to metabolic disease has a polygenic basis. The absence of large single gene effects and the detection of multiple small effects support this notion, which suggests that only in combination with other predisposing variants does a sizeable phenotypic effect arise. In addition, this implies that certain sets of polygenic variants relevant to these conditions in one individual may not be the same in another individual. Identifying genes associated with any complex trait involves a range of experimental strategies including positional cloning using genome-wide linkage analysis, candidate gene association and more recently genome-wide association studies (GWAS). Genome scanning in several different ethnic groups has identified a number of chromosome regions harboring T2DM and obesity susceptibility genes. Calpain 10 (CAPN10), which encodes cysteine protease calpain 10, was the first T2DM susceptibility gene identified through a genome-wide scan followed by positional cloning [42]. Genetic and functional data indicate that CAPN10 plays an important role in insulin resistance and intermediate phenotypes [43]. CAPN10 variants have been linked with several MetS phenotypes including hypertriglyceridaemia, BMI and hypertension [44–46]. With the advent of high-throughput genetic analysis and the completion of the Human Genome Project our understanding of the genetic architecture and biology of obesity, MetS, insulin resistance and T2DM is improving. GWAS represent a powerful approach to the identification of genes involved in common polygenic diseases. Typically, these studies, which are performed without any prior knowledge regarding the nature or location of the causative genes, analyze thousands of SNPs across the entire genome and involve very large subject numbers. Patterns of association between genotypes and disease status are then identified and evaluated statistically.

1.3.2 GWAS AND METABOLIC HEALTH RELEVANT LOCI

In the last six years GWAS have identified more than 50 loci, many of which are novel, relevant to obesity and diabetes. 2006 heralded the identification of the most important T2DM susceptibility gene known so far, transcription factor 7-like 2 (TCF7L2). Two novel single nucleotide polymorphisms (SNP) in the Wnt signaling regulated TCF7L2 gene were associated with increased T2DM risk [47,48], most likely through defective beta-cell function and impaired insulin secretion [47]. Several large studies subsequently replicated and confirmed the association with T2DM risk in various populations and the TCF7L2 rs7903146 SNP has emerged as one of the most important T2DM susceptibility gene variants known to date [49–54]. TCF7L2 polymorphisms have also been associated with MetS components such as dyslipidemia and waist circumference [55,56]. However, prospective and population-based MetS association studies have produced conflicting results; with some reporting an association with MetS, hyperglycaemia, impaired insulin secretion and hypertriglyceridaemia [56,57], whilst others found no association with MetS or insulin resistance [58,59]. More recently, TCF7L2 rs7903146 was associated with increased MetS risk, arising from their impaired insulin sensitivity, greater insulin resistance, increased abdominal obesity and hypertension [60]. This association has also been confirmed in a recent systematic review [61]. Following the identification of TCF7L2, previously unknown genetic variants in the fat mass and obesity-associated (FTO) gene on chromosome 16 were also linked to T2DM risk through an effect on BMI [62], such that the 16% of adults homozygous for the rs9939609 "risk" A allele were 3 kg heavier and had 1.7 fold increased risk of obesity relative to the homozygous non-risk allele carriers. It has been suggested that the impaired satiety, greater food intake and more frequent loss of eating control reported by individuals with at least one risk allele may account for the observed increased obesity

risk [63–65]. Importantly, researchers have demonstrated increased obesity risk associated with this polymorphism from childhood into old age
[62]. A number of large studies replicated and confirmed the association
with obesity risk in European populations [66–68] and the FTO rs9939609
SNP is now recognized as one of the most important gene variants predisposing to obesity. A recent systematic review on the genetics of the MetS
confirmed increased MetS risk with FTO [61]. Other obesity susceptibility
loci identified by GWAS include SHRB1 and BDNF [69,70]. Meta-analysis of 21 GWAS cohorts (Meta-Analyses of Glucose and Insulin-related
traits Consortium [MAGIC]), identified associations between a number
of SNPs in 8 loci (including the candidate genes ADCY5, FADS1, and
GLIS3) and fasting glucose concentrations, and also between SNPs in one
locus (with the candidate gene IGF1) with both fasting insulin levels and
insulin resistance [71]. More recently, a joint meta-analysis investigating
whether genes involved in insulin resistance pathways could be discovered by accounting for differences in BMI and interactions between BMI
and genetic variants identified eight SNPs in six loci (including the candidate genes COBLL1/GRB14, IRS1, PPP1R3B, PDGFC, UHRF1BP1,
and LYPLAL1) that were associated with fasting insulin, high triglyceride
and low high-density lipoprotein levels, suggesting a new series of pathways to identify genes with contributions to multiple phenotypes [72]. It is
clear that the contribution of GWAS cohorts to such consortia will provide
increased power to detect variants associated with measures of glucose
homeostasis, insulin resistance, obesity, T2DM and MetS phenotypes.

1.3.3 LIPID METABOLISM GENES

Dyslipidaemia is one of the very early features of an obese or MetS phenotype and frequently precedes metabolic disturbances in insulin and
glucose homeostasis. Both fasting and postprandial lipid metabolism are
disturbed in the MetS [73] and in T2DM, particularly among individuals
with poor metabolic control [74]. A number of recent reviews present the
evidence linking candidate genes with modulation of postprandial lipid
metabolism [75,76]. To summarize, genes of note include the apolipoprotein APOA1/C3/A4/A5 cluster, APOB, APOE, cholesterol ester transfer

protein, hepatic and lipoprotein lipases, TCF7L2, glucokinase regulatory protein, fatty acid binding proteins, microsomal triglyceride transfer protein, peroxisome proliferator-activated receptor-γ (PPARγ), scavenger receptor class B type 1 (SCARB1) and perilipin (PLIN). Prior to the GWAS era the lipid sensitive transcription factor PPARγ Pro12Ala polymorphism was identified as the most widely reproduced genetic variation for T2DM risk [39]. The original study investigating the polymorphism in T2DM demonstrated that the alanine allele of this polymorphism was associated with lower BMI, improved insulin sensitivity and thus reduced diabetes risk by 75% [77]. Inconsistent results from subsequent association studies prompted a meta-analysis which confirmed a modest (1.25-fold) but significant increase in diabetes risk for the Pro12Pro genotype [78]. The Pro12Pro genotype also predicts conversion from impaired glucose tolerance to T2DM, with a 3 fold increased T2DM risk in the Pro12 homozygotes compared to 12Ala carriers [79]. In obese subjects diabetes risk is almost doubled among the Pro12 allele carriers. Interestingly the obese phenotype seems to exacerbate the detrimental effect of the PPARγ Pro12 allele on insulin sensitivity [80]. Perhaps more importantly is that this variant is very common in most populations. Approximately 98% of Europeans carry at least one copy of the Pro allele, so it is reasonable to speculate that this SNP contributes to a considerable proportion of T2DM risk. In the LIPGENE-SU.VI.MAX study we identified a number of novel associations between SNPs of genes involved in fatty acid and lipoprotein metabolism, including long-chain acyl CoA synthetase 1 (ACSL1), acetyl-CoA carboxylase β (ACC2), apolipoprotein A-I (APOA1) and apolipoprotein B (APOB) and lipoprotein lipase (LPL) with risk of the MetS or its phenotypes [81–84].

1.3.4 PRO-INFLAMMATORY CYTOKINES AND OTHER INFLAMMATORY MEDIATORS

Obesity is a chronic low-grade inflammatory state that is associated with increased risk for the MetS, diabetes and CVD. Indeed the NCEP ATP III identified pro-inflammatory status as being a key MetS characteristic [4]. Previous studies have demonstrated the influence of a variety of pro-inflammatory cytokine polymorphisms, including tumor necrosis factor alpha (TNFα) (rs1800629) and interleukin 6 (IL-6) (rs1800795 and rs1800797) in the risk of central obesity, diabetes and MetS phenotypes [85–89]. Although results have been inconsistent [90,91], which may be, in part, explained by the fact that as cytokines act in a complex network, single gene activity may not provide full insight into the role of cytokine genes in the MetS. In a recent MetS case control study, Phillips et al. examined the relationship between lymphotoxin-α (LTA), IL-6, and TNF-α genetic variants with MetS risk in the LIPGENE-SU.VI.MAX cohort [92]. TNF-α rs1800629 major G allele homozygotes and LTA rs915654 minor A allele carriers had 20%–40% higher MetS risk [92]. The combined effect of carrying both risk genotypes, which represent half of this population, further increased MetS risk probably attributable to their greater risk of abdominal obesity. Interestingly, this additive effect was further influenced by the presence of IL-6 rs1800797 G allele. Risk genotype carriers who were also homozygous for the IL-6 rs1800797 G allele were most at risk of developing the MetS (OR 2.10), fasting hyperglycaemia (OR 2.65) and abdominal obesity (OR 1.52).

Interestingly, IL-6 activates signal transducer and activator of transcription 3 (STAT3), a transcription factor released during the acute-phase response [93]. Common genetic variants at the STAT3 locus have recently been associated with increased risk of abdominal obesity [94]. As individual genetic variants generally confer only a moderate risk to a trait, analyzing multiple risk alleles simultaneously can be more informative and enhance predictive power, particularly in polygenic conditions. A significant genotype effect between the number of risk alleles and risk of abdominal obesity was identified in the LIPGENE-SU.VI.MAX study, with approximately a 2.5 fold increased risk in individuals carrying two or more risk alleles compared to individuals carrying 1 or no risk alleles [95]. Complement component 3 (C3) is another potential candidate gene with respect to inflammation and diabesity. Elevated concentrations of C3, a protein with a central role in the innate immune system, have been associated with insulin resistance, obesity, the MetS and diabetes [96–99]. In keeping with previous findings, a dose-dependent relationship between C3 concentrations and the number of MetS components were identified in the LIPGENE-SU.VI.MAX study [100]. Interestingly the increased

MetS risk conferred by elevated C3 concentrations (OR 3.11) was abolished in abdominally lean individuals. Furthermore, examination of associations between C3 polymorphisms and MetS risk demonstrated that common genetic variants at the C3 locus were associated with risk of the MetS and its phenotypes including dyslipidaemia, abdominal obesity and insulin sensitivity [101]. In recent years adipose tissue has been recognized as an important hormonally active organ. The adipocytokines adiponectin and leptin are thought to play important roles in cardiovascular and metabolic homeostasis. Polymorphisms in the adiponectin (ADIPOQ) gene and its receptors (ADIPOR1) have been associated with adiponectin levels, insulin resistance, and MetS phenotypes [102,103]. Homozygosity for the leptin receptor (LEPR) rs3790433 G allele was also associated with increased MetS risk, which may be accounted for by increased risk of elevated insulin concentrations and insulin resistance [104].

In summary, a major role for genetic susceptibility to obesity, insulin resistance, the MetS and T2DM has been identified. There is no doubt that GWAS in very large populations will rapidly advance our understanding of the genetic basis of these conditions. Indeed the discovery of associated variants in unsuspected genes and outside coding regions illustrates the ability of GWAS to provide potentially important clues into the pathogenesis of these common conditions. Notwithstanding the accumulating evidence regards genetic susceptibility to obesity, T2DM and the MetS, regardless of the approach used, gene-disease-association studies are fraught with difficulties including lack of replication, inadequate statistical power, multiple hypothesis testing, population stratification, publication bias and phenotypic differences. Despite numerous successful discoveries, the effect sizes are small and only explain a fraction of inter-individual variation. Considering the limited success in confirming polygenic variants for diabesity to date it is obvious that more polygenic variants await discovery. Furthermore, in order to further improve our understanding of how these genetic variants interact with environmental factors to modulate disease risk more extensive phenotypic data are required, especially in relation to diet and lifestyle factors.

1.4 NUTRIGENETICS AND METABOLIC DISEASE

According to the "thrifty genotype" hypothesis [105], evolutionary selection of genes (obesity genes) that were originally beneficial for energy storage which conferred a protective effect in times of food deprivation by promoting fat deposition, might, at least in part, explain the current escalating incidence of obesity in a modern Westernized environment of physical inactivity and excessive caloric consumption. This is supported by recent findings that obesity and T2DM reaches epidemic proportions in certain ethnic groups such as Pima Indians, Pacific Islanders, Afro-Americans and Hispanic-Americans [106]. Indeed, the much lower prevalence of metabolic disease observed in Pima Indians in Mexico compared to their counterparts in America illustrates that even in populations who are genetically predisposed to these conditions, their development is largely determined by environmental factors [107,108]. Such data add to the growing body of evidence which suggests that individual's phenotype represents a complex interaction between genetic and environmental factors over their life course. Nutrition is a key environmental factor in the pathogenesis and progression of common polygenic, diet-related metabolic conditions. The concept of gene-diet interaction describes dietary modulation of the effect of genotype on a particular phenotype (for example obesity, insulin resistance and dyslipidemia) and/or modulation of the effect of a dietary factor on a particular phenotype by a genetic variant. It is generally accepted that the effect of dietary changes on plasma biomarker concentrations differs significantly between individuals. Such inter-individual variability in response to dietary modification is, to a large extent, determined by genetic factors. As discussed already, dietary fat is an important environmental factor and current evidence to support the nutrigenetics concept with respect to obesity, the MetS and T2DM is largely based on data relating to dietary fat [15,32,33]. Other food components such as carbohydrate or fiber can play a role in the development of these conditions. Nevertheless these nutrigenetic investigations provide proof of concept. The PPARγ Pro12Ala polymorphism provides an excellent example of the relevance of gene-nutrient interactions in the development of obesity, the MetS and T2DM. In a prospective population-based cohort study, researchers demonstrated an important interaction between habitual dietary fat composition and this SNP [109]. As the ratio of total PUFA to SFA increased a significant inverse relationship was shown for both fasting insulin concentrations and BMI in the Ala carriers, suggesting that the potential protective effect of the Ala allele may be lost in the presence of a high SFA diet. More recently an inverse relationship between Ala frequency and T2DM prevalence has been observed in populations where energy from lipids exceeded 30% of the total energy intake [110]. In recent years a number of well-designed studies (LIPGENE MetS case dietary intervention, LIPGENE-SU.VI.MAX MetS case-control study, Genetics of Lipid Lowering Drugs and Diet Network (GOLDN)) have examined interactions between dietary and/or plasma fatty acid composition and genotype in these diet-related conditions. Some of the key findings from these studies are presented below and summarized in Table 1.

1.4.1 TCF7L2

Data from the Diabetes Prevention Program and the Diabetes Prevention Study indicate that lifestyle or environmental factors can modulate the genetic effects of TCF7L2 polymorphisms [47,54]. In the Diabetes Prevention Study overweight individuals with impaired glucose tolerance were allocated to an intensive diet and lifestyle intervention group or a control group. After a mean 4-year follow-up period they found that TCF7L2 polymorphisms were associated with the incidence of diabetes in the control group, but not the intervention group, suggesting that environmental factors can reduce genetic susceptibility even when risk genotypes are related to impaired insulin secretion. In the GOLDN study total dietary PUFA modulated the genetic effects of the TCF7L2 rs7903146 polymorphism on postprandial lipemia [56]. In the LIPGENE dietary intervention study, TCF7L2 SNPs were associated with plasma lipid concentrations, carbohydrate metabolism, blood pressure and inflammatory markers. Interactions with total SFA were noted. For example, among the rs11196224 major homozygotes elevated plasma SFA was associated with increased insulin resistance [112]. Similarly in the LIPGENE-SU.VI.MAX study rs7903146 was associated with increased MetS risk, arising from their impaired insulin sensitivity, greater insulin resistance, increased abdominal obesity and

TABLE 1: Gene-nutrient interactions which modulate metabolic syndrome risk.

Gene Locus	Polymorphism	Dietary Factors	Odds Ratio	Conclusions	Reference Number
Acetyl-CoA carboxylase β (ACC2)	rs4766587	n-6 PUFA	1.82	Risk conferred by the A allele was exacerbated among individuals with a high-fat intake (>35% energy) (OR 1.62), particularly a high intake (>5.5% energy) of n-6 PUFA (OR 1.82 for gene-nutrient interaction).	[83]
Apolipoprotein A-I (APOA1)	rs670	MUFA	1.57	MetS risk was exacerbated among the habitual high-fat consumers (>35% energy, OR 1.58). In addition a high MUFA fat increased MetS risk (OR 1.57).	[84]
Apolipoprotein B (APOB)	rs512535	MUFA	1.89	MetS risk was increased among the habitual high-fat consumers (>35% energy, OR 2.00). Moreover a high MUFA intake increased MetS risk (OR 1.89).	[84]
Complement component 3 (C3)	rs2250656 rs11569562	n-6 PUFA	2.2 (rs2250656) 0.32 (rs11569562)	AA genotype for rs2250656 had increased MetS risk relative to minor G subjects. GG genotype for rs11569562 had decreased MetS risk compared with minor A allele carriers.	[101]
Interleukin 1 beta (IL-1β)	6054 G	n-3 PUFA	3.29 (GG) 1.95 (GA)	Low n-3-PUFA intake (below the median) among the 6054 G allele carriers was associated with increased MetS risk (OR 3.29, for GG and OR 1.95, for GA) compared with the AA genotype.	[88]

Metabolic Syndrome (MetS); monounsaturated fatty acids (MUFA); polyunsaturated fatty acids (PUFA); saturated fatty acids (SFA) Adapted from Perez-Martinez et al. [31].

hypertension [60]. Interestingly dietary fat intake, recorded 7.5 years prior to MetS case/control selection, modulated the genetic influence on MetS risk. In particular high dietary SFA intake (≥15.5% of energy) accentuated the deleterious effects of rs7903146 on MetS risk, suggesting that the long-term effect of dietary fatty acid composition and consumption may have the potential to modify the genetic susceptibility of developing the MetS.

1.4.2 FTO

Limited cross-sectional analysis of the influence of dietary factors on BMI according to FTO rs9939609 genotype indicates that high-fat diets increase obesity risk [113,114]. However, these studies did not investigate specific effects of dietary fat type or fatty acid composition. Recent data from a study of 354 children identified an interaction between dietary SFA and the ratio of total PUFA to SFA and obesity associated with FTO rs9939609 [115]. In the LIPGENE-SU.VI.MAX study FTO rs9939609 was associated with increased risk of having a BMI in the overweight or obese category and of being abdominally obese [116]. Increased obesity risk was maintained over the 7.5 year follow-up period and while rs9939609 was not associated with MetS risk, risk of these obesity related measures was higher in the risk allele carrying MetS cases relative to their non-risk allele carrying counterparts. A novel finding in that study was that high habitual dietary SFA consumption (≥15.5% of energy) and low PUFA:SFA accentuates obesity risk in the A allele carriers but not in the TT homozygotes in this adult European population, suggesting that genetic predisposition to obesity may be modulated by dietary SFA intake. This may be particularly relevant to individuals with diet-related metabolic disease who are at increased cardiometabolic risk. Recent data from the GOLDN study also identified an interaction between rs9939609 and SFA intake, whereby homozygous participants in this American population had a higher BMI only when they had a high SFA intake (>mean) [117]. Interestingly, hypothalamic FTO over-expression has been shown to result in a 4-fold increase in STAT3 expression [118]. Given the importance of STAT3 in the leptin signaling pathway these data suggest a potential mechanism for mediating FTO's actions and potential modulation by SFA.

1.4.3 FATTY ACID AND LIPID METABOLISM GENES

Acetyl-CoA carboxylase β (ACC2) plays a key role in fatty acid synthesis and oxidation pathways, disturbance of which is associated with impaired insulin sensitivity and MetS. The LIPGENE-SU.VI.MAX study examined whether several ACC2 polymorphisms (rs2075263, rs2268387, rs2284685, rs2284689, rs2300453, rs3742023, rs3742026, rs4766587, and rs6606697) influence MetS risk, and whether dietary fatty acids modulate this interaction [83]. Minor A allele carriers of rs4766587 had increased MetS risk (OR 1.29) compared with the GG homozygotes, which may in part be explained by their increased BMI, abdominal obesity, and impaired insulin sensitivity. Dietary fat intake modulated MetS risk such that risk conferred by the A allele was exacerbated among individuals with a high-fat intake (>35% energy) (OR 1.62). Conversely MetS risk was abolished among individuals with a low-fat intake. Examination of individual fatty acid classes identified a gene-nutrient interaction with PUFA intake, whereby A allele carriers with high PUFA intake (>5.5% energy) had increased MetS risk (OR 1.53). This gene-nutrient interaction was reflected by both n-6 (OR 1.80) and n-3 PUFA (OR 1.75). Importantly, some of these findings were replicated in an independent cohort (LIPGENE MetS only dietary intervention cohort). Thus, the genetic associations with increased BMI, body-weight, waist circumference, and insulin resistance were confirmed. Consistent with the original findings, these genetic differences persisted in the high-fat but not among the low-fat consumers. In summary, genetic variation at the ACC2 gene locus influences MetS risk, which was modulated by dietary fat.

Long-chain acyl CoA synthetase 1 (ACSL1) is important for mitochondrial beta-oxidation of long chain fatty acids and plays an important role in fatty acid metabolism. Disturbance of these pathways may result in insulin resistance and dyslipidemia, key MetS features [119–121]. Examination of the relationship between ACSL1 polymorphisms (rs4862417, rs6552828, rs13120078, rs9997745, and rs12503643) and MetS risk and potential interactions with dietary fat conducted in the LIPGENE-SU. VI.MAX study [82]. Subjects with the GG genotype for rs9997745 SNP had increased MetS risk (OR 1.90), displayed elevated fasting glucose and insulin concentrations and increased insulin resistance compared with subjects carrying the A allele. Moreover MetS risk was modulated by dietary fat, whereby the risk conferred by GG homozygosity was effectively abolished among those subjects consuming either a low-fat or when total dietary PUFA intake was in the top 50th percentile. Examination of the HAPMAP data for this SNP indicates allele frequency differences between ethnic groups. Whereas the allele frequency in the LIPGENE-SU. VI.MAX study is not far from that in the European HAPMAP population, the opposite is true in Africans where the G allele is the minor allele. It is quite interesting to note that this SNP is not polymorphic in Asians. This implies that in an Asian population everybody will carry the "risk" allele in the presence of a total PUFA-poor diet (which has been the traditional diet in Asian countries) and may be particularly at risk of MetS. In conclusion, ACSL1 rs9997745 influences MetS risk, most likely via disturbances in fatty acid metabolism, which was modulated by dietary fat consumption.

APOA1 is the major protein component of HDL and also is an activator of the enzyme lecithin-cholesterol acyltransferase (LCAT), a key component of reverse cholesterol transport. In contrast, plasma APOB, the main component of LDL, is essential for the assembly and secretion of the triglyceride-rich lipoproteins. Several SNPs at these loci have been proposed to influence MetS risk. In the LIPGENE-SU.VI.MAX study, the ApoB rs512535 and ApoA1 rs670 major G allele homozygotes had increased MetS risk (OR 1.65 and OR 1.42, respectively), which may be explained by their increased abdominal obesity and impaired insulin sensitivity but not dyslipidemia [84]. These associations derived primarily from the male GG homozygotes (ApoB rs512535 OR 1.92 and ApoA1 rs670 OR 1.50). On the other hand, MetS risk was exacerbated among the habitual high-fat consumers (>35% energy) (ApoB rs512535 OR 2.00 and OR 1.58 for ApoA1 rs670). In addition a high dietary MUFA intake increased MetS risk (OR 1.89 and OR 1.57 for ApoB rs512535 and ApoA1 rs670, respectively). MetS risk was diminished among the habitual low-fat consumers (<35% energy). In summary ApoB and ApoA1 polymorphisms may influence MetS risk. Modulation of these associations by gender and dietary fat suggests novel gene-gender-diet interactions. As already alluded to in section 3.4, the postprandial period, which is the most metabolically abnormal, is of particular importance as humans spend most of their lives in this phase. During which circulating lipoproteins are involved in a cascade of changes to their composition and concentration. It is generally accepted that the impact of dietary fat on postprandial lipoprotein response differs between individuals, and genetic factors are thought to be one of the key determinants of such inter-individual variability. Nutrigenetics of postprandial lipid metabolism including evidence from human dietary interventions has recently been reviewed [122,123].

1.4.4 PRO-INFLAMMATORY CYTOKINES, ADIPOCYTOKINES AND INFLAMMATORY MEDIATORS

As already eluded to, inflammation plays a key role in insulin resistance. We have shown that total plasma PUFA/SFA levels modified the observed additive genetic effects of IL-6, TNFα and LTA [92]. When stratified according to median plasma PUFA/SFA levels, MetS risk was four-fold higher in the 3 SNP risk genotype carriers with the lowest PUFA/SFA levels compared to non-carriers and was thought to be driven by the SFA content, with high SFA levels alone accounting for 5-fold increased MetS risk. A low PUFA/SFA ratio also exacerbated their increased risk of several phenotypes (abdominal obesity, fasting hyperglycaemia, hypertension and pro-inflammatory profile). Interestingly, when risk genotype carriers with the lowest PUFA/SFA levels were compared with their risk genotype carriers with the highest PUFA/SFA levels significant improvements to their metabolic profile were noted. Most importantly a high PUFA/SFA ratio attenuated genetic predisposition to MetS risk. Moreover, risk genotype carriers with the highest PUFA/SFA levels had reduced pro-inflammatory status, lower TAG levels and HOMA-IR values than risk genotype carriers with the lowest PUFA/SFA levels. These findings support current guidelines to reduce dietary SFA intake and increase PUFA consumption.

Investigation of the effect of ADIPOQ SNPs, rs266729 and rs17300539, on metabolic-related traits, and their modulation by dietary fat in white Americans in the GOLDN study revealed significant interaction with MUFA intake [124]. In subjects with high MUFA intake (>median) rs17300539 A allele carriers had lower BMI and decreased obesity risk. Dietary intervention analysis has demonstrated that CC homozygotes for rs266729 were less insulin resistant after consumption of MUFA and carbohydrate rich diets compared to the SFA rich diet [125]. Furthermore in the LIPGENE dietary intervention study two SNPs (rs266729 in ADI-POQ, and rs10920533 in ADIPOR1) interacted with plasma SFAs to alter insulin and HOMA-IR [126]. This study demonstrated that a reduction in plasma SFA decreased insulin resistance in carriers of the minor allele of rs266729 ADIPOQ and rs10920533 ADIPOR1. Personalised nutrition advice based on this data would recommend a decrease in SFA consumption in the diet of MetS subjects carrying the minor allele of rs266729 ADIPOQ and/or rs10920533 ADIPOR1.

The LEPR is an adipocytokine receptor that is involved in the regulation of fat metabolism and has been associated with insulin resistance, a key feature of MetS. In the LIPGENE-SU.VI.MAX study GG homozygotes of rs3790433 SNP at the LEPR gene had increased MetS risk compared with the minor A allele carriers (OR 1.65), which may be accounted for by their increased risk of elevated insulin concentrations and insulin resistance [104]. Low (less than median) plasma n-3 and high n-6 PUFA status exacerbated the genetic risk conferred by GG homozygosity to hyperinsulinemia (OR 2.92–2.94) and insulin resistance (OR 3.40–3.47). Interestingly, these associations were abolished against a high n-3 or low n-6 PUFA background. Importantly, these findings were replicated in an independent MetS cohort. Homozygosity for the LEPR rs3790433 G allele was associated with insulin resistance, which may predispose to increased MetS risk. Of note, significant improvements to indices of insulin sensitivity and insulin resistance by the GG homozygotes were identified following a 12 weeks low-fat dietary intervention supplemented with n-3 long chain (LC) PUFA. These data suggest that genetic influences associated with this LEPR polymorphism may be selectively modulated by n-3 LC-PUFA. In conclusion these data from the LIPGENE project suggest novel gene-nutrient interactions whereby the deleterious effects associated with LEPR rs3790433 GG homozygosity were more evident against a background of low n-3 or high n-6 PUFA, and to a lesser extent with high SFA status. As LEPR rs3790433 GG homozygotes appear to be sensitive to plasma and dietary fatty acid composition, these individuals may derive the most benefit from dietary manipulation and current guidelines to reduce dietary SFA and increase n-3 PUFA intake.

The LIPGENE-SU.VI.MAX study also identified a gene-nutrient interaction between STAT3 polymorphisms with SFA. High dietary SFA intake ($\geq$15.5% of energy) modulated the genetic association between STAT3 polymorphisms with obesity [94]; carriers of more than 2 risk alleles with the highest SFA consumption further increased their risk of abdominal obesity by 32% compared to those carrying one or fewer risk alleles. This data suggests that individuals with certain STAT3 genotypes are more sensitive to SFA and that these individuals may derive the most benefit from dietary manipulation and current guidelines to reduce dietary SFA intake. While the mechanisms underlying these findings are unknown, it is possible that toll-like receptor-4, the molecular link between fatty acids, obesity, inflammation and insulin resistance [127] may play a role. Interestingly, hypothalamic FTO over-expression has been shown to result in a 4-fold increase in STAT3 expression [118]. Given the importance of STAT3 in the leptin signaling pathway these data suggest a potential mechanism for mediating FTO's actions and potential modulation by SFA.

The increased MetS risk associated with the C3 rs2250656 A allele in the LIPGENE-SU.VI.MAX study may be explained by their classic MetS profile and raised inflammatory status [101]. Interestingly plasma PUFA modified MetS risk whereby the combination of carrying two "risk" A alleles and having low n-6 or total PUFA (below the median) exacerbated MetS risk, suggesting that these individuals, who represent approximately half of the population and who are genetically predisposed to the MetS, are also more sensitive to PUFA. Similarly reduced MetS risk associated with the rs11569562 polymorphism was subject to a significant effect modification by PUFA, with the greatest protection from the MetS being achieved by GG homozygotes with the highest total PUFA status [101]. Likewise GG homozygotes with the highest LC n-3 PUFAs had the lowest risk of hypertriglyceridaemia. In keeping with these findings, in the LIPGENE dietary intervention study the "protective" rs11569562 GG genotype was associated with enhanced insulin sensitivity and these individuals were more responsive to LC n-3 PUFA, compared to the A allele carriers. Following a 12 weeks low-fat (28% energy), high-complex carbohydrate diet intervention supplemented with 1.24 g/day LC n-3 PUFA, GG homozygotes displayed beneficial changes to their lipid profile (10% reduction in NEFA, 8% non-significant reduction in TAG, 5% reduction in total cholesterol and 17% reduction in LDL concentrations), compared to the A allele carriers. No changes were observed between genotypes when subjects on the same diet received a 1 g/day high oleic acid control supplement. In addition, the "at risk" rs2250656 A allele carriers had reduced insulin sensitivity and increased BMI, relative to the GG homozygotes. Again, genetic influences were modified by LC n-3 PUFA supplementation, whereby the A allele carriers achieved a 35% improvement in insulin sensitivity following intervention whereas no changes were noted between genotypes following oleic acid supplementation. Interestingly PUFAs are ligands of FXR [128], a nuclear receptor which regulates C3 expression [129]. Thus it is possible that alteration of C3 expression via modulation of FXR is a potential mechanism by which gene-nutrient interaction of C3 genotype and dietary PUFAs could influence C3 levels and thus MetS risk. Although speculative, it may be worthy of further investigation to help elucidate the molecular basis of such gene-nutrient interactions and their impact on markers of inflammation and insulin resistance.

1.5 NUTRIGENETICS AND PERSONALISED NUTRITION

Phenylketonuria (PKU) was the first genetic disease in which a gene-diet interaction was described. This condition is a good example of how a single nutrient can be used to manage genetic predisposition to a monogenic disease. People with PKU lack the enzyme required to metabolize phenylalanine, an essential amino acid found in dairy, meat, fish, nuts and pulses, with the result that dangerous levels of phenylpyruvic acid may build up which are toxic to the brain Thus, individuals with PKU need to stick to a low phenylalanine diet for life to avoid PKU symptoms. Coeliac disease, an inflammatory condition which results from intolerance to dietary gluten, is an example of how personalised nutrition can potentially work. High concordance rates from twin studies indicate a strong genetic influence, but it seems that carrying certain genes reveals a genetic predisposition to dietary factors rather than disease development [130]. Obesity is another example of how nutrigenetics can be used to personalise an individual's diet with a view to improving long term weight management. An interesting study by Arkadianos et al. [131] examined weight loss and weight loss maintenance following a personalised calorie-controlled diet and exercise program, based on 24 SNPs in 19 genes involved in metabolism, in subjects with a history of weight loss failure compared to control subjects who just received generic dietary and exercise advice. This study showed that the nutrigenetically tailored diet achieved better compliance, improvements in glucose levels and BMI reduction not only during the weight loss period but importantly also over the following year. Another personalised dietary intervention, based on 4 SNPs in four genes, with stratification of overweight/obese and control subjects into diet or diet and exercise groups, demonstrated that individuals were slow to take optimal health advice, particularly in the combined diet and exercise group [132].

While this was a small study based on a limited number of genetic variants it raises the issue of negative consumer opinion, which poses a potential barrier to the application of nutrigenetic based intervention. A recent pan-European study investigated the attitudes of consumers towards genetic testing and personalised nutrition [133]. The results of this study were encouraging, with 66% of respondents willing to undergo genetic testing and 27% willing to adhere to a personalised diet [133]. Interestingly individuals with MetS and T2DM related health conditions were particularly positive toward nutrigenetic intervention. These findings are encouraging for the future application of genome-customized diets for obesity, MetS and T2DM prevention and therapy following personalised approaches. However, as success or failure of any new technology is consumer driven, consumer research in the application of personalised nutrition is essential.

1.6 CONCLUSIONS

In this review, some recent novel nutrigenetic data in the context of metabolic disease have been presented, which suggest that certain nutrients, in particular dietary fatty acids, may have the potential to modify the genetic predisposition to these diet-related conditions. While this review has focused on dietary fat, more holistic methods which incorporate an individuals' diet or dietary patterns, rather than selecting individual dietary components, need to be developed to advance the state of the art. Moreover, other modifiable environmental factors which interact with the diet should be considered in gene-environment studies (i.e., physical activity, alcohol intake, smoking status) across a range of metabolic conditions. Nevertheless current data provides proof of concept. The shift towards "personalised" nutritional advice is an attractive proposition. Nutrigenetics has the potential to change diet-related disease prevention and therapy. While recent advances in high-throughput genetic analysis have improved our understanding of the contribution of genetics to metabolic health and disease, the molecular mechanisms underlying many of these gene-nutrient interactions remain unclear. Functional studies are needed to ascertain their biological significance and potential clinical utility. Nutrigenetics is just one piece in a very complex jigsaw, which needs to move forward with nutritional science in order to translate observational findings into molecular mechanisms. The combined application of nutritional and genetic epidemiology with metabolite and molecular profiling at the gene, transcriptome, proteome and metabolome level to define an individuals' metabotype will be crucial in this regard. Such concerted actions, using larger study cohorts and collaborative research efforts across different disciplines may lead to the identification of sensitive/responsive metabotypes (i.e., modifiable by dietary fatty acids or other nutrients). The challenge for current and future research is validation and translation of nutrigenetic findings, which may provide the basis for successful personalised and public health approaches for metabolic disease prevention.

REFERENCES

1. Moller, D.E.; Kaufman, K.D. Metabolic syndrome: A clinical and molecular perspective. Annu. Rev. Med. 2005, 56, 45–62.
2. Alberti, K.G.; Zimmet, P.Z. Definition, diagnosis and classification of diabetes mellitus and its complications. Part 1: Diagnosis and classification of diabetes mellitus provisional report of a who consultation. Diabet. Med. 1998, 15, 539–553.
3. Balkau, B.; Charles, M.A. Comment on the provisional report from the WHO consultation. European Group for the Study of Insulin Resistance (EGIR). Diabet. Med. 1999, 16, 442–443.
4. National Cholesterol Education Program; National Heart, Lung, and Blood Institute; National Institutes of Health. Third Report of the National Cholesterol Education Program (NCEP) Expert Panel on Detection, Evaluation, and Treatment of High

Blood Cholesterol in Adults (Adult Treatment Panel III) final report. Circulation 2002, 106, 3143–3421.

5. Alberti, K.G.; Zimmet, P.; Shaw, J. The metabolic syndrome—A new worldwide definition. Lancet 2005, 366, 1059–1062.

6. Alberti, K.G.; Eckel, R.H.; Grundy, S.M.; Zimmet, P.Z.; Cleeman, J.I.; Donato, K.A.; Fruchart, J.C.; James, W.P.; Loria, C.M.; Smith, S.C., Jr. Harmonizing the metabolic syndrome: A joint interim statement of the international diabetes federation task force on epidemiology and prevention; national heart, lung, and blood institute; american heart association; world heart federation; international atherosclerosis society; and international association for the study of obesity. Circulation 2009, 120, 1640–1645.

7. Ford, E.S.; Li, C.; Zhao, G. Prevalence and correlates of metabolic syndrome based on a harmonious definition among adults in the US. J. Diabetes 2010, 2, 180–193.

8. Flier, J.S. Obesity wars: Molecular progress confronts an expanding epidemic. Cell 2004, 116, 337–350.

9. Anderson, J.W.; Kendall, C.W.; Jenkins, D.J. Importance of weight management in type 2 diabetes: Review with meta-analysis of clinical studies. J. Am. Coll. Nutr. 2003, 22, 331–339.

10. Shaw, J.E.; Sicree, R.A.; Zimmet, P.Z. Global estimates of the prevalence of diabetes for 2010 and 2030. Diabetes Res. Clin. Pract. 2010, 87, 4–14.

11. Bellia, A.; Giardina, E.; Lauro, D.; Tesauro, M.; Di Fede, G.; Cusumano, G.; Federici, M.; Rini, G.B.; Novelli, G.; Lauro, R.; et al. "The linosa study": Epidemiological and heritability data of the metabolic syndrome in a caucasian genetic isolate. Nutr. Metab. Cardiovasc. Dis. 2009, 19, 455–461.

12. Henneman, P.; Aulchenko, Y.S.; Frants, R.R.; van Dijk, K.W.; Oostra, B.A.; van Duijn, C.M. Prevalence and heritability of the metabolic syndrome and its individual components in a dutch isolate: The erasmus rucphen family study. J. Med. Genet. 2008, 45, 572–577.

13. Eckel, R.H.; Grundy, S.M.; Zimmet, P.Z. The metabolic syndrome. Lancet 2005, 365, 1415–1428.

14. Lottenberg, A.M.; Afonso Mda, S.; Lavrador, M.S.; Machado, R.M.; Nakandakare, E.R. The role of dietary fatty acids in the pathology of metabolic syndrome. J. Nutr. Biochem. 2012, 23, 1027–1040.

15. Phillips, C.; Lopez-Miranda, J.; Perez-Jimenez, F.; McManus, R.; Roche, H.M. Genetic and nutrient determinants of the metabolic syndrome. Curr. Opin. Cardiol. 2006, 21, 185–193.

16. Szabo de Edelenyi, F.; Goumidi, L.; Bertrais, S.; Phillips, C.; Macmanus, R.; Roche, H.; Planells, R.; Lairon, D. Prediction of the metabolic syndrome status based on dietary and genetic parameters, using random forest. Genes Nutr. 2008, 3, 173–176.

17. Vessby, B. Dietary fat, fatty acid composition in plasma and the metabolic syndrome. Curr. Opin. Lipidol. 2003, 14, 15–19.

18. Warensjo, E.; Sundstrom, J.; Lind, L.; Vessby, B. Factor analysis of fatty acids in serum lipids as a measure of dietary fat quality in relation to the metabolic syndrome in men. Am. J. Clin. Nutr. 2006, 84, 442–448.

19. Hu, F.B.; van Dam, R.M.; Liu, S. Diet and risk of Type II diabetes: The role of types of fat and carbohydrate. Diabetologia 2001, 44, 805–817.

20. Melanson, E.L.; Astrup, A.; Donahoo, W.T. The relationship between dietary fat and fatty acid intake and body weight, diabetes, and the metabolic syndrome. Ann. Nutr. Metab. 2009, 55, 229–243.
21. Meyer, K.A.; Kushi, L.H.; Jacobs, D.R., Jr.; Folsom, A.R. Dietary fat and incidence of type 2 diabetes in older iowa women. Diabetes Care 2001, 24, 1528–1535.
22. Perez-Jimenez, F.; Lopez-Miranda, J.; Pinillos, M.D.; Gomez, P.; Paz-Rojas, E.; Montilla, P.; Marin, C.; Velasco, M.J.; Blanco-Molina, A.; Jimenez Pereperez, J.A.; et al. A mediterranean and a high-carbohydrate diet improve glucose metabolism in healthy young persons. Diabetologia 2001, 44, 2038–2043.
23. Tierney, A.C.; McMonagle, J.; Shaw, D.I.; Gulseth, H.L.; Helal, O.; Saris, W.H.; Paniagua, J.A.; Golabek-Leszczynska, I.; Defoort, C.; Williams, C.M.; et al. Effects of dietary fat modification on insulin sensitivity and on other risk factors of the metabolic syndrome—LIPGENE: A european randomized dietary intervention study. Int. J. Obes. (Lond.) 2011, 35, 800–809.
24. Due, A.; Larsen, T.M.; Mu, H.; Hermansen, K.; Stender, S.; Astrup, A. Comparison of 3 ad libitum diets for weight-loss maintenance, risk of cardiovascular disease, and diabetes: A 6-mo randomized, controlled trial. Am. J. Clin. Nutr. 2008, 88, 1232–1241.
25. Due, A.; Larsen, T.M.; Hermansen, K.; Stender, S.; Holst, J.J.; Toubro, S.; Martinussen, T.; Astrup, A. Comparison of the effects on insulin resistance and glucose tolerance of 6-mo high-monounsaturated-fat, low-fat, and control diets. Am. J. Clin. Nutr. 2008, 87, 855–862.
26. Vessby, B.; Unsitupa, M.; Hermansen, K.; Riccardi, G.; Rivellese, A.A.; Tapsell, L.C.; Nalsen, C.; Berglund, L.; Louheranta, A.; Rasmussen, B.M.; et al. Substituting dietary saturated for monounsaturated fat impairs insulin sensitivity in healthy men and women: The kanwu study. Diabetologia 2001, 44, 312–319.
27. Oliver, E.; McGillicuddy, F.C.; Harford, K.A.; Reynolds, C.M.; Phillips, C.M.; Ferguson, J.F.; Roche, H.M. Docosahexaenoic acid attenuates macrophage-induced inflammation and improves insulin sensitivity in adipocytes-specific differential effects between LC n-3 PUFA. J. Nutr. Biochem. 2011, 23, 1192–1200.
28. Storlien, L.H.; Baur, L.A.; Kriketos, A.D.; Pan, D.A.; Cooney, G.J.; Jenkins, A.B.; Calvert, G.D.; Campbell, L.V. Dietary fats and insulin action. Diabetologia 1996, 39, 621–631.
29. Storlien, L.H.; Jenkins, A.B.; Chisholm, D.J.; Pascoe, W.S.; Khouri, S.; Kraegen, E.W. Influence of dietary fat composition on development of insulin resistance in rats. Relationship to muscle triglyceride and omega-3 fatty acids in muscle phospholipid. Diabetes 1991, 40, 280–289.
30. Willett, W. Nutritional Epidemiology, 2nd ed.; Oxford University Press: New York, NY, USA, 1998.
31. Perez-Martinez, P.; Phillips, C.M.; Delgado-Lista, J.; Garcia-Rios, A.; Lopez-Miranda; Francisco Perez-Jimenez, J. Nutrigenetics, metabolic syndrome risk and personalized nutrition. Curr. Vasc. Pharmacol. 2012, in press.
32. Phillips, C.M.; Tierney, A.C.; Roche, H.M. Gene-nutrient interactions in the metabolic syndrome. J. Nutrigenet. Nutrigenomics 2008, 1, 136–151.
33. Roche, H.M.; Phillips, C.; Gibney, M.J. The metabolic syndrome: The crossroads of diet and genetics. Proc. Nutr. Soc. 2005, 64, 371–377.

34. Allison, D.B.; Faith, M.S.; Nathan, J.S. Risch's lambda values for human obesity. Int. J. Obes. Relat. Metab. Disord. 1996, 20, 990–999.

35. Maes, H.H.; Neale, M.C.; Eaves, L.J. Genetic and environmental factors in relative body weight and human adiposity. Behav. Genet. 1997, 27, 325–351.

36. Stunkard, A.J.; Sorensen, T.I.; Hanis, C.; Teasdale, T.W.; Chakraborty, R.; Schull, W.J.; Schulsinger, F. An adoption study of human obesity. N. Engl. J. Med. 1986, 314, 193–198.

37. Bouchard, C.; Perusse, L.; Leblanc, C.; Tremblay, A.; Theriault, G. Inheritance of the amount and distribution of human body fat. Int. J. Obes. 1988, 12, 205–215.

38. Vogler, G.P.; Sorensen, T.I.; Stunkard, A.J.; Srinivasan, M.R.; Rao, D.C. Influences of genes and shared family environment on adult body mass index assessed in an adoption study by a comprehensive path model. Int. J. Obes. Relat. Metab. Disord. 1995, 19, 40–45.

39. Florez, J.C.; Hirschhorn, J.; Altshuler, D. The inherited basis of diabetes mellitus: Implications for the genetic analysis of complex traits. Annu. Rev. Genomics. Hum. Genet. 2003, 4, 257–291.

40. Gloyn, A.L. The search for type 2 diabetes genes. Ageing Res. Rev. 2003, 2, 111–127.

41. Andreasen, C.H.; Andersen, G. Gene-environment interactions and obesity—Further aspects of genomewide association studies. Nutrition 2009, 25, 998–1003.

42. Horikawa, Y.; Oda, N.; Cox, N.J.; Li, X.; Orho-Melander, M.; Hara, M.; Hinokio, Y.; Lindner, T.H.; Mashima, H.; Schwarz, P.E.; et al. Genetic variation in the gene encoding calpain-10 is associated with type 2 diabetes mellitus. Nat. Genet. 2000, 26, 163–175.

43. Saez, M.E.; Gonzalez-Sanchez, J.L.; Ramirez-Lorca, R.; Martinez-Larrad, M.T.; Zabena, C.; Gonzalez, A.; Moron, F.J.; Ruiz, A.; Serrano-Rios, M. The CAPN10 gene is associated with insulin resistance phenotypes in the Spanish population. PLoS One 2008, 3, e2953.

44. Carlsson, E.; Fredriksson, J.; Groop, L.; Ridderstrale, M. Variation in the calpain-10 gene is associated with elevated triglyceride levels and reduced adipose tissue messenger ribonucleic acid expression in obese Swedish subjects. J. Clin. Endocrinol. Metab. 2004, 89, 3601–3605.

45. Garant, M.J.; Kao, W.H.; Brancati, F.; Coresh, J.; Rami, T.M.; Hanis, C.L.; Boerwinkle, E.; Shuldiner, A.R. SNP43 of CAPN10 and the risk of type 2 diabetes in African-Americans: The atherosclerosis risk in communities study. Diabetes 2002, 51, 231–237.

46. Shima, Y.; Nakanishi, K.; Odawara, M.; Kobayashi, T.; Ohta, H. Association of the SNP-19 genotype 22 in the calpain-10 gene with elevated body mass index and hemoglobin A1c levels in Japanese. Clin. Chim. Acta 2003, 336, 89–96.

47. Florez, J.C.; Jablonski, K.A.; Bayley, N.; Pollin, T.I.; de Bakker, P.I.; Shuldiner, A.R.; Knowler, W.C.; Nathan, D.M.; Altshuler, D. TCF7L2 polymorphisms and progression to diabetes in the Diabetes Prevention Program. N. Engl. J. Med. 2006, 355, 241–250.

48. Grant, S.F.; Thorleifsson, G.; Reynisdottir, I.; Benediktsson, R.; Manolescu, A.; Sainz, J.; Helgason, A.; Stefansson, H.; Emilsson, V.; Helgadottir, A.; et al. Variant

of transcription factor 7-like 2 (TCF7L2) gene confers risk of type 2 diabetes. Nat. Genet. 2006, 38, 320–323.

49. Cauchi, S.; El Achhab, Y.; Choquet, H.; Dina, C.; Krempler, F.; Weitgasser, R.; Nejjari, C.; Patsch, W.; Chikri, M.; Meyre, D.; et al. TCF7L2 is reproducibly associated with type 2 diabetes in various ethnic groups: A global meta-analysis. J. Mol. Med. (Berl.) 2007, 85, 777–782.

50. Cauchi, S.; Meyre, D.; Dina, C.; Choquet, H.; Samson, C.; Gallina, S.; Balkau, B.; Charpentier, G.; Pattou, F.; Stetsyuk, V.; et al. Transcription factor TCF7L2 genetic study in the french population: Expression in human beta-cells and adipose tissue and strong association with type 2 diabetes. Diabetes 2006, 55, 2903–2908.

51. Chandak, G.R.; Janipalli, C.S.; Bhaskar, S.; Kulkarni, S.R.; Mohankrishna, P.; Hattersley, A.T.; Frayling, T.M.; Yajnik, C.S. Common variants in the TCF7L2 gene are strongly associated with type 2 diabetes mellitus in the indian population. Diabetologia 2007, 50, 63–67.

52. Hayashi, T.; Iwamoto, Y.; Kaku, K.; Hirose, H.; Maeda, S. Replication study for the association of TCF7L2 with susceptibility to type 2 diabetes in a japanese population. Diabetologia 2007, 50, 980–984.

53. Humphries, S.E.; Gable, D.; Cooper, J.A.; Ireland, H.; Stephens, J.W.; Hurel, S.J.; Li, K.W.; Palmen, J.; Miller, M.A.; Cappuccio, F.P.; et al. Common variants in the TCF7L2 gene and predisposition to type 2 diabetes in UK European whites, Indian Asians and Afro-Caribbean men and women. J. Mol. Med. (Berl.) 2006, 84, 1005–1014.

54. Wang, J.; Kuusisto, J.; Vanttinen, M.; Kuulasmaa, T.; Lindstrom, J.; Tuomilehto, J.; Uusitupa, M.; Laakso, M. Variants of transcription factor 7-like 2 (TCF7L2) gene predict conversion to type 2 diabetes in the finnish diabetes prevention study and are associated with impaired glucose regulation and impaired insulin secretion. Diabetologia 2007, 50, 1192–1200.

55. Melzer, D.; Murray, A.; Hurst, A.J.; Weedon, M.N.; Bandinelli, S.; Corsi, A.M.; Ferrucci, L.; Paolisso, G.; Guralnik, J.M.; Frayling, T.M. Effects of the diabetes linked TCF7L2 polymorphism in a representative older population. BMC Med. 2006, 4, 34.

56. Warodomwichit, D.; Arnett, D.K.; Kabagambe, E.K.; Tsai, M.Y.; Hixson, J.E.; Straka, R.J.; Province, M.; An, P.; Lai, C.Q.; Borecki, I.; et al. Polyunsaturated fatty acids modulate the effect of TCF7L2 gene variants on postprandial lipemia. J. Nutr. 2009, 139, 439–446.

57. Sjogren, M.; Lyssenko, V.; Jonsson, A.; Berglund, G.; Nilsson, P.; Groop, L.; Orho-Melander, M. The search for putative unifying genetic factors for components of the metabolic syndrome. Diabetologia 2008, 51, 2242–2251.

58. Marzi, C.; Huth, C.; Kolz, M.; Grallert, H.; Meisinger, C.; Wichmann, H.E.; Rathmann, W.; Herder, C.; Illig, T. Variants of the transcription factor 7-like 2 gene (TCF7L2) are strongly associated with type 2 diabetes but not with the metabolic syndrome in the MONICA/KORA surveys. Horm. Metab. Res. 2007, 39, 46–52.

59. Saadi, H.; Nagelkerke, N.; Carruthers, S.G.; Benedict, S.; Abdulkhalek, S.; Reed, R.; Lukic, M.; Nicholls, M.G. Association of TCF7L2 polymorphism with diabetes mellitus, metabolic syndrome, and markers of beta cell function and insulin resistance in a population-based sample of emirati subjects. Diabetes Res. Clin. Pract. 2008, 80, 392–398.

60. Phillips, C.M.; Goumidi, L.; Bertrais, S.; Field, M.R.; McManus, R.; Hercberg, S.; Lairon, D.; Planells, R.; Roche, H.M. Dietary saturated fat, gender and genetic variation at the TCF7L2 locus predict the development of metabolic syndrome. J. Nutr. Biochem. 2012, 23, 239–244.

61. Povel, C.M.; Boer, J.M.; Reiling, E.; Feskens, E.J. Genetic variants and the metabolic syndrome: A systematic review. Obes. Rev. 2011, 12, 952–967.

62. Frayling, T.M.; Timpson, N.J.; Weedon, M.N.; Zeggini, E.; Freathy, R.M.; Lindgren, C.M.; Perry, J.R.; Elliott, K.S.; Lango, H.; Rayner, N.W.; et al. A common variant in the FTO gene is associated with body mass index and predisposes to childhood and adult obesity. Science 2007, 316, 889–894.

63. Haupt, A.; Thamer, C.; Staiger, H.; Tschritter, O.; Kirchhoff, K.; Machicao, F.; Haring, H.U.; Stefan, N.; Fritsche, A. Variation in the FTO gene influences food intake but not energy expenditure. Exp. Clin. Endocrinol. Diabetes 2009, 117, 194–197.

64. Speakman, J.R.; Rance, K.A.; Johnstone, A.M. Polymorphisms of the fto gene are associated with variation in energy intake, but not energy expenditure. Obesity (Silver Spring) 2008, 16, 1961–1965.

65. Tanofsky-Kraff, M.; Han, J.C.; Anandalingam, K.; Shomaker, L.B.; Columbo, K.M.; Wolkoff, L.E.; Kozlosky, M.; Elliott, C.; Ranzenhofer, L.M.; Roza, C.A.; et al. The FTO gene rs9939609 obesity-risk allele and loss of control over eating. Am. J. Clin. Nutr. 2009, 90, 1483–1488.

66. Dina, C.; Meyre, D.; Gallina, S.; Durand, E.; Korner, A.; Jacobson, P.; Carlsson, L.M.; Kiess, W.; Vatin, V.; Lecoeur, C.; et al. Variation in FTO contributes to childhood obesity and severe adult obesity. Nat. Genet. 2007, 39, 724–726.

67. Hinney, A.; Nguyen, T.T.; Scherag, A.; Friedel, S.; Bronner, G.; Muller, T.D.; Grallert, H.; Illig, T.; Wichmann, H.E.; Rief, W.; et al. Genome wide association (GWA) study for early onset extreme obesity supports the role of fat mass and obesity associated gene (FTO) variants. PLoS One 2007, 2, e1361.

68. Hunt, S.C.; Stone, S.; Xin, Y.; Scherer, C.A.; Magness, C.L.; Iadonato, S.P.; Hopkins, P.N.; Adams, T.D. Association of the FTO gene with BMI. Obesity (Silver Spring) 2008, 16, 902–904.

69. Thorleifsson, G.; Walters, G.B.; Gudbjartsson, D.F.; Steinthorsdottir, V.; Sulem, P.; Helgadottir, A.; Styrkarsdottir, U.; Gretarsdottir, S.; Thorlacius, S.; Jonsdottir, I.; et al. Genome-wide association yields new sequence variants at seven loci that associate with measures of obesity. Nat. Genet. 2009, 41, 18–24.

70. Willer, C.J.; Speliotes, E.K.; Loos, R.J.; Li, S.; Lindgren, C.M.; Heid, I.M.; Berndt, S.I.; Elliott, A.L.; Jackson, A.U.; Lamina, C.; et al. Six new loci associated with body mass index highlight a neuronal influence on body weight regulation. Nat. Genet. 2009, 41, 25–34.

71. Dupuis, J.; Langenberg, C.; Prokopenko, I.; Saxena, R.; Soranzo, N.; Jackson, A.U.; Wheeler, E.; Glazer, N.L.; Bouatia-Naji, N.; Gloyn, A.L.; et al. New genetic loci implicated in fasting glucose homeostasis and their impact on type 2 diabetes risk. Nat. Genet. 2010, 42, 105–116.

72. Manning, A.K.; Hivert, M.F.; Scott, R.A.; Grimsby, J.L.; Bouatia-Naji, N.; Chen, H.; Rybin, D.; Liu, C.T.; Bielak, L.F.; Prokopenko, I.; et al. A genome-wide approach accounting for body mass index identifies genetic variants influencing fasting glycemic traits and insulin resistance. Nat. Genet. 2012, 44, 659–669.

73. Perez-Caballero, A.I.; Alcala-Diaz, J.F.; Perez-Martinez, P.; Garcia-Rios, A.; Delgado-Casado, N.; Marin, C.; Yubero-Serrano, E.; Camargo, A.; Caballero, J.; Malagon, M.M.; et al. Lipid metabolism after an oral fat test meal is affected by age-associated features of metabolic syndrome, but not by age. Atherosclerosis 2013, 226, 258–262.

74. Phillips, C.; Murugasu, G.; Owens, D.; Collins, P.; Johnson, A.; Tomkin, G.H. Improved metabolic control reduces the number of postprandial apolipoprotein B-48-containing particles in Type 2 diabetes. Atherosclerosis 2000, 148, 283–291.

75. Perez-Martinez, P.; Delgado-Lista, J.; Perez-Jimenez, F.; Lopez-Miranda, J. Update on genetics of postprandial lipemia. Atheroscler. Suppl. 2010, 11, 39–43.

76. Perez-Martinez, P.; Lopez-Miranda, J.; Perez-Jimenez, F.; Ordovas, J.M. Influence of genetic factors in the modulation of postprandial lipemia. Atheroscler. Suppl. 2008, 9, 49–55.

77. Deeb, S.S.; Fajas, L.; Nemoto, M.; Pihlajamaki, J.; Mykkanen, L.; Kuusisto, J.; Laakso, M.; Fujimoto, W.; Auwerx, J. A Pro12Ala substitution in PPARgamma2 associated with decreased receptor activity, lower body mass index and improved insulin sensitivity. Nat. Genet. 1998, 20, 284–287.

78. Altshuler, D.; Hirschhorn, J.N.; Klannemark, M.; Lindgren, C.M.; Vohl, M.C.; Nemesh, J.; Lane, C.R.; Schaffner, S.F.; Bolk, S.; Brewer, C.; et al. The common PPARgamma Pro12Ala polymorphism is associated with decreased risk of type 2 diabetes. Nat. Genet. 2000, 26, 76–80.

79. Andrulionyte, L.; Zacharova, J.; Chiasson, J.L.; Laakso, M. Common polymorphisms of the PPAR-gamma2 (Pro12Ala) and PGC-1alpha (Gly482Ser) genes are associated with the conversion from impaired glucose tolerance to type 2 diabetes in the STOP-NIDDM trial. Diabetologia 2004, 47, 2176–2184.

80. Ghoussaini, M.; Meyre, D.; Lobbens, S.; Charpentier, G.; Clement, K.; Charles, M.A.; Tauber, M.; Weill, J.; Froguel, P. Implication of the Pro12Ala polymorphism of the PPAR-gamma 2 gene in type 2 diabetes and obesity in the French population. BMC Med. Genet. 2005, 6, doi:10.1186/1471-2350-6-11.

81. Garcia-Rios, A.; Delgado-Lista, J.; Perez-Martinez, P.; Phillips, C.M.; Ferguson, J.F.; Gjelstad, I.M.; Williams, C.M.; Karlstrom, B.; Kiec-Wilk, B.; Blaak, E.E.; et al. Genetic variations at the lipoprotein lipase gene influence plasma lipid concentrations and interact with plasma n-6 polyunsaturated fatty acids to modulate lipid metabolism. Atherosclerosis 2011, 218, 416–422.

82. Phillips, C.M.; Goumidi, L.; Bertrais, S.; Field, M.R.; Cupples, L.A.; Ordovas, J.M.; Defoort, C.; Lovegrove, J.A.; Drevon, C.A.; Gibney, M.J.; et al. Gene-nutrient interactions with dietary fat modulate the association between genetic variation of the ACSL1 gene and metabolic syndrome. J. Lipid Res. 2010, 51, 1793–1800.

83. Phillips, C.M.; Goumidi, L.; Bertrais, S.; Field, M.R.; Cupples, L.A.; Ordovas, J.M.; McMonagle, J.; Defoort, C.; Lovegrove, J.A.; Drevon, C.A.; et al. ACC2 gene polymorphisms, metabolic syndrome, and gene-nutrient interactions with dietary fat. J. Lipid Res. 2010, 51, 3500–3507.

84. Phillips, C.M.; Goumidi, L.; Bertrais, S.; Field, M.R.; McManus, R.; Hercberg, S.; Lairon, D.; Planells, R.; Roche, H.M. Gene-nutrient interactions and gender may modulate the association between Apoa1 and Apob gene polymorphisms and metabolic syndrome risk. Atherosclerosis 2011, 214, 408–414.

85. Dalziel, B.; Gosby, A.K.; Richman, R.M.; Bryson, J.M.; Caterson, I.D. Association of the TNF-alpha-308 G/A promoter polymorphism with insulin resistance in obesity. Obes. Res. 2002, 10, 401–407.

86. Hamid, Y.H.; Rose, C.S.; Urhammer, S.A.; Glumer, C.; Nolsoe, R.; Kristiansen, O.P.; Mandrup-Poulsen, T.; Borch-Johnsen, K.; Jorgensen, T.; Hansen, T.; et al. Variations of the interleukin-6 promoter are associated with features of the metabolic syndrome in caucasian danes. Diabetologia 2005, 48, 251–260.

87. Huth, C.; Heid, I.M.; Vollmert, C.; Gieger, C.; Grallert, H.; Wolford, J.K.; Langer, B.; Thorand, B.; Klopp, N.; Hamid, Y.H.; et al. Il6 gene promoter polymorphisms and type 2 diabetes: Joint analysis of individual participants' data from 21 studies. Diabetes 2006, 55, 2915–2921.

88. Shen, J.; Arnett, D.K.; Perez-Martinez, P.; Parnell, L.D.; Lai, C.Q.; Peacock, J.M.; Hixson, J.E.; Tsai, M.Y.; Straka, R.J.; Hopkins, P.N.; et al. The effect of IL6–174C/G polymorphism on postprandial triglyceride metabolism in the GOLDN studyboxs. J. Lipid Res. 2008, 49, 1839–1845.

89. Sookoian, S.; Garcia, S.I.; Gianotti, T.F.; Dieuzeide, G.; Gonzalez, C.D.; Pirola, C.J. The G-308A promoter variant of the tumor necrosis factor-alpha gene is associated with hypertension in adolescents harboring the metabolic syndrome. Am. J. Hypertens. 2005, 18, 1271–1275.

90. Meirhaeghe, A.; Cottel, D.; Amouyel, P.; Dallongeville, J. Lack of association between certain candidate gene polymorphisms and the metabolic syndrome. Mol. Genet. Metab. 2005, 86, 293–299.

91. Qi, L.; Zhang, C.; van Dam, R.M.; Hu, F.B. Interleukin-6 genetic variability and adiposity: Associations in two prospective cohorts and systematic review in 26,944 individuals. J. Clin. Endocrinol. Metab. 2007, 92, 3618–3625.

92. Phillips, C.M.; Goumidi, L.; Bertrais, S.; Ferguson, J.F.; Field, M.R.; Kelly, E.D.; Mehegan, J.; Peloso, G.M.; Cupples, L.A.; Shen, J.; et al. Additive effect of polymorphisms in the IL-6, LTA, and TNF-{alpha} genes and plasma fatty acid level modulate risk for the metabolic syndrome and its components. J. Clin. Endocrinol. Metab. 2010, 95, 1386–1394.

93. Aaronson, D.S.; Horvath, C.M. A road map for those who don't know JAK-STAT. Science 2002, 296, 1653–1655.

94. Phillips, C.M.; Goumidi, L.; Bertrais, S.; Field, M.R.; Peloso, G.M.; Shen, J.; McManus, R.; Hercberg, S.; Lairon, D.; Planells, R.; et al. Dietary saturated fat modulates the association between STAT3 polymorphisms and abdominal obesity in adults. J. Nutr. 2009, 139, 2011–2017.

95. Weedon, M.N.; McCarthy, M.I.; Hitman, G.; Walker, M.; Groves, C.J.; Zeggini, E.; Rayner, N.W.; Shields, B.; Owen, K.R.; Hattersley, A.T.; et al. Combining information from common type 2 diabetes risk polymorphisms improves disease prediction. PLoS Med. 2006, 3, e374.

96. Engstršm, G.; Hedblad, B.; Eriksson, K.F.; Janzon, L.; LindgŠrde, F. Complement C3 is a risk factor for the development of diabetes: A population-based cohort study. Diabetes 2005, 54, 570–575.

97. Halkes, C.J.; van Dijk, H.; de Jaegere, P.P.; Plokker, H.W.; van Der Helm, Y.; Erkelens, D.W.; Castro Cabezas, M. Postprandial increase of complement component 3

in normolipidemic patients with coronary artery disease: Effects of expanded-dose simvastatin. Arterioscler. Thromb. Vasc. Biol. 2001, 21, 1526–1530.

98. Muscari, A.; Massarelli, G.; Bastagli, L.; Poggiopollini, G.; Tomassetti, V.; Drago, G.; Martignani, C.; Pacilli, P.; Boni, P.; Puddu, P. Relationship of serum C3 to fasting insulin, risk factors and previous ischaemic events in middle-aged men. Eur. Heart J. 2000, 21, 1081–1090.

99. Van Oostrom, A.J.; Alipour, A.; Plokker, T.W.; Sniderman, A.D.; Cabezas, M.C. The metabolic syndrome in relation to complement component 3 and postprandial lipemia in patients from an outpatient lipid clinic and healthy volunteers. Atherosclerosis 2007, 190, 167–173.

100. Phillips, C.M.; Kesse-Guyot, E.; Ahluwalia, N.; McManus, R.; Hercberg, S.; Lairon, D.; Planells, R.; Roche, H.M. Dietary fat, abdominal obesity and smoking modulate the relationship between plasma complement component 3 concentrations and metabolic syndrome risk. Atherosclerosis 2012, 220, 513–519.

101. Phillips, C.M.; Goumidi, L.; Bertrais, S.; Ferguson, J.F.; Field, M.R.; Kelly, E.D.; Peloso, G.M.; Cupples, L.A.; Shen, J.; Ordovas, J.M.; et al. Complement component 3 polymorphisms interact with polyunsaturated fatty acids to modulate risk of metabolic syndrome. Am. J. Clin. Nutr. 2009, 90, 1665–1673.

102. Menzaghi, C.; Trischitta, V.; Doria, A. Genetic influences of adiponectin on insulin resistance, type 2 diabetes, and cardiovascular disease. Diabetes 2007, 56, 1198–1209.

103. Sheng, T.; Yang, K. Adiponectin and its association with insulin resistance and type 2 diabetes. J. Genet. Genomics 2008, 35, 321–326.

104. Phillips, C.M.; Goumidi, L.; Bertrais, S.; Field, M.R.; Ordovas, J.M.; Cupples, L.A.; Defoort, C.; Lovegrove, J.A.; Drevon, C.A.; Blaak, E.E.; et al. Leptin receptor polymorphisms interact with polyunsaturated fatty acids to augment risk of insulin resistance and metabolic syndrome in adults. J. Nutr. 2010, 140, 238–244.

105. Neel, J.V. Diabetes mellitus: A "thrifty" genotype rendered detrimental by "progress"? Am. J. Hum. Genet. 1962, 14, 353–362.

106. Wild, S.; Roglic, G.; Green, A.; Sicree, R.; King, H. Global prevalence of diabetes: Estimates for the year 2000 and projections for 2030. Diabetes Care 2004, 27, 1047–1053.

107. Esparza-Romero, J.; Valencia, M.E.; Martinez, M.E.; Ravussin, E.; Schulz, L.O.; Bennett, P.H. Differences in insulin resistance in Mexican and U.S. Pima Indians with normal glucose tolerance. J. Clin. Endocrinol. Metab. 2010, 95, E358–E362.

108. Schulz, L.O.; Bennett, P.H.; Ravussin, E.; Kidd, J.R.; Kidd, K.K.; Esparza, J.; Valencia, M.E. Effects of traditional and western environments on prevalence of type 2 diabetes in Pima indians in Mexico and the U.S. Diabetes Care 2006, 29, 1866–1871.

109. Luan, J.; Browne, P.O.; Harding, A.H.; Halsall, D.J.; O'Rahilly, S.; Chatterjee, V.K.; Wareham, N.J. Evidence for gene-nutrient interaction at the PPARgamma locus. Diabetes 2001, 50, 686–689.

110. Scacchi, R.; Pinto, A.; Rickards, O.; Pacella, A.; de Stefano, G.F.; Cannella, C.; Corbo, R.M. An analysis of peroxisome proliferator-activated receptor gamma (PPAR-γ2) Pro12Ala polymorphism distribution and prevalence of type 2 diabetes mellitus (T2DM) in world populations in relation to dietary habits. Nutr. Metab. Cardiovasc. Dis. 2007, 17, 632–641.

111. Robitaille, J.; Gaudet, D.; Perusse, L.; Vohl, M.C. Features of the metabolic syndrome are modulated by an interaction between the peroxisome proliferator-activated receptor-delta -87T>C polymorphism and dietary fat in French-Canadians. Int. J. Obes. (Lond.) 2007, 31, 411–417.

112. Delgado-Lista, J.; Perez-Martinez, P.; Garcia-Rios, A.; Phillips, C.M.; Williams, C.M.; Gulseth, H.L.; Helal, O.; Blaak, E.E.; Kiec-Wilk, B.; Basu, S.; et al. Pleiotropic effects of TCF7L2 gene variants and its modulation in the metabolic syndrome: From the lipgene study. Atherosclerosis 2011, 214, 110–116.

113. Lee, H.J.; Kim, I.K.; Kang, J.H.; Ahn, Y.; Han, B.G.; Lee, J.Y.; Song, J. Effects of common FTO gene variants associated with BMI on dietary intake and physical activity in Koreans. Clin. Chim. Acta 2010, 411, 1716–1722.

114. Sonestedt, E.; Roos, C.; Gullberg, B.; Ericson, U.; Wirfalt, E.; Orho-Melander, M. Fat and carbohydrate intake modify the association between genetic variation in the fto genotype and obesity. Am. J. Clin. Nutr. 2009, 90, 1418–1425.

115. Moleres, A.; Ochoa, M.C.; Rendo-Urteaga, T.; Martinez-Gonzalez, M.A.; Azcona San Julian, M.C.; Martinez, J.A.; Marti, A. Dietary fatty acid distribution modifies obesity risk linked to the rs9939609 polymorphism of the fat mass and obesity-associated gene in a Spanish case-control study of children. Br. J. Nutr. 2012, 107, 533–538.

116. Garcia-Rios, A.; Perez-Martinez, P.; Delgado-Lista, J.; Phillips, C.M.; Gjelstad, I.M.; Wright, J.W.; Karlstrom, B.; Kiec-Wilk, B.; van Hees, A.M.; Helal, O.; et al. A Period 2 genetic variant interacts with plasma SFA to modify plasma lipid concentrations in adults with metabolic syndrome. J. Nutr. 2012, 142, 1213–1218.

117. Corella, D.; Arnett, D.K.; Tucker, K.L.; Kabagambe, E.K.; Tsai, M.; Parnell, L.D.; Lai, C.Q.; Lee, Y.C.; Warodomwichit, D.; Hopkins, P.N.; et al. A high intake of saturated fatty acids strengthens the association between the fat mass and obesity-associated gene and bmi. J. Nutr. 2011, 141, 2219–2225.

118. Gutierrez-Aguilar, R.; Kim, D.H.; Woods, S.C.; Seeley, R.J. Expression of new loci associated with obesity in diet-induced obese rats: From genetics to physiology. Obesity (Silver Spring) 2011, 20, 306–312.

119. Coleman, R.A.; Lewin, T.M.; Muoio, D.M. Physiological and nutritional regulation of enzymes of triacylglycerol synthesis. Annu. Rev. Nutr. 2000, 20, 77–103.

120. McGarry, J.D. Banting lecture 2001: Dysregulation of fatty acid metabolism in the etiology of type 2 diabetes. Diabetes 2002, 51, 7–18.

121. Shimabukuro, M.; Zhou, Y.T.; Levi, M.; Unger, R.H. Fatty acid-induced beta cell apoptosis: A link between obesity and diabetes. Proc. Natl. Acad. Sci. USA 1998, 95, 2498–2502.

122. Garcia-Rios, A.; Perez-Martinez, P.; Delgado-Lista, J.; Lopez-Miranda, J.; Perez-Jimenez, F. Nutrigenetics of the lipoprotein metabolism. Mol. Nutr. Food Res. 2012, 56, 171–183.

123. Perez-Martinez, P.; Garcia-Rios, A.; Delgado-Lista, J.; Perez-Jimenez, F.; Lopez-Miranda, J. Nutrigenetics of the postprandial lipoprotein metabolism: Evidences from human intervention studies. Curr. Vasc. Pharmacol. 2011, 9, 287–291.

124. Warodomwichit, D.; Shen, J.; Arnett, D.K.; Tsai, M.Y.; Kabagambe, E.K.; Peacock, J.M.; Hixson, J.E.; Straka, R.J.; Province, M.A.; An, P.; et al. ADIPOQ polymor-

phisms, monounsaturated fatty acids, and obesity risk: The GOLDN study. Obesity (Silver Spring) 2009, 17, 510–517.

125. Perez-Martinez, P.; Lopez-Miranda, J.; Cruz-Teno, C.; Delgado-Lista, J.; Jimenez-Gomez, Y.; Fernandez, J.M.; Gomez, M.J.; Marin, C.; Perez-Jimenez, F.; Ordovas, J.M. Adiponectin gene variants are associated with insulin sensitivity in response to dietary fat consumption in Caucasian men. J. Nutr. 2008, 138, 1609–1614.

126. Ferguson, J.F.; Phillips, C.M.; Tierney, A.C.; Perez-Martinez, P.; Defoort, C.; Helal, O.; Lairon, D.; Planells, R.; Shaw, D.I.; Lovegrove, J.A.; et al. Gene-nutrient interactions in the metabolic syndrome: Single nucleotide polymorphisms in ADIPOQ and ADIPOR1 interact with plasma saturated fatty acids to modulate insulin resistance. Am. J. Clin. Nutr. 2010, 91, 794–801.

127. Shi, H.; Kokoeva, M.V.; Inouye, K.; Tzameli, I.; Yin, H.; Flier, J.S. TLR4 links innate immunity and fatty acid-induced insulin resistance. J. Clin. Invest. 2006, 116, 3015–3025.

128. Zhao, A.; Yu, J.; Lew, J.L.; Huang, L.; Wright, S.D.; Cui, J. Polyunsaturated fatty acids are FXR ligands and differentially regulate expression of FXR targets. DNA Cell Biol. 2004, 23, 519–526.

129. Fraga, M.F.; Ballestar, E.; Paz, M.F.; Ropero, S.; Setien, F.; Ballestar, M.L.; Heine-Suner, D.; Cigudosa, J.C.; Urioste, M.; Benitez, J.; et al. Epigenetic differences arise during the lifetime of monozygotic twins. Proc. Natl. Acad. Sci. USA 2005, 102, 10604–10609.

130. Greco, L.; Romino, R.; Coto, I.; Di Cosmo, N.; Percopo, S.; Maglio, M.; Paparo, F.; Gasperi, V.; Limongelli, M.G.; Cotichini, R.; et al. The first large population based twin study of coeliac disease. Gut 2002, 50, 624–628.

131. Arkadianos, I.; Valdes, A.M.; Marinos, E.; Florou, A.; Gill, R.D.; Grimaldi, K.A. Improved weight management using genetic information to personalize a calorie controlled diet. Nutr. J. 2007, 6, 29.

132. Tapueru-French, C. Can the Use of Genetics Benefit Weight Loss IN A New Zealand Setting? M.Sc. Thesis, University of Auckland, Auckland, New Zealand, 2009.

133. Stewart-Knox, B.J.; Bunting, B.P.; Gilpin, S.; Parr, H.J.; Pinhao, S.; Strain, J.J.; de Almeida, M.D.; Gibney, M. Attitudes toward genetic testing and personalised nutrition in a representative sample of European consumers. Br. J. Nutr. 2009, 101, 982–989.

This chapter was originally published under the Creative Commons Attribution License. Phillips, C. M. Nutrigenetics and Metabolic Disease: Current Status and Implications for Personalised Nutrition. Nutrients 2013, 5, 32-57. doi:10.3390/nu5010032.

CHAPTER 2

A REVIEW OF THE NATURE AND EFFECTIVENESS OF NUTRITION INTERVENTIONS IN ADULT MALES: A GUIDE FOR INTERVENTION STRATEGIES

PENNIE J. TAYLOR, GREGORY S .KOLT, CORNEEL VANDELANOTTE, CRISTINA M. CAPERCHIONE, W. KERRY MUMMERY, EMMA S. GEORGE, MOHANRAJ KARUNANITHI, and MANNY J NOAKES

2.1 INTRODUCTION

In terms of dietary behaviours, males are less likely to meet the recommended intakes of fruit and vegetables compared to women [1-5]. According to the 'Health of Australia's Males' report [1] 68% of Australian males are classified as overweight or obese [1,6].

Despite males having a shorter life expectancy [1,7] and being more susceptible to the medical consequences of chronic disease compared to their female counterparts [2,8], participation in preventive health services are lower amongst males [1,2]. Based on current literature, males are less likely to attend face-to-face dietary counselling sessions and tend to be more apprehensive of health-related initiatives, media advertising campaigns, and scientific studies on healthy eating [9]. "Healthy eating" messages are presented by a range of media sources, especially internet-based media, which can provide contradictory nutritional messages to readers or messages based on evidence with unknown scientific quality [2]. For males, who most often do not seek professional consultation [1,2,10], this can lead to self-monitoring of their current health status based on instinctive assumptions from questionable evidence or utilising uninformed partners and friends as a source of advice regarding diet and lifestyle be-haviour changes [1,2,10]. A lack of willpower, motivation [11-14] and time [9,15,16] has also been identified by men as the primary barriers that hinder the adoption of healthy dietary behaviours.

Despite research recognising that poor dietary patterns, such as low fruit and vegetable intake and energy excess, as being associated with a greater risk of developing non communicable diseases including diabetes mellitus, heart disease and some cancers [4,6-8], scientific literature spe-cific to lifestyle programs that effectively target engagement by males, however, is very limited. A recent review by Pagoto and collegues [17] noted that only 5% of lifestyle intervention studies targeting weight loss were exclusive to males compared to 32% exclusively targeting females [17]. With recent publications identifying males as a hard-to-reach popula-tion and at greater risk of premature death compared to females, interven-tions should consider appropriate strategies to engage males in improving lifestyle and behaviour outcomes [1,2,8].

Therefore the purpose of this review is to provide a critical evaluation of the nature and effectiveness of nutrition interventions that target the male population. Also to identify strategies that are likely to be effective in improving program engagement by adult males.

2.2 METHODS

2.2.1 LITERATURE SEARCH

An individualised search for full-text publications was conducted using the following electronic databases: The Cochrane Library; Web of Sci-ence; SCOPUS; MEDLINE and CINAHL. Reference lists of retrieved ar-ticles and other relevant systematic reviews [18-20] were also reviewed. The following search terms were used: men OR male; diet OR nutrition OR dietary intake; lifestyle; intervention OR program; fruit and vegeta-ble intake. The search was limited to the adult population (over 18 years of age). Two reviewers independently assessed studies identified in the search for relevance from the title, abstract and key words. Those meeting the inclusion criteria were retrieved and further assessed for relevance. In the event where agreement was not met for studies inclusion/exclusion a third reviewer was engaged.

2.2.2 STUDY INCLUSION CRITERIA

To identify the most effective strategies for improving nutrition and supporting lifestyle risk modification in adult males, this review considers and reports outcomes from studies that have delivered lifestyle interventions aimed at improving dietary intake and/or behaviour in adult males. Nutrition research has grown significantly over time, and more recently to encompass web-based intervention modalities. To ensure relevant current technology and intervention design, dates of publication were restricted to January 1990 to August 2011.

Publications were selected for this review if they met the following criteria:

(i) included adult males only, OR both males and females where data on male participants was reported exclusively AND
(ii) intervention delivered assessed changes in weight and dietary intakes and/or dietary behaviours as a primary or secondary outcome.

Studies were excluded if:

(i) participants recruited had special dietary requirements and dietary interventions targeted these conditions (e.g. diverticular disease; diabetes, heart disease; renal disease; all types of cancers and gastrointestinal disorders) AND
(ii) if studies examined the effect of different diets on weight loss only.

Randomised Control Trials (RCT) with appropriate control group comparisons were the priority for inclusion; quasi-experimental trials (non random allocation) with a comparison group were also considered. If multiple publications presented data from the same population in the same intervention study, only the most relevant publication was included. Although priority was given to nutrition-based interventions, due to the limited body of literature available on nutrition only in males, studies that investigated nutrition in combination with physical activity were also included in the review. Combined studies were excluded in situations where limited data was reported on dietary intake and/or dietary behaviour in combination with physical activity. During the review process a consensus was drawn to include RCT's where males contributed > 90% of the active cohort and where studies provided substantial insight to nutrition intervention design.

2.2.3 DATA EXTRACTION

Data from the included studies, such as methodological quality, intervention and intervention comparison details were independently extracted and quality graded by two reviewers.

Methodological quality of each study reviewed was scored using the McMasters University quality assessment tool for quantitative studies developed by the Effective Public Health Practice Project quality assessment tool [21] . Factors assessed and rated were: selection bias, study design, confounders, blinding, data collection methods and withdrawals/dropouts. A standardised dictionary developed for the Effective Public Health Practice Project quality assessment tool, was used to classify the factors as strong, moderate or weak (see Table 1).

Individual ratings contributed to a global rating for each study assigning a weak (where 2 or more factors were rated as WEAK); moderate (where less than 4 factors were rated as STRONG and one WEAK rating) and strong (where 4 or more factors were rated as STRONG with NO WEAK ratings). In the event of a discrepancy in interpretation of the findings, all queries were resolved through discussion amongst authors during the review process.

To determine intervention effectiveness, a study must have described, (i) a significant change over the intervention period ($p<0.05$) in an objective measure of body weight, expressed in kilograms (kg) AND/OR Body Mass Index (BMI) expressed as kilograms per metre squared (Kg/m^2) in studies with prescribed caloric restriction OR (ii) at least one significant change ($p<0.05$) in a dietary intake measure including, fruit and vegetable, total energy and total fat intakes.

TABLE 1: Effective Public Health Practice Project Quality Assessment Components and Ratings

Components	Strong	Moderate	Weak
Selection Bias	Very Likely to be representative of the target population and greater than 80% participation rate	Somewhat likely to pre representative of the target population and 60–79% participation rate	All other responses or not stated
Design	RCT or CCT	Cohort analytic, case control, cohort or an interrupted time series	All other designs of design not stated
Confounders	Controlled for at least 80% of confounders	Controlled for at least 60–79% of confounders	Confounders not controlled for or not stated
Blinding	Blinding on outcome assessor and study participants to intervention status and/or research question	Blinding of either outcome assessor or study participants	Outcome assessor and study participants are aware of intervention status and/or research question
Data Collection Methods	Tools are valid and reliable	Tools are valid but reliability not described	No evidenced of validity or reliability
Withdrawals and Droupouts	Follow-up rate of >80% of participants	Follow-up rate of 60–79% of participants	Follow-up rate of <60% of participants or withdrawals and dropouts not described

2.3 RESULTS

2.3.1 RESULTS OF SEARCH STRATEGY

Individual data-base searches resulted in 1381 titles and abstracts being identified for this review. Duplicate articles and those that did not meet the specified inclusion criteria, within the title or abstract were removed, leaving a total of 89 publications for consideration. Of these, 18 were review articles and 71 intervention studies. Full text articles were retrieved and assessed against the inclusion criteria. Of the 71 intervention studies, nine met the inclusion criteria for this review (Table 2). The majority of the studies excluded from this review included both male and female participants where data was unable to be extracted for males exclusively. All studies targeted overweight and obese males (BMI >25 kg/m^2) with one study [10] reporting mean baseline BMI <25 kgm^2. Of the nine studies, eight were RCTs [3,22-28] and one was quasi-experimental [10]. Four (40%) of the included studies were based solely on nutrition interventions in males [3,25,27,28] whilst the remainder combined nutrition and physical activity interventions [10,22-24,26]. All but one study [2] included male participants only. A consensus between the primary and co-authors resulted in this study being included. The basis for its inclusion being that 97% of the participants in the study were male and outcome measures were commensurate with the inclusion criteria.

2.3.2 DESCRIPTION OF INCLUDED STUDIES AND PARTICIPANTS

A total of 6,167 overweight/obese males were represented in these nine studies. Sample sizes of the included studies ranged from 53 to 5,042 male participants and were aged 18 years and over. Studies were based in Australia [3,18,23,24,26], Belgium [27], Japan [10], United Kingdom [25] and the United States of America [28]. Six studies were worksite interventions [10,24-28] and three were community-based interventions [3,22,23] of which two involved family members [22,23]. Study duration ranged from 12 weeks to 24 months, of these seven were considered short-term (<6 months) [3,10,22-25,27] and two long-term (>6 months) [26,28].

Eight studies observed weight changes [3,10,22-27], three studies [22,25,26] had the primary aim of achieving weight loss through caloric restriction whilst five studies [3,10,24,26-28] focused on achieving changes in dietary intake patterns relating to modifications in fruit, vegetable, dairy and total fat intakes. Four studies reported the social cognitive theory as supporting the intervention [22-24,28].

The characteristics of each study including participants, the intervention, comparison, duration and outcomes that reported changes in dietary behaviours and/or weight changes were stratified. Additional features of the intervention were further tabulated by personnel delivering the intervention (e.g. dietician), mode of delivery (e.g. print, internet or face-face) and form of nutrition information provided (e.g. generic or tailored).

TABLE 2: Summary of selected intervention studies

Study	Participants	Measures	Comparisons Intervention/s	Comparison	Duration	Outcomes
Arao et al. (2007) Non-randomised Control trial	177 men with risk factors for chronic disease	Primary:	Intervention - LiSM-PAN Group (individual counselling based on stages of change and environmental and social support, work- and home-based)	Control group (Standard Conventional Health-care (SCH))	6 mths	· LiSM PAN group: showed significant positive changes in leisure time exercise energy expenditure (LEEE) (mean inter-group difference: 400.6 kcal/ week, 95% CI: 126.1, 675.0 kcal/week).
Japan	Wt 68.3 kg (+/- 10.1)	· leisure time exercise energy expenditure (LEEE)	· Individual counselling: 6 month program based on stages of change. 15 mins of one-on-onecounselling on PA and dietary goals. PA goal was specific number of steps/ day based on stage of change. Those already active/maintaining given additional PA goals. Structured counselling given by trained professionals	· Generic printed materials on exercise, diet and cooking provided		· No mean inter-group differences reported for dietary habits. (p= 0.432)
	· Ht 166.2 cm (+/- 6.1)	· Secondary:	· Environmental and social support: walking course and exercise facilities installed at workplace, caloric content of lunch menu displayed at workplace café, providing better nutrition through consultation with café manger.	· no counselling nor environmental/social support		· No significant between group difference in changes in dietary fat; fruit and vegetables intakes (0.071-0.238)

TABLE 2: *Cont.*

Study	Participants	Measures	Comparisons		Dura-tion	Outcomes
			Intervention/s	Comparison		
	· 40-59 years	· VO2max	· Increasing support from family/at home by encouraging participants to discuss health and strategies to improve health with family, men asked to participate in PA with family/spouse, spouse given printed materials on healthy diet/cooking	· Participants given written feedback and recommendations from results of medical check-up and baseline data.		· Greater decreases in BMI, SBP, LDL in intervention vs. control. [<0.001]
		· Intake of fats, fruits and vegetables assessed by FFQ).		Occupational nurse encouraged participants to follow recommendations		· Compliance
		·BMI				· Retention rate for LiSM-PAN group program = 95.2%.
		· BP				· average rate of compliance was 97.1% in monthly counselling
		· Blood glucose				· average achievements of basic target were 86.7% for self-monitoring on the walking steps and 54.7% for controlling dietary targeted activities
		· Lipid				

TABLE 2: *Cont.*

Study	Participants	Measures	Comparisons		Dura-tion	Outcomes
			Intervention/s	Comparison		
Booth et al. (2008) Randomised con-trolled trial	· 54 free-living, overweight or obese males · Mean age 48 years	Primary: · Changes in dietary intake of: i) fruit, ii) vegetables, iii) dairy	WELL intervention (Weight-loss; exercise; lower blood pressure and longevity intervention group) delivered face to face + 2 2 telephone calls by trained research staff overseen by dietician · Print based material provided on DASH diet with a weight loss focus.	Low Fat group (Based on the healthy weight guide by the Na-tional Heart Foun-dation (2002) – no prescribed food volume given) and delivered face to face + 2 telephone calls by trained research staff overseen by dietician · Generalised written informa-tion in the booklet recommended: a) limit high and full-fat foods, b) consume more fruit, vegetables and other plant based products, c) consume fish and legumes at least twice a week.	12 wks	Overall: · 86% retention rate

TABLE 2: *Cont.*

Study	Participants	Measures	Comparisons		Dura-tion	Outcomes
			Intervention/s	Comparison		
Australia		· Other measures		· Other recommendations: limit high fat foods, choose low fat or reduced fat products, and use a variety of plant based oils for cooking.		· No Difference in mean weight loss between groups 5-6% of TBW lost.
			· Daily targets set: Participants required to consume at least 4 serves of vegetables, at least 4 serves of fruit, at least 3 serves of dairy and a maximum of 4 serves (4tspn) MUFA.			
		· BP (taken daily by volunteer)	· Weekly targets set: Participants required to consume 4 serves of nuts and seeds, at least 3 serves of fish, 1 serve of legumes, max 2 serves of red meat.	· Self monitoring through 3-day food diaries completed weekly – each day for 3 consecutive days. Diaries reviewed		· WELL diet achieved a greater Fruit; Vegetable and Dairy intake compared to LG Diet group measured by food group diaries (p<0.01)
		· Weight (taken at each face-to-face visit)	· No restriction on rice/pasta/ wholegrain bread and lower-salt cereals as long as they were consuming the volume of other foods listed above.			· (-7.67.7 mmHg SBP and -5.44.9 mmHg DBP) than LF group (-2.1 6.4 mmHg SBP and
		· Height - baseline	· Self monitoring through 3-day food diaries completed weekly –each day for 3 consecutive days. Diaries reviewed by study staff.			· 1.0 4.1 mmHg DBP (difference in BP change between groups P = 0.001).
		· BMI				

TABLE 2: *Cont.*

Study	Participants	Measures	Comparisons		Dura-tion	Outcomes
			Intervention/s	Comparison		
Brackman et al (1999) Randomised controlled trial (4 worksites randomised) Belgium	· 638 middle aged men · Mean age 43.7 (+/-6.6) · Mean BMI 26.5 kg/m2	Primary: · Dietary Habits (24hr food record) · Serum Lipid levels · Second-ary:	Low-Fat Dietary Intervention · Participants informed of baseline screening measures at 2 weeks through individual counselling session and informed of personal risk factor profiles · Mass media used within Intervention sites to stress the link between Cholesterol and heart disease and the role of a low fat diet. · Poster displays and leaflets providing strategies on how to reduce dietary fat provided at intervention sites	Control sites (no access to support) · Provided written summary of risk factor profile with nil dietary education/information provided. Exception for those with abnormal values who were referred to their GP.	3 mths	· 82% retention rate at 3 months · Significant reduction in total energy and total fat intake in the intervention group (p<0.05) but no difference for percent of energy from types of fat. · Intervention group increased protein and carbohydrate more than the control (p<0.05) · BMI increased by 0.3 kg/m2 in the intervention group vs. controls (p<0.001)

TABLE 2: *Cont.*

Study	Participants	Measures	Comparisons		Dura-tion	Outcomes
			Intervention/s	Comparison		
		· (Self ad-ministered Health Question-naire (smoking; PAL and medical history) · Nutrition knowledge (10-item question-naire) · WHR · BMI	· Video outlining importance of reducing blood cholesterol by reducing dietary fat intake presented with question and answer time at a worksite safety meeting · Participants offered several non-compulsory dietician-led 2 hour dietary group education sessions at the worksite out of work hours · Summary newsletter pro-vided at the end of the study to reinforce dietary messages			· Nutrition knowledge significantly greater in the intervention groups (p<0.001) No significant effect for to-tal cholesterol between groups · HDL cholesterol increased in the control group compared to intervention group (p<0.001)
Leslie et al. (2002)	· 122 over-weight/obese males	Primary	Energy Deficit diet (ED) (a 2512 kJ (600 kcal deficit) with individualized energy prescriptions	Generalised low calorie diet (6279 kJ=1500 kcal)	24 wks (12 wks inter-vention + 12 wk mainte-nance)	· Weight loss sig-nificant in both ED and GLC groups but no difference between groups in weight loss or maintenance.

TABLE 2: *Cont.*

Study	Participants	Comparisons			Outcomes
		Measures	Intervention/s	Comparison	Duration	

Study	Participants	Measures	Intervention/s	Comparison	Duration	Outcomes
Randomised controlled trial	· 18-55 years	· Weight loss	1. ED with meat	3. GLC (general low calorie) meat		· No effect of meat vs no meat on weight loss or biochemical measure between groups
United Kingdom		· Weight loss maintenance	2. ED no meat	4. GLC no meat		· Significantly more attrition from the GLC group than the ED group.
			All attended initial dietary consult (60 minute) delivered by dietician and face-to-face reviews every 2 weeks for 20 minutes for first 12 weeks.			· 69% Retention at 24 weeks.
			All groups underwent 12 weeks weight loss followed by 12 week maintenance phase			
			All contacted by email at 2 week intervals and self reported anthropometric and dietary information			
		· Secondary requested.				
		· Lipids (plasma)				
		· Dietary Habits				

2.3.3 PERSONNEL

Interventions were dietician-led [3,25,27] dietician and exercise physiologist-led (sex not defined) [26] male researcher-led [22-24] or researcher-led (sex non-defined) [10,29].

2.3.4 QUALITY OF INCLUDED STUDIES

Assessment of study quality identified four studies as having a strong quality rating [10,22,24,26], four rated as moderate [3,23,25,28] and one rated as weak [27]. Quality Assessment results can be seen in Table 3.

Those that were rated moderate and weak performed poorly in the quality assessment criteria for selection bias, confounders and data collection methods. Study designs differed substantially in the nature of the interventions and the nature of the control group. For example, some studies compared the intervention to a wait-listed, no intervention control group [23,24,27,28] whilst others compared two different forms of intervention [3,10,22,25,26]. Several modes of delivery within studies were common such as combining print-based materials (n=8) [10,22-24,26-29], face-to-face (n=9) [3,10,22-29], or phone counselling (n=2) [3,28] and use of internet-based tools (n=3) [23-25].

2.3.5 INTERVENTION EFFECTIVENESS

Intervention 'effectiveness', as defined previously, was found for seven of the nine studies and is shown in Table 4. Five effective studies were in the worksite setting [10,24,26-28] and two were community based [3,23]. One community study focussed on overweight males and their primary school aged children [23]. Four studies reported using a theoretical basis for the intervention utilising social cognitive theories and readiness to change models [10,23,24,28]. The duration of study interventions were short-term, 12 weeks to six-months [3,23,24,27] and long-term 12–24 months [26,28]. Interventions targeted nutritional intake and dietary behaviours [3,27,28] and dietary intake in combination with physical activity

TABLE 3: Quality assessment results against the Effective Public Health Practice Project Quality Assessment Tool for all included studies

Author/Date	Selection Bias	Design	Confounders	Blinding	Data Collection Methods	Withdrawals/ Dropouts	Global rat- ings
Arao et al. (2007)	Moderate	Strong	Strong	Moderate	Strong	Strong	Strong
Booth et al. (2008)	Moderate	Strong	Moderate	Moderate	Strong	Moderate	Moderate
Brackman et al. (1999)	Weak	Moderate	Moderate	Weak	Moderate	Moderate	Weak
Leslie et al. (2002)	Strong	Strong	Moderate	Strong	Moderate	Moderate	Moderate
Morgan (2009, 2011c) SHED-IT	Moderate	Strong	Strong	Moderate	Strong	strong	Strong
Morgan et al. (2011a) HDHK	Moderate	Strong	Moderate	Weak	Strong	strong	Moderate
Morgan et al. (2011b) POWER	Moderate	Strong	Moderate	Moderate	Strong	Strong	Moderate
Pritchard et al. (1997)	Strong	Strong	Moderate	Moderate	Strong	Strong	Strong
Tilley et al. (1999, 1997)	Strong	Strong	Moderate	Weak	Moderate	Moderate	Moderate

[10,23-25]. Pritchard et al. was the only study that compared weight loss through caloric restriction against weight loss through exercise only to a control group whereby participants attended weight monitoring sessions only [26].

2.3.6 EFFECTIVENESS OF INTERVENTIONS AIMING TO ACHIEVE WEIGHT LOSS

Of the effective interventions, four studies [10,23,24,26] reported significant positive changes in weight (kg) and/or BMI (kg/m^2) changes (p ≤0.05). Study sample sizes ranged from 53 to 177.

Two studies involved internet usage [23,24] including the use of a commercially available self-monitoring tool in combination with face-to-face group interaction [23,24] and an independent study website [24]. All other studies provided face-to-face, individual intervention delivery [10,23,26]. All studies used combinations of nutrition and physical activity strategies to promote weight-loss [10,23,24,26] and two used caloric restriction through personalised low-fat dietary plans aimed at reducing baseline intake by 500kcal per day [23,26]. Other nutrition strategies included reduction of total fat intake, portion serves and building awareness of targets for healthier eating practices [10,23,24,26].

2.3.7 EFFECTIVENESS OF INTERVENTIONS AIMING TO ACHIEVE CHANGE IN DIETARY INTAKE AND FOOD BEHAVIORS

There were four effective interventions (p <0.05) targeting determinants of dietary intake and dietary behaviours and/or nutritional intake [3,24,27,28]. Study sample sizes ranged from 53 to 5042. One study included single face-to-face interaction, use of a commercially available self-monitoring tool and access as needed to an independent study website [23]. Other modes of delivery included multiple face-to-face interactions combined with phone support [3] or non-compulsory group education sessions [27]. Tilley et al. conducted a long-term study, duration of 2 years, incorporating multiple group face-to-face interactions with phone support

TABLE 4: Dietary evaluation summary of interventions to improve Men's nutritional and weight-loss outcomes - Effectiveness table

Study	Weight status/lipid studies/BP		Nutritional determinants		Nutrition Intake		Intervention Effectiveness
Arao et al. (2007)	BMI	X	Dietary control	×	Fat intake (serves)	×	YES
	Lipid studies	X			Fruit intake (serves)	×	
	Blood Pressure(mmHg)	X			Vegetable intake (serves)	×	
	Blood glucose	X					
Booth et al. (2008)	Body Weight (kg)	×	-		Fat intake (serves)	X	YES
	BMI	×			Fruit intake (serves)	X	
	Blood Pressure(mmHg)	X			Dairy intake (serves)	X	
Braekman et al. (1999)	BMI	×	Nutrition knowledge	X	Total Energy and macro-nutrient intake	X	YES
	Lipid studies (HDL)	X					
Leslie et al. (2002)	Height (cm)	×	Dietary practices monitored through dietary	×	Total energy intake	×	NO
	Body Weight (kg)	×	targets monitor				
	Waist Circumference (cm)	×					

TABLE 4: *Cont.*

Study	Weight status/lipid studies/BP		Nutritional determinants		Nutrition Intake		Intervention Effectiveness
Morgan (2009, 2011c) SHED-IT	Lipid studies	×					
	Body Weight (kg)	×	Dietary knowledge and belief cognitions	×	Total energy intake	×	NO
	BMI	×	Quality of Life & general health measures	×			
	WC (CM)	×	Frequency of take-away food consumption	×			
	BP (mmHg)	×	& eating while watching TV				
Morgan et al. (2011a) HDHK	[fathers] Weight (kg)	X	Social support	na	Total energy intake	×	YES
	BMI	X	Intentions				
	WC	X	Self-efficacy				
	BP	X	Outcome expectations				
Morgan et al. (2011b) POWER	Body Weight (kg)	X	Dietary patterns	X	Dietary behaviours - Sweetened beverages (Serves).	X	YES
	BMI	X	Dietary cognitions.	X			

TABLE 4: *Cont.*

Study	Weight status/lipid studies/BP	Nutritional determinants	Nutrition Intake	Intervention Effectiveness
	WC	X		
	BP	X		
Pritchard et al. (1997)	Body Weight (kg)	X	-	Energy intake (kcal) ×	YES
Tilley et al. (1999, 1997)	Fat Mass	X		Percentage dietary fat ×	YES
	-	-	-	Total Fat (% Energy) X	
				Fibre intake (g/1000kj) X	
				Fruit & Vegetable intakes X (Serves)	

Note: X = Significant between group difference identified (See table one) ; × = NO statistically significant change between intervention vs control identified.

[28]. A range of dietary behaviours were targeted, most commonly a combination of total fat reduction and increasing daily fruit and vegetable intake [3,27,28], dietary patterns and/or nutrition knowledge [3,24,27].

2.3.8 MODE OF INTERVENTION DELIVERY

2.3.8.1 GROUP-BASED DELIVERY

Three group-based intervention studies were included in this review [23,27,28]. All three delivered face-to-face group sessions and print-based materials in a group-based setting to convey information on healthy eating [27,28] and physical activity [23]. One of these studies included telephone delivery of motivational tips and feedback on diet quality during the intervention [28]. Two studies were worksite interventions [27,28] and one study accessed families in the community [23]. Interventions were dietician-led (sex non-defined) [27], male researcher-led [23] or researcher-led (sex non-defined) [28]. The "Healthy Dads, Healthy Kids" study included the use of a publically available website, CalorieKing™, for self-monitoring dietary intake and physical activity habits. Fifty-three obese males (mean age 40.6 +/− 7.1 years) and their primary school aged children (n=71) participated in the study which compared a structured, 8 week face-to-face group program known as the "Healthy Dads, Healthy Kids" program, against a wait-list control group [23]. The program followed the constructs of Social Cognitive Theory and Family Systems Theory and delivered dietary information based on the Australian Guide to Healthy Eating's recommended daily intakes, meal planning, food label reading and goal setting in addition to physical activity messages [23]. There was a significant between-group difference in weight loss at 6 months (7.6 kg vs 0.0 kg, P<0.001 for treatment effect), and a significant correlation was identified between percentage weight loss and number of diet entries, daily physical activity entries and weekly weight check-ins. There was no measurable effect for changes in total energy and dietary intake [23]. The authors noted that targeting fathers for lifestyle interventions improved the health behaviours of their children [23].

The worksite interventions included greater numbers of participants (n=638) [27], (n=5,042) [28] compared to the community run intervention (n=53) [23]. The study by Braeckman and colleagues [27] recruited 638 males (mean age 43.7 +/− 6.6) from 4 worksites to participate in the study comparing a 3-month generic low-fat dietary intervention with control group sites, where no dietary education or information was provided [27]. The intervention sites were exposed to a range of media, including posters, leaflets, newsletters and videos outlining the importance of a low-fat diet in reducing the risk of heart disease and provided optional 2-hour dietician-led dietary group education sessions which were run out of business hours [27]. At 3 months there was a significant between-group difference for total energy and total fat, with the intervention groups reporting a reduction of 8.6% and 7.2% respectively (p<0.05) and a significant improvement of 44.5% in nutrition knowledge (p<0.001) [28]. Body mass index (BMI) for the intervention group increased by 0.3 kg/m^2 compared to the control group and the authors suggest an increase in BMI was observed in the intervention group as a result of participants overeating "healthier" foods and under-reporting post intervention dietary intakes [27].. There was no measurable effect for percent of energy from types of fat or total cholesterol between groups [27].

The final study by Tilley and colleagues [28] recruited 5042 participants from 28 male dominant worksites (mean age 56.5 +/− 12.2 years) into the Next Step Trial; >90% of this cohort were males [28,29].

The Next Step Trial was a multi-component, cancer control program, which compared a generic 2-year face-to-face group program against control groups [28,29]. The Next Step Trial followed the constructs of Social Cognitive Theory and the transtheoretical model of change to deliver dietary information that promoted a reduced fat dietary intake and an increased fruit and vegetable intake according to the USDA food guide pyramid [28,29]. Five nutrition education groups, provided during paid working hours, delivered information regarding diet quality, behaviour change, goal setting, tips on how to obtain family and co-workers support and information regarding the importance of cancer screening [28,29]. One motivational telephone call was delivered to participants, and group attendance records were maintained [28,29]. Computer generated person-alised feedback was provided to participants in the form of graphs com-paring employee's dietary intake against the USDA food guide pyramid [28,29]. The intervention sites were exposed to a range of media including posters, leaflets and newsletters on how to reduce dietary fat and increase fruit and vegetable intake [28,29]. There was a significant between-group effect for all outcome measures at 12-months (p<0.006), with total energy reducing by 1.5%, dietary fibre increasing by 0.5 g/1000Kcal and daily fruit and vegetable servings increasing by 0.13 per day, with a significant effect observed for fibre only at 24-months (p =0.002) [28,29]. A signifi-cant dose–response was observed for group attendance rates, with those attending all 5 group sessions (19%) reducing their total energy from di-etary fat by 3%, increasing dietary fibre by 1 g/1000Kcal and fruit and vegetable intakes increasing between 0.3 to 0.7 serves per day [28,29].

This study identifies that participation in worksite health promotion initiatives where employers enable employees paid release time from work to attend education sessions may be an effective strategy to enhance intervention effectiveness [28,29].

2.3.8.2 FACE-TO-FACE DELIVERY

Face-to-face methods featured in all nine interventions [3,10,22-28]. All studies combined face-to-face delivery with at least one of the following additional modes of delivery including internet [3,22-25] or print-based material [10,22-24,26-28], telephone contact [3,28] and group-based con-tact [23,27,28]. Participants from six of the studies were recruited from worksite environments [10,24-28] whilst three studies recruited from the general community [3,22,23]. The frequency of face-to-face contact var-ied between one session only [19]; fortnightly [3,22,24-27] and monthly [10,28]; with nutrition education sessions led by a dietician [3,25-27], male researcher [22-24], or researcher (sex not defined) [10,28]; with time for dietary intervention delivery ranging from 15 minutes to two hours [3,10,22-28]. Of the nine studies, four had developed the behaviour change intervention based on Social Cognitive Theory [10,22,23,28] to guide goal setting and self-monitoring behaviours and to guide researcher feedback on participant performance [10,22,23,28]. All face-to-face studies included the provision of individualised feedback on dietary intake patterns, adherence to targets and weight loss goals to participants. Of the nine studies, only three reported significant positive between-group effects for changes in dietary intake patterns [3,27,28]. The comparison groups in these studies were generic low-fat dietary intervention [3], no access to information or support [27,28].

Booth and colleagues [3] studied the impact of prescribed dietary targets vs. generalised information for fruit, vegetable, and dairy consumption in a cohort of 63 males (mean age 48 years) over a 12-week period [3]. Dietician-led face-to-face sessions were conducted fortnightly and supplemented with two phone counselling sessions over the study duration [3]. Findings by Booth et al. suggested that the setting of quantitative-based dietary targets performed better than generic information in improving self-reported dietary behaviours [3].

2.3.8.3 INTERNET-BASED DELIVERY

Three studies reported utilising internet technology; three were work-site based programs utilising on-site intranet access including email [24,25] and one was a community based program [23]. The common internet-based features utilised in the abovementioned studies were: interactivity with the user [22-24]; personalised feedback based on user entered food data [23-25]; and use of transtheoretical behaviour change constructs within nutrition messages [22-24]. The Transtheoretical model assesses an individual's readiness to act on a new healthier behaviour and provides strategies or processes of change to guide the individual through the stages of change to action and maintenance [30].

Leslie et al. conducted a study where a dietician delivered generic dietary information via email in combination with face-to-face sessions [25]. Although weight loss was significant in both intervention groups compared to the control group, adherence with dietary advice was limited [25].

2.4 DISCUSSION

This review evaluated recent literature where interventions targeted males in community and worksite settings as a means to improve nutrition and dietary behaviours. From the nine studies included in this review, seven were considered effective in achieving satisfactory outcomes [3,10,22-24,26-28]. Few male-based dietary intervention studies have focussed primarily on weight management [10,22-24,26], and to a lesser extent on diet quality changes [3,25,27,28]. Many lacked sufficient detail on intervention description and few described any behavioural theory underpinning the approach used. This makes interpretation of the effectiveness of different delivery modes difficult to determine, with this review suggesting that intervention effectiveness was not associated with mode of intervention delivery.

Little information exists for why males are under-represented in dietary modification interventions [17], suggesting that future investigations need to consider the barriers preventing males from engaging in dietary modification strategies. This was reinforced by Robertson et al. who were unable to determine any effective approaches to enhance the uptake of health services by males [18].

This review identified an increasing trend towards recruitment of male participants within the workplace. Six of the nine studies included in this review recruited male participants from worksite environments [10,24-28], two incorporated an internet-based methodology for delivering interventions [24,25]. This may suggest that utilising worksites where the internet is readily available to employees may be an effective delivery mechanism with the potential to be a wide reaching and cost-effective option for targeting the male population [18-20,31].

An important issue for interventions is getting participants to effectively and frequently engage in the provided programs. Previous studies have shown that higher use of internet-based features combined with telephone interaction proved to be effective in changing dietary behaviours [19,32]. The features of successful internet-based lifestyle change programs remain to be identified, and as noted by Neve and colleagues [19], it is not yet possible to attribute the effectiveness of internet-based interventions to specific intervention components due to heterogeneity of study designs. Manzoni et al. supported this view but noted greater efficacy of behavioural internet-based programs which included tailored feedback on self-monitoring of weight, eating and activity over education only internet-based interventions [33]. Qualitative feedback received through semi-structured interviews from participants of the SHED-IT trial strengthen the comments made by Manzoni et al. reporting that most respondents found that Calorie King™ was a invaluable component of the SHED-IT program, reported to aid in their understanding of the effects of diet and exercise on weight loss due to the instant and visual feedback gained online [34]. However, limitations for the use of Calorie king™ existed including time needed to enter daily intake, navigation of the website was not intuitive and respondents suggesting that the foods contained within Calorie King™ did not cater for those who did not dependant of fast foods, pre-prepared or standard foods [34].

To enhance effectiveness of dietary interventions for males, literature suggests that nutrition messages need to remain clear, concise and achievable [3,35]. Messages are also well received when they are based on the participants' identified needs and presented frequently in an engaging and fun manner [3,10,23,28,35]. This result is consistent with Morgan et al.'s evaluation of the SHED-IT intervention, which highlighted that males were attracted to programmes that do not require extensive time commitment but present key nutrition messages in a thoughtful but comical manner [34]. Practical advice, such as simple weekly meal plans that are inclusive of "treat" foods and beverages recipes, and the engagement of wives or partners, has been deemed helpful to overcome the barriers to healthy eating in males [14,34].

Several studies demonstrated the importance of self-monitoring (e.g. keeping dietary intake records, tracking/monitoring weight online) to enhance adherence and encourage long-term behaviour change [3,22-25,28]. Self-monitoring is strongly associated with program adherence and positive long-term behaviour change. Previous studies have recognised that more frequent self-monitoring of dietary behaviour is positively correlated with weight loss independent of delivery mode [36-39].

Although excluded from this review due to the lack of any control group comparison, Auon et al. provide valuable insight on how collaboration between health care service providers and existing community groups/clubs can lead to sustainable lifestyle strategies for males [40].

Auon et al. [40] delivered a community-based lifestyle intervention program recruiting 750 overweight males from 23 Rotary Groups. The program was titled the "Waist Disposal Challenge" and was delivered by health service dieticians and exercise physiologists. These included face-to-face group educational presentations (up to three over 12 months), a face-to-face BMI monitoring/competition - weight and height measured and recorded monthly with competition between Rotary clubs and telephone lifestyle coaching through four calls to each participant [40]. This study found significant decreases in BMI; increased awareness of lifestyle changes that promote weight loss; and increased motivation through friendly club competition, suggesting that face-to-face group presentations on nutrition and physical activity may help males to facilitate lifestyle changes [40]. This perception was also observed by Morgan et al.'s findings whereby participants of the SHED-IT trial suggested improvements for future trials be to consider increasing face-to-face contact in the form of small groups and meetings [34].

Also highlighted was that similar community-based projects that utilise existing community groups/clubs in collaboration with health care service providers, potentially enable interventions to be self-managed and sustainable [40]. The study undertaken by Morgan et al. [24] with male shift workers clearly demonstrates that information sessions, program booklets, group-based financial incentives and an online component in the Workplace Power Program was able to achieve reductions in weight and soft drink consumption [41]. However, Collins et al. [42] noted that whilst males were able to reduce portion size in the SHED-IT Program, reductions in alcohol or increases in fruit and vegetable intake were not achieved, recommending that specific food based guidelines be included in future programs for males [41,42].

2.4.1 STRENGTHS AND LIMITATIONS

The strengths of this review are that it draws on evidence from randomised control trials targeting males and systematically assesses the methodological quality of studies. However, it is not without limitations. Although the majority of the studies in our review reported effective interventions, the findings from this review are based on a small number of studies of which trials were limited by short intervention periods and absence of long-term follow-up. Additionally publication bias could be a result of the broad search focus that may have resulted in relevant papers being missed; this is in addition to unpublished (grey literature) not being sought.

2.5 CONCLUSION

The area of male health has been recently ignited on the health agenda [1,2] but remains under-developed in exploring effective dietary interventions. Although seven of the nine reviewed studies were effective in achieving weight-loss outcomes and/or change in dietary practices in the short-term, uncertainty remains as to the primary features of successful dietary interventions for males due to limited details provided for nutrition intervention protocols, variability in mode of delivery and comparisons as well as content of information provided to participants between studies. However, the reviewed studies do provide valuable insight to potential recruitment and intervention strategies for this hard to reach male population, favouring worksite recruitment practices where employer engagement and intervention support is obtained.

To enable evaluation of intervention effectiveness between studies it is recommended that larger scale effectiveness studies that report nutrition intervention protocols and content of information provided in males (or gender stratified) are needed.

Although outside the scope of this review, studies exploring the cost-effectiveness of worksite vs community interventions targeting eating behaviour and energy balance would be valuable to explore.

REFERENCES

1. Australian Institute of Health and Welfare: The Health of Australia's Males. Canberra: AIHW; 2011. AIHW Cat. no. PHE 141

2. Department of Health and Ageing: National Male Health Policy: Building on the Strengths of Australian Males. Canberra: Department of Health and Ageing; 2010.

3. Booth A, Nowson C, Worsley A, Margerison C, Jorna M: Dietary approaches for weight loss with increased intakes of fruit, vegetables and dairy products. Nutr Diet 2008, 65:115-120.

4. Australian Institute of Health and Welfare: Indicators for Chronic Diseases and Their Determinants, 2008. Canberra: AIHW, Cat. no. PHE 75; 2008. AIHW Cat. no. PHE 75

5. National Health and Medical Research Council: Dietary Guidelines for Australian Adults. Canberra: NH&MRC; 2003.

6. Australian Institute of Health and Welfare: Australia's Health 2010. Australia's Health Series no. 12. Cat no. AUS 122. Canberra: AIHW; 2010.

7. World Health Organisation: World Health Statistics 2010. Geneva: World Health Organisation; 2010.

8. Williams DR: The health of men: structured inequalities and opportunities. Am J Public Health 2003, 93(5):724-31.

9. Gough B, Conner M: Barriers to healthy eating amongst men: a qualitative analysis. Soc Sci Med 2006, 62:387-395.

10. Arao T, Oida Y, Maruyama C, Mutou T, Sawada S, Matsuzuki H, et al.: Impact of lifestyle intervention on physical activity and diet of Japanese workers. Prev Med 2007, 45(2–3):146-52.

11. Crawford D, Baghurst K: Diet and health: a national survey of beliefs, behaviours and barriers to change in the community. Aust J Nutr Diet 1990, 47:97-104.

12. López-Azpiazu I, Martínez-González M, Kearney J, Gibney M, Martínez J: Perceived barriers of, and benefits to, healthy eating reported by a Spanish national sample. Public Health Nutr 2007, 2:209-215.

13. Milligan R, Burke V, Beilin L, et al.: Health-related behaviours and psycho-social characteristics of 18 year-old Australians. Soc Sci Med 1997, 45:1549-1562.

14. O'Kane G, Craig P, Black D, Thorpe C: The riverina men's study: rural australian men's barriers to healthy lifestyle habits. Int J Men's Health 2008, 7:237-254.

15. Lappalainen R, Saba A, Holm L, Mykkanen H, Gibney M, Moles A: Difficulties in trying to eat healthier: descriptive analysis of perceived barriers for healthy eating. Eur J Clin Nutr 1997, 51:S36.

16. Newsom J, Kaplan M, Huguet N, McFarland B: Health behaviors in a representative sample of older Canadians: prevalences, reported change, motivation to change, and perceived barriers. Gerontologist 2004, 44:193-205.

17. Pagoto SL, Schneider K, Oleski JL, Luciani JM, Bodenlos JS, Whited MC: Male inclusion in randomised controlled trials of lifestyle weight loss interventions. Obesity 2011, 20(6):1234-1239.

18. Robertson LM, Douglas F, Ludbrook A, Reid G, van Teijlingen E: What works with men? A systematic review of health promoting interventions targeting men. BMC Health Serv Res 2008, 8:141.

19. Neve M, Morgan PJ, Jones PR, Collins CE: Effectiveness of web-based interventions in achieving weight loss and weight loss maintenance in overweight and obese adults: a systematic review with meta-analysis. Obes Rev 2010, 11:306-321.

20. Norman GJ, Zabinski MF, Adams MA, Rosenberg DE, Yaroch AL, Atienza AA: A review of eHealth interventions for physical activity and dietary behavior change. Am J Prev Med 2007, 33(4):336-45.
21. Thomas B, Ciliska D, Dobbins M, Micucci S: A process for systematically reviewing the literature: providing the research evidence for public health nursing interventions. Worldviews Evid Based Nurs 2004, 1:176-184.
22. Morgan PJ, Lubans DR, Collins CE, Warren JA, Callister R: The SHED-IT randomized controlled trial: evaluation of an internet-based weight-loss program for men. Obesity 2009, 17(11):2025-32.
23. Morgan P, Lubans D, Callister R, Okely A, Burrows T, Fletcher R, et al.: The 'Healthy Dads, Healthy Kids' randomized controlled trial: Efficacy of a healthy lifestyle program for overweight fathers and their children. Int J Obes 2011, 35:436-447.
24. Morgan PJ, Collins CE, Plotnikoff RC, Cook AT, Berthon B, Mitchell S, Callister R: Efficacy of a workplace-based weight loss program for overweight male shift workers: the Workplace POWER (Preventing Obesity Without Eating like a Rabbit) randomized controlled trial. Prev Med 2011 May 1, 52(5):317-25.
25. Leslie W, Lean M, Baillie H, Hankey C: Weight management: a comparison of existing dietary approaches in a work-site setting. Int J Obes 2002, 26:1469-1475.
26. Pritchard J, Nowson C, Wark J: A worksite program for overweight middle-aged men achieves lesser weight loss with exercise than with dietary change. J Am Diet Assoc 1997, 97(1):37-42.
27. Braeckman L, De Bacquer D, Maes L, De Backer G: Effects of a low-intensity worksite-based nutrition intervention. Occup Med 1999, 49(8):549.
28. Tilley BC, Glanz K, Kristal AR, Hirst K, Shuhui L, Vernon SW, Myers R: Nutrition Intervention for high-risk auto workers: results of the next step trial. Prev Med 1999, 28:284-92.
29. Tilley BC, Vernon SW, Glanz K, Myers R, Sanders K, Lu M, Hirst K, Kristal AR, Smereka C, Sowers MF: Worksite cancer screening and nuririotn interventio fro high-risk auto workers: design and baseline findings of the Next Step trial. Prev Med 1997, 26:227-35.
30. Prochaska J, Velicer W: The transtheoretical model of health behavior change. Am J Health Promot 1997, 12:38-48.
31. Leslie W, Tan F, Brug J: Short-term efficacy of a web-based computer-tailored nutrition intervention: main effects and mediators. Ann Behav Med 2005, 29:54-63.
32. Wyatt HR, Ogden LG, Cassic KS, Hoagland EA, McKinnon T, Eich N, Chernyshev BS, Wood T, Cuomo J, Hill JO: Successful internet-based lifestyle change program on body weight and markers of metabolic health. Obesity and Weight Management 2009, 5(4):167-173.
33. Manzoni GM, Pagnini F, Corti S, Molinari E, Castelnuovo G: Internet-based behavioral interventions for obesity: an updated systematic review. Clin Pract Epidemiol Ment Health. 2011 Mar 4, 7:19-28.
34. Morgan PJ, Warren JM, Lubans DR, Collins CE, Callister R: Engaging men in weight loss: experiences of men who participated in the male only SHED-IT pilot study. Obesity Research and Clinical Practice. 2011 March, 5:e239-e248.
35. De Bourdeaudhuij I, Stevens V, Vandelanotte C, Brug J: Evaluation of an interactive computer-tailored nutrition intervention in a real-life setting. Ann Behav Med 2007, 33:39-48.

36. Burke LE, Warziski M, Starrett T, Choo J, Music E, Sereika S, Stark S, Sevick MA: Self-monitoring dietary intake: current and future practices. J Renal Nutr 2005, 15(3):281-90.

37. Yon BA, Johnson RK, Harvey-Berino J, Gold BC, Howard AB: Personal digital assistants are comparable to traditional diaries for dietary self-monitoring during a weight loss program. J Behav Med 2007 Apr, 30(2):165-75.

38. Butryn M, Phelan S, Hill J, Wing R: Consistent self-monitoring of weight: a key component of successful weight loss maintenance. Obesity 2007, 15:3091-3096.

39. Paul-Ebhohimhen V, Avenell A: A systematic review of the effectiveness of group versus individual treatments for adult obesity. Obes Facts 2009, 2:17-24.

40. Aoun S, Osseiran-Moisson R, Collins F, Newton R, Newton MA: self-management concept for men at the community level: The `Waist' Disposal Challenge. J Health Psychol 2009, 14(5):663-74.

41. Morgan PJ, Lubans DR, Collins CE, Warren JM, Callister R: 12-month outcomes and process evaluation of the SHED-IT RCT: an internet-based weight loss program targeting men. Obesity 2011 Jan, 19(1):142-51.

42. Collins CE, Morgan PJ, Warren JM, Lubans DR, Callister R: Men participating in a weight-loss intervention are able to implement key dietary messages, but not those relating to vegetables or alcohol: the Self-Help, Exercise and Diet using Internet Technology (SHED-IT) study. Public Health Nutr 2011 Jan, 14(1):168-75.

CHAPTER 3

NUTRIENT TIMING REVISITED: IS THERE A POST-EXERCISE ANABOLIC WINDOW?

ALAN ALBERT ARAGON and BRAD JON SCHOENFELD

3.1 INTRODUCTION

Over the past two decades, nutrient timing has been the subject of numerous research studies and reviews. The basis of nutrient timing involves the consumption of combinations of nutrients--primarily protein and carbohydrate--in and around an exercise session. The strategy is designed to maximize exercise-induced muscular adaptations and facilitate repair of damaged tissue [1]. Some have claimed that such timing strategies can produce dramatic improvements in body composition, particularly with respect to increases in fat-free mass [2]. It has even been postulated that the timing of nutritional consumption may be more important than the absolute daily intake of nutrients [3].

The post-exercise period is often considered the most critical part of nutrient timing. An intense resistance training workout results in the depletion of a significant proportion of stored fuels (including glycogen and amino acids) as well as causing damage to muscle fibers. Theoretically, consuming the proper ratio of nutrients during this time not only initiates the rebuilding of damaged tissue and restoration of energy reserves, but it does so in a supercompensated fashion that enhances both body composition and exercise performance. Several researchers have made reference to an "anabolic window of opportunity" whereby a limited time exists after training to optimize training-related muscular adaptations [3-5].

However, the importance – and even the existence – of a post-exercise 'window' can vary according to a number of factors. Not only is nutrient timing research open to question in terms of applicability, but recent evidence has directly challenged the classical view of the relevance of post-exercise nutritional intake on anabolism. Therefore, the purpose of this paper will be twofold: 1) to review the existing literature on the effects of nutrient timing with respect to post-exercise muscular adaptations, and; 2) to draw relevant conclusions that allow evidence-based nutritional recommendations to be made for maximizing the anabolic response to exercise.

3.1.1 GLYCOGEN REPLETION

A primary goal of traditional post-workout nutrient timing recommendations is to replenish glycogen stores. Glycogen is considered essential to optimal resistance training performance, with as much as 80% of ATP production during such training derived from glycolysis [6]. MacDougall et al. [7] demonstrated that a single set of elbow flexion at 80% of 1 repetition maximum (RM) performed to muscular failure caused a 12% reduction in mixed-muscle glycogen concentration, while three sets at this intensity resulted in a 24% decrease. Similarly, Robergs et al. [8] reported that 3 sets of 12 RM performed to muscular failure resulted in a 26.1% reduction of glycogen stores in the vastus lateralis while six sets at this intensity led to a 38% decrease, primarily resulting from glycogen depletion in type II fibers compared to type I fibers. It therefore stands to reason that typical high volume bodybuilding-style workouts involving multiple exercises and sets for the same muscle group would deplete the majority of local glycogen stores.

In addition, there is evidence that glycogen serves to mediate intracellular signaling. This appears to be due, at least in part, to its negative regulatory effects on AMP-activated protein kinase (AMPK). Muscle anabolism and catabolism are regulated by a complex cascade of signaling pathways. Several pathways that have been identified as particularly important to muscle anabolism include mammalian target of rapamycin (mTOR), mitogen-activated protein kinase (MAPK), and various calcium-(Ca^{2+}) dependent pathways. AMPK, on the other hand, is a cellular energy sensor that serves to enhance energy availability. As such, it blunts energy-consuming processes including the activation of mTORC1 mediated by insulin and mechanical tension, as well as heightening catabolic processes such as glycolysis, beta-oxidation, and protein degradation [9]. mTOR is considered a master network in the regulation of skeletal muscle growth [10,11], and its inhibition has a decidedly negative effect on anabolic processes [12]. Glycogen has been shown to inhibit purified AMPK in cell-free assays [13], and low glycogen levels are associated with an enhanced AMPK activity in humans in vivo[14].

Creer et al. [15] demonstrated that changes in the phosphorylation of protein kinase B (Akt) are dependent on pre-exercise muscle glycogen content. After performing 3 sets of 10 repetitions of knee extensions with a load equating to 70% of 1 repetition maximum, early phase post-exercise Akt phosphorylation was increased only in the glycogen-loaded muscle, with no effect seen in the glycogen-depleted contralateral muscle. Glycogen inhibition also has been shown to blunt S6K activation, impair translation, and reduce the amount of mRNA of genes responsible for regulating muscle hypertrophy [16,17]. In contrast to these findings, a recent study by Camera et al. [18] found that high-intensity resistance training with low muscle glycogen levels did not impair anabolic signaling or muscle protein synthesis (MPS) during the early (4 h) postexercise recovery period. The discrepancy between studies is not clear at this time.

Glycogen availability also has been shown to mediate muscle protein breakdown. Lemon and Mullin [19] found that nitrogen losses more than doubled following a bout of exercise in a glycogen-depleted versus glycogen-loaded state. Other researchers have displayed a similar inverse relationship between glycogen levels and proteolysis [20]. Considering the totality of evidence, maintaining a high intramuscular glycogen content at the onset of training appears beneficial to desired resistance training outcomes.

Studies show a supercompensation of glycogen stores when carbohydrate is consumed immediately post-exercise, and delaying consumption by just 2 hours attenuates the rate of muscle glycogen re-synthesis by as much as 50% [21]. Exercise enhances insulin-stimulated glucose uptake following a workout with a strong correlation noted between the amount of uptake and the magnitude of glycogen utilization [22]. This is in part due to an increase in the translocation of GLUT4 during glycogen depletion [23,24] thereby facilitating entry of glucose into the cell. In addition, there is an exercise-induced increase in the activity of glycogen synthase—the principle enzyme involved in promoting glycogen storage [25]. The combination of these factors facilitates the rapid uptake of glucose following an exercise bout, allowing glycogen to be replenished at an accelerated rate.

There is evidence that adding protein to a post-workout carbohydrate meal can enhance glycogen re-synthesis. Berardi et al. [26] demonstrated that consuming a protein-carbohydrate supplement in the 2-hour period following a 60-minute cycling bout resulted in significantly greater glycogen resynthesis compared to ingesting a calorie-equated carbohydrate solution alone. Similarly, Ivy et al. [27] found that consumption of a combination of protein and carbohydrate after a 2+ hour bout of cycling and sprinting increased muscle glycogen content significantly more than either a carbohydrate-only supplement of equal carbohydrate or caloric equivalency. The synergistic effects of protein-carbohydrate have been attributed to a more pronounced insulin response [28], although it should be noted that not all studies support these findings [29]. Jentjens et al. [30] found that given ample carbohydrate dosing (1.2 g/kg/hr), the addition of a protein and amino acid mixture (0.4 g/kg/hr) did not increase glycogen synthesis during a 3-hour post-depletion recovery period.

Despite a sound theoretical basis, the practical significance of expeditiously repleting glycogen stores remains dubious. Without question, expediting glycogen resynthesis is important for a narrow subset of endurance sports where the duration between glycogen-depleting events is limited to less than approximately 8 hours [31]. Similar benefits could potentially be obtained by those who perform two-a-day split resistance training bouts (i.e. morning and evening) provided the same muscles will be worked during the respective sessions. However, for goals that are not specifically focused on the performance of multiple exercise bouts in the same day, the urgency of glycogen resynthesis is greatly diminished. High-intensity resistance training with moderate volume (6-9 sets per muscle group) has only been shown to reduce glycogen stores by 36-39% [8,32]. Certain athletes are prone to performing significantly more volume than this (i.e., competitive bodybuilders), but increased volume typically accompanies decreased frequency. For example, training a muscle group with 16-20 sets in a single session is done roughly once per week, whereas routines with 8-10 sets are done twice per week. In scenarios of higher volume and frequency of resistance training, incomplete resynthesis of pre-training glycogen levels would not be a concern aside from the far-fetched scenario where exhaustive training bouts of the same muscles occur after recovery intervals shorter than 24 hours. However, even in the event of complete glycogen depletion, replenishment to pre-training levels occurs well-within this timeframe, regardless of a significantly delayed post-exercise carbohydrate intake. For example, Parkin et al [33] compared the immediate post-exercise ingestion of 5 high-glycemic carbohydrate meals with a 2-hour wait before beginning the recovery feedings. No significant between-group differences were seen in glycogen levels at 8 hours and 24 hours post-exercise. In further support of this point, Fox et al. [34] saw no significant reduction in glycogen content 24 hours after depletion despite adding 165 g fat collectively to the post-exercise recovery meals and thus removing any potential advantage of high-glycemic conditions.

3.1.2 PROTEIN BREAKDOWN

Another purported benefit of post-workout nutrient timing is an attenuation of muscle protein breakdown. This is primarily achieved by spiking insulin levels, as opposed to increasing amino acid availability [35,36]. Studies show that muscle protein breakdown is only slightly elevated immediately post-exercise and then rapidly rises thereafter [36]. In the fasted state, muscle protein breakdown is significantly heightened at 195 minutes following resistance exercise, resulting in a net negative protein balance [37]. These values are increased as much as 50% at the 3 hour mark, and elevated proteolysis can persist for up to 24 hours of the post-workout period [36].

Although insulin has known anabolic properties [38,39], its primary impact post-exercise is believed to be anti-catabolic [40-43]. The mechanisms by which insulin reduces proteolysis are not well understood at this time. It has been theorized that insulin-mediated phosphorylation of PI3K/Akt inhibits transcriptional activity of the proteolytic Forkhead family of transcription factors, resulting in their sequestration in the sarcoplasm away from their target genes [44]. Down-regulation of other aspects of the ubiquitin-proteasome pathway are also believed to play a role in the process [45]. Given that muscle hypertrophy represents the difference between myofibrillar protein synthesis and proteolysis, a decrease in protein breakdown would conceivably enhance accretion of contractile proteins and thus facilitate greater hypertrophy. Accordingly, it seems logical to conclude that consuming a protein-carbohydrate supplement following exercise would promote the greatest reduction in proteolysis since the combination of the two nutrients has been shown to elevate insulin levels to a greater extent than carbohydrate alone [28].

However, while the theoretical basis behind spiking insulin post-workout is inherently sound, it remains questionable as to whether benefits extend into practice. First and foremost, research has consistently shown that, in the presence of elevated plasma amino acids, the effect of insulin elevation on net muscle protein balance plateaus within a range of 15–30 mU/L [45,46]; roughly 3–4 times normal fasting levels. This insulinogenic effect is easily accomplished with typical mixed meals, considering that it takes approximately 1–2 hours for circulating substrate levels to peak, and 3–6 hours (or more) for a complete return to basal levels depending on the size of a meal. For example, Capaldo et al. [47] examined various metabolic effects during a 5-hour period after ingesting a solid meal comprised of 75 g carbohydrate 37 g protein, and 17 g fat. This meal was able to raise insulin 3 times above fasting levels within 30 minutes of consumption. At the 1-hour mark, insulin was 5 times greater than fasting. At the 5-hour mark, insulin was still double the fasting levels. In another example, Power et al. [48] showed that a 45g dose of whey protein isolate takes approximately 50 minutes to cause blood amino acid levels to peak. Insulin concentrations peaked 40 minutes after ingestion, and remained at elevations seen to maximize net muscle protein balance (15-30 mU/L, or 104-208 pmol/L) for approximately 2 hours. The inclusion of carbohydrate to this protein dose would cause insulin levels to peak higher and stay elevated even longer. Therefore, the recommendation for lifters to spike insulin post-exercise is somewhat trivial. The classical post-exercise objective to quickly reverse catabolic processes to promote recovery and growth may only be applicable in the absence of a properly constructed pre-exercise meal.

Moreover, there is evidence that the effect of protein breakdown on muscle protein accretion may be overstated. Glynn et al. [49] found that the post-exercise anabolic response associated with combined protein and carbohydrate consumption was largely due to an elevation in muscle protein synthesis with only a minor influence from reduced muscle protein breakdown. These results were seen regardless of the extent of circulating insulin levels. Thus, it remains questionable as to what, if any, positive effects are realized with respect to muscle growth from spiking insulin after resistance training.

3.1.3 PROTEIN SYNTHESIS

Perhaps the most touted benefit of post-workout nutrient timing is that it potentiates increases in MPS. Resistance training alone has been shown to promote a twofold increase in protein synthesis following exercise, which is counterbalanced by the accelerated rate of proteolysis [36]. It appears that the stimulatory effects of hyperaminoacidemia on muscle protein synthesis, especially from essential amino acids, are potentiated by previous exercise [35,50]. There is some evidence that carbohydrate has an additive effect on enhancing post-exercise muscle protein synthesis when combined with amino acid ingestion [51], but others have failed to find such a benefit [52,53].

Several studies have investigated whether an "anabolic window" exists in the immediate post-exercise period with respect to protein synthesis. For maximizing MPS, the evidence supports the superiority of post-exercise free amino acids and/or protein (in various permutations with or without carbohydrate) compared to solely carbohydrate or non-caloric placebo [50,51,54-59]. However, despite the common recommendation to consume protein as soon as possible post-exercise [60,61], evidence-based support for this practice is currently lacking. Levenhagen et al. [62] demonstrated a clear benefit to consuming nutrients as soon as possible after exercise as opposed to delaying consumption. Employing a within-subject design,10 volunteers (5 men, 5 women) consumed an oral supplement containing 10 g protein, 8 g carbohydrate and 3 g fat either immediately following or three hours post-exercise. Protein synthesis of the legs and whole body was increased threefold when the supplement was ingested immediately after exercise, as compared to just 12% when consumption was delayed. A limitation of the study was that training involved moderate intensity, long duration aerobic exercise. Thus, the increased fractional synthetic rate was likely due to greater mitochondrial and/or sarcoplasmic protein fractions, as opposed to synthesis of contractile elements [36]. In contrast to the timing effects shown by Levenhagen et al. [62], previous work by Rasmussen et al. [56] showed no significant difference in leg net amino acid balance between 6 g essential amino acids (EAA) co-ingested with 35 g carbohydrate taken 1 hour versus 3 hours post-exercise. Compounding the unreliability of the post-exercise 'window' is the finding by Tipton et al. [63] that immediate pre-exercise ingestion of the same EAA-carbohydrate solution resulted in a significantly greater and more sustained MPS response compared to the immediate post-exercise ingestion, although the validity of these findings have been disputed based on flawed methodology [36]. Notably, Fujita et al [64] saw opposite results using a similar design, except the EAA-carbohydrate was ingested 1 hour prior to exercise compared to ingestion immediately pre-exercise in Tipton et al. [63]. Adding yet more incongruity to the evidence, Tipton et al. [65] found no significant difference in net MPS between the ingestion of 20 g whey immediately pre- versus the same solution consumed 1 hour post-exercise. Collectively, the available data lack any consistent indication of an ideal post-exercise timing scheme for maximizing MPS.

It also should be noted that measures of MPS assessed following an acute bout of resistance exercise do not always occur in parallel with chronic upregulation of causative myogenic signals [66] and are not necessarily predictive of long-term hypertrophic responses to regimented resistance training [67]. Moreover, the post-exercise rise in MPS in untrained subjects is not recapitulated in the trained state [68], further confounding practical relevance. Thus, the utility of acute studies is limited to providing clues and generating hypotheses regarding hypertrophic adaptations; any attempt to extrapolate findings from such data to changes in lean body mass is speculative, at best.

3.1.4 MUSCLE HYPERTROPHY

A number of studies have directly investigated the long-term hypertrophic effects of post-exercise protein consumption. The results of these trials are curiously conflicting, seemingly because of varied study design and methodology. Moreover, a majority of studies employed both pre- and post-workout supplementation, making it impossible to tease out the impact of consuming nutrients after exercise. These confounding issues highlight the difficulty in attempting to draw relevant conclusions as to the validity of an "anabolic window." What follows is an overview of the current research on the topic. Only those studies that specifically evaluated immediate (≤ 1 hour) post-workout nutrient provision are discussed (see Table 1 for a summary of data).

Esmarck et al. [69] provided the first experimental evidence that consuming protein immediately after training enhanced muscular growth compared to delayed protein intake. Thirteen untrained elderly male volunteers were matched in pairs based on body composition and daily protein intake and divided into two groups: P0 or P2. Subjects performed a progressive resistance training program of multiple sets for the upper and lower body. P0 received an oral protein/carbohydrate supplement immediately post-exercise while P2 received the same supplement 2 hours following the exercise bout. Training was carried out 3 days a week for 12 weeks. At the end of the study period, cross-sectional area (CSA) of the quadriceps femoris and mean fiber area were significantly increased in the P0 group while no significant increase was seen in P2. These results support the presence of a post-exercise window and suggest that delaying post-workout nutrient intake may impede muscular gains.

TABLE 1: Post-exercise nutrition and muscle hypertrophy

Study	Subjects	Supplementation	Protein matched with Control?	Measurement instrument	Training protocol	Results
Esmarck et al.[69]	13 untrained elderly males	10 g milk/soy protein combo consumed either immediately or 2 hours after exercise	Yes	MRI and muscle biopsy	Progressive resistance training consisting of multiple sets of lat pulldown, leg press and knee extension performed 3 days/wk for 12 wk	Significant increase in muscle CSA with immediate vs. delayed supplementation
Cribb and Hayes [70]	23 young recreational male bodybuilders	1 g/kg of a supplement containing 40 g whey isolate, 43 g glucose, and 7 g creatine monohydrate consumed either immediately before and after exercise or in the early morning and late evening	Yes	DXA and muscle biopsy	Progressive resistance training consisting of exercises for the major muscle groups performed 3 days/wk for 10 wks	Significant increases in lean body mass and muscle CSA of type II fibers in immediate vs. delayed supplementation
Willoughby et al. [71]	19 untrained young males	20 g protein or 20 g dextrose consumed 1 hour before and after exercise	No	Hydrostatic weighing, muscle biopsy, surface measurements	Progressive resistance training consisting of 3 sets of 6–8 repetitions for all the major muscles performed 4 days/wk for 10 wks	Significant increase in total body mass, fat-free mass, and thigh mass with protein vs. carb supplementation
Hulmi et al.[72]	31 untrained young males	15 g whey isolate or placebo consumed immediately before and after exercise	No	MRI, muscle biopsy	Progressive, periodized total body resistance training consisting of 2–5 sets of 5–20 repetitions performed 2 days/ wk for 21 wks.	Significant increase in CSA of the vastus lateralis but not of the other quadriceps muscles in supplemented group versus placebo.

TABLE 1: *Cont...*

Study	Subjects	Supplementation	Protein matched with Control?	Measurement instrument	Training protocol	Results
Verdijk et al.[73]	28 untrained elderly males	10 g casein hydrolysate or placebo consumed immediately before and after exercise	No	DXA, CT, and muscle biopsy	Progressive resistance training consisting of multiple sets of leg press and knee extension performed 3 days/wk for 12 wks	No significant differences in muscle CSA between groups
Hoffman et al.[74]	33 well-trained young males	Supplement containing 42 g protein (milk/collagen blend) and 2 g carbohydrate consumed either immediately before and after exercise or in the early morning and late evening	Yes	DXA	Progressive resistance training consisting of 3–4 sets of 6–10 repetitions of multiple exercises for the entire body performed 4 days/wk for 10 weeks.	No significant differences in total body mass or lean body mass between groups.
Erskine et al.[75]	33 untrained young males	20 g high quality protein or placebo consumed immediately before and after exercise	No	MRI	4-6 sets of elbow flexion performed 3 days/wk for 12 weeks	No significant differences in muscle CSA between groups

In contrast to these findings, Verdijk et al. [73] failed to detect any increases in skeletal muscle mass from consuming a post-exercise protein supplement in a similar population of elderly men. Twenty-eight untrained subjects were randomly assigned to receive either a protein or placebo supplement consumed immediately before and immediately following the exercise session. Subjects performed multiple sets of leg press and knee extension 3 days per week, with the intensity of exercise progressively increased over the course of the 12 week training period. No significant differences in muscle strength or hypertrophy were noted between groups at the end of the study period indicating that post exercise nutrient timing strategies do not enhance training-related adaptation. It should be noted that, as opposed to the study by Esmark et al. [69] this study only investigated adaptive responses of supplementation on the thigh musculature; it therefore is not clear based on these results whether the upper body might respond differently to post-exercise supplementation than the lower body.

In an elegant single-blinded design, Cribb and Hayes [70] found a significant benefit to post-exercise protein consumption in 23 recreational male bodybuilders. Subjects were randomly divided into either a PRE-POST group that consumed a supplement containing protein, carbohydrate and creatine immediately before and after training or a MOR-EVE group that consumed the same supplement in the morning and evening at least 5 hours outside the workout. Both groups performed regimented resistance training that progressively increased intensity from 70% 1RM to 95% 1RM over the course of 10 weeks. Results showed that the PRE-POST group achieved a significantly greater increase in lean body mass and increased type II fiber area compared to MOR-EVE. Findings support the benefits of nutrient timing on training-induced muscular adaptations. The study was limited by the addition of creatine monohydrate to the supplement, which may have facilitated increased uptake following training. Moreover, the fact that the supplement was taken both pre- and post-workout confounds whether an anabolic window mediated results.

Willoughby et al. [71] also found that nutrient timing resulted in positive muscular adaptations. Nineteen untrained male subjects were randomly assigned to either receive 20 g of protein or 20 grams dextrose administered 1 hour before and after resistance exercise. Training consisted of 3 sets of 6–8 repetitions at 85%–90% intensity. Training was performed 4 times a week over the course of 10 weeks. At the end of the study period, total body mass, fat-free mass, and thigh mass was significantly greater in the protein-supplemented group compared to the group that received dextrose. Given that the group receiving the protein supplement consumed an additional 40 grams of protein on training days, it is difficult to discern whether results were due to the increased protein intake or the timing of the supplement.

In a comprehensive study of well-trained subjects, Hoffman et al. [74] randomly assigned 33 well-trained males to receive a protein supplement either in the morning and evening (n = 13) or immediately before and immediately after resistance exercise (n = 13). Seven participants served as unsupplemented controls. Workouts consisted of 3–4 sets of 6–10 repetitions of multiple exercises for the entire body. Training was carried out on 4 day-a-week split routine with intensity progressively increased over the course of the study period. After 10 weeks, no significant differences were noted between groups with respect to body mass and lean body mass. The study was limited by its use of DXA to assess body composition, which lacks the sensitivity to detect small changes in muscle mass compared to other imaging modalities such as MRI and CT [76].

Hulmi et al. [72] randomized 31 young untrained male subjects into 1 of 3 groups: protein supplement (n = 11), non-caloric placebo (n = 10) or control (n = 10). High-intensity resistance training was carried out over 21 weeks. Supplementation was provided before and after exercise. At the end of the study period, muscle CSA was significantly greater in the protein-supplemented group compared to placebo or control. A strength of the study was its long-term training period, providing support for the beneficial effects of nutrient timing on chronic hypertrophic gains. Again, however, it is unclear whether enhanced results associated with protein supplementation were due to timing or increased protein consumption.

Most recently, Erskine et al. [75] failed to show a hypertrophic benefit from post-workout nutrient timing. Subjects were 33 untrained young males, pair-matched for habitual protein intake and strength response to a 3-week pre-study resistance training program. After a 6-week washout period where no training was performed, subjects were then randomly assigned to receive either a protein supplement or a placebo immediately before and after resistance exercise. Training consisted of 6– 8 sets of elbow flexion carried out 3 days a week for 12 weeks. No significant differences were found in muscle volume or anatomical cross-sectional area between groups.

3.2 DISCUSSION

Despite claims that immediate post-exercise nutritional intake is essential to maximize hypertrophic gains, evidence-based support for such an "anabolic window of opportunity" is far from definitive. The hypothesis is based largely on the pre-supposition that training is carried out in a fasted state. During fasted exercise, a concomitant increase in muscle protein breakdown causes the pre-exercise net negative amino acid balance to persist in the post-exercise period despite training-induced increases in muscle protein synthesis [36]. Thus, in the case of resistance training after an overnight fast, it would make sense to provide immediate nutritional intervention--ideally in the form of a combination of protein and carbohydrate--for the purposes of promoting muscle protein synthesis and reducing proteolysis, thereby switching a net catabolic state into an anabolic one. Over a chronic period, this tactic could conceivably lead cumulatively to an increased rate of gains in muscle mass.

This inevitably begs the question of how pre-exercise nutrition might influence the urgency or effectiveness of post-exercise nutrition, since not everyone engages in fasted training. In practice, it is common for those with the primary goal of increasing muscular size and/or strength to make a concerted effort to consume a pre-exercise meal within 1-2 hours prior to the bout in attempt to maximize training performance. Depending on its size and composition, this meal can conceivably function as both a pre- and an immediate post-exercise meal, since the time course of its digestion/absorption can persist well into the recovery period. Tipton et al. [63] observed that a relatively small dose of EAA (6 g) taken immediately pre-exercise was able to elevate blood and muscle amino acid levels by roughly 130%, and these levels remained elevated for 2 hours after the exercise bout. Although this finding was subsequently challenged by Fujita et al. [64], other research by Tipton et al. [65] showed that the ingestion of 20 g whey taken immediately pre-exercise elevated muscular uptake of amino acids to 4.4 times pre-exercise resting levels during exercise, and did not return to baseline levels until 3 hours post-exercise. These data indicate that even minimal-to-moderate pre-exercise EAA or high-quality protein taken immediately before resistance training is capable of sustaining amino acid delivery into the post-exercise period. Given this scenario, immediate post-exercise protein dosing for the aim of mitigating catabolism seems redundant. The next scheduled protein-rich meal (whether it occurs immediately or 1–2 hours post-exercise) is likely sufficient for maximizing recovery and anabolism.

On the other hand, there are others who might train before lunch or after work, where the previous meal was finished 4–6 hours prior to commencing exercise. This lag in nutrient consumption can be considered significant enough to warrant post-exercise intervention if muscle retention or growth is the primary goal. Layman [77] estimated that the anabolic effect of a meal lasts 5-6 hours based on the rate of postprandial amino acid metabolism. However, infusion-based studies in rats [78,79] and humans [80,81] indicate that the postprandial rise in MPS from ingesting amino acids or a protein-rich meal is more transient, returning to baseline within 3 hours despite sustained elevations in amino acid availability. It thus has been hypothesized that a "muscle full" status can be reached where MPS becomes refractory, and circulating amino acids are shunted toward oxidation or fates other than MPS. In light of these findings, when training is initiated more than ~3–4 hours after the preceding meal, the classical recommendation to consume protein (at least 25 g) as soon as possible seems warranted in order to reverse the catabolic state, which in turn could expedite muscular recovery and growth. However, as illustrated previously, minor pre-exercise nutritional interventions can be undertaken if a significant delay in the post-exercise meal is anticipated.

An interesting area of speculation is the generalizability of these recommendations across training statuses and age groups. Burd et al. [82] reported that an acute bout of resistance training in untrained subjects stimulates both mitochondrial and myofibrillar protein synthesis, whereas in trained subjects, protein synthesis becomes more preferential toward the myofibrillar component. This suggests a less global response in advanced trainees that potentially warrants closer attention to protein timing and type (e.g., high-leucine sources such as dairy proteins) in order to optimize rates of muscular adaptation. In addition to training status, age can influence training adaptations. Elderly subjects exhibit what has been termed "anabolic resistance," characterized by a lower receptivity to amino acids and resistance training [83]. The mechanisms underlying this phenomenon are not clear, but there is evidence that in younger adults, the acute anabolic response to protein feeding appears to plateau at a lower dose than in elderly subjects. Illustrating this point, Moore et al. [84] found that 20 g whole egg protein maximally stimulated post-exercise MPS, while 40 g increased leucine oxidation without any further increase in MPS in young men. In contrast, Yang et al. [85] found that elderly subjects displayed greater increases in MPS when consuming a post-exercise dose of 40 g whey protein compared to 20 g. These findings suggest that older subjects require higher individual protein doses for the purpose of optimizing the anabolic response to training. Further research is needed to better assess post-workout nutrient timing response across various populations, particularly with respect to trained/untrained and young/elderly subjects.

The body of research in this area has several limitations. First, while there is an abundance of acute data, controlled, long-term trials that systematically compare the effects of various post-exercise timing schemes are lacking. The majority of chronic studies have examined pre- and post-exercise supplementation simultaneously, as opposed to comparing the two treatments against each other. This prevents the possibility of isolating the effects of either treatment. That is, we cannot know whether pre- or post-exercise supplementation was the critical contributor to the outcomes (or lack thereof). Another important limitation is that the majority of chronic studies neglect to match total protein intake between the conditions compared. As such, it's not possible to ascertain whether positive outcomes were influenced by timing relative to the training bout, or simply by a greater protein intake overall. Further, dosing strategies employed in the preponderance of chronic nutrient timing studies have been overly conservative, providing only 10–20 g protein near the exercise bout. More research is needed using protein doses known to maximize acute anabolic response, which has been shown to be approximately 20–40 g, depending on age [84,85]. There is also a lack of chronic studies examining the co-ingestion of protein and carbohydrate near training. Thus far, chronic studies have yielded equivocal results. On the whole, they have not corroborated the consistency of positive outcomes seen in acute studies examining post-exercise nutrition.

Another limitation is that the majority of studies on the topic have been carried out in untrained individuals. Muscular adaptations in those without resistance training experience tend to be robust, and do not necessarily reflect gains experienced in trained subjects. It therefore remains to be determined whether training status influences the hypertrophic response to post-exercise nutritional supplementation.

A final limitation of the available research is that current methods used to assess muscle hypertrophy are widely disparate, and the accuracy of the measures obtained are inexact [68]. As such, it is questionable whether these tools are sensitive enough to detect small differences in muscular hypertrophy. Although minor variances in muscle mass would be of little relevance to the general population, they could be very meaningful for elite athletes and bodybuilders. Thus, despite conflicting evidence, the potential benefits of post-exercise supplementation cannot be readily dismissed for those seeking to optimize a hypertrophic response. By the same token, widely varying feeding patterns among individuals challenge the common assumption that the post-exercise "anabolic window of opportunity" is universally narrow and urgent.

3.2.1 PRACTICAL APPLICATIONS

Distilling the data into firm, specific recommendations is difficult due to the inconsistency of findings and scarcity of systematic investigations seeking to optimize pre- and/or post-exercise protein dosage and timing. Practical nutrient timing applications for the goal of muscle hypertrophy inevitably must be tempered with field observations and experience in order to bridge gaps in the scientific literature. With that said, high-quality protein dosed at 0.4–0.5 g/kg of LBM at both pre- and post-exercise is a simple, relatively fail-safe general guideline that reflects the current evidence showing a maximal acute anabolic effect of 20–40 g [53,84,85]. For example, someone with 70 kg of LBM would consume roughly 28–35 g protein in both the pre- and post exercise meal. Exceeding this would be have minimal detriment if any, whereas significantly under-shooting or neglecting it altogether would not maximize the anabolic response.

Due to the transient anabolic impact of a protein-rich meal and its potential synergy with the trained state, pre- and post-exercise meals should not be separated by more than approximately 3–4 hours, given a typical resistance training bout lasting 45–90 minutes. If protein is delivered within particularly large mixed-meals (which are inherently more anticatabolic), a case can be made for lengthening the interval to 5–6 hours. This strategy covers the hypothetical timing benefits while allowing significant flexibility in the length of the feeding windows before and after training. Specific timing within this general framework would vary depending on individual preference and tolerance, as well as exercise duration. One of many possible examples involving a 60-minute resistance training bout could have up to 90-minute feeding windows on both sides of the bout, given central placement between the meals. In contrast, bouts exceeding typical duration would default to shorter feeding windows if the 3–4 hour pre- to post-exercise meal interval is maintained. Shifting the training session closer to the pre- or post-exercise meal should be dictated by personal preference, tolerance, and lifestyle/scheduling constraints.

Even more so than with protein, carbohydrate dosage and timing relative to resistance training is a gray area lacking cohesive data to form concrete recommendations. It is tempting to recommend pre- and post-exercise carbohydrate doses that at least match or exceed the amounts of protein consumed in these meals. However, carbohydrate availability during and after exercise is of greater concern for endurance as opposed to strength or hypertrophy goals. Furthermore, the importance of co-ingesting post-exercise protein and carbohydrate has recently been challenged by studies examining the early recovery period, particularly when sufficient protein is provided. Koopman et al [52] found that after full-body resistance training, adding carbohydrate (0.15, or 0.6 g/kg/hr) to amply dosed casein hydrolysate (0.3 g/kg/hr) did not increase whole body protein balance during a 6-hour post-exercise recovery period compared to the protein-only treatment. Subsequently, Staples et al [53] reported that after lower-body resistance exercise (leg extensions), the increase in post-exercise muscle protein balance from ingesting 25 g whey isolate was not improved by an additional 50 g maltodextrin during a 3-hour recovery period. For the goal of maximizing rates of muscle gain, these findings support the broader objective of meeting total daily carbohydrate need instead of specifically timing its constituent doses. Collectively, these data indicate an increased potential for dietary flexibility while maintaining the pursuit of optimal timing.

REFERENCES

1. Kerksick C, Harvey T, Stout J, Campbell B, Wilborn C, Kreider R, Kalman D, Ziegenfuss T, Lopez H, Landis J, Ivy JL, Antonio J: International Society of Sports Nutrition position stand: nutrient timing. J Int Soc Sports Nutr. 2008, 5:17.

2. Ivy J, Portman R: Nutrient Timing: The Future of Sports Nutrition. North Bergen, NJ: Basic Health Publications; 2004.

3. Candow DG, Chilibeck PD: Timing of creatine or protein supplementation and re-sistance training in the elderly. Appl Physiol Nutr Metab 2008, 33(1):184-90.

4. Hulmi JJ, Lockwood CM, Stout JR: Effect of protein/essential amino acids and resistance training on skeletal muscle hypertrophy: A case for whey protein. Nutr Metab (Lond). 2010, 7:51. BioMed Central Full Text

5. Kukuljan S, Nowson CA, Sanders K, Daly RM: Effects of resistance exercise and fortified milk on skeletal muscle mass, muscle size, and functional performance in middle-aged and older men: an 18-mo randomized controlled trial. J Appl Physiol 2009, 107(6):1864-73.

6. Lambert CP, Flynn MG: Fatigue during high-intensity intermittent exercise: applica-tion to bodybuilding. Sports Med. 2002, 32(8):511-22.

7. MacDougall JD, Ray S, Sale DG, McCartney N, Lee P, Garner S: Muscle substrate utilization and lactate production. Can J Appl Physiol 1999, 24(3):209-15.

8. Robergs RA, Pearson DR, Costill DL, Fink WJ, Pascoe DD, Benedict MA, Lam-bert CP, Zachweija JJ: Muscle glycogenolysis during differing intensities of weight-resistance exercise. J Appl Physiol 1991, 70(4):1700-6.

9. Goodman CA, Mayhew DL, Hornberger TA: Recent progress toward understand-ing the molecular mechanisms that regulate skeletal muscle mass. Cell Signal 2011, 23(12):1896-906.

10. Bodine SC, Stitt TN, Gonzalez M, Kline WO, Stover GL, Bauerlein R, Zlotchenko E, Scrimgeour A, Lawrence JC, Glass DJ, Yancopoulos GD: Akt/mTOR pathway is a crucial regulator of skeletal muscle hypertrophy and can prevent muscle atrophy in vivo. Nat Cell Biol. 2001, 3(11):1014-9.

11. Jacinto E, Hall MN: Tor signalling in bugs, brain and brawn. Nat Rev Mol Cell Biol 2003, 4(2):117-26.

12. Izumiya Y, Hopkins T, Morris C, Sato K, Zeng L, Viereck J, Hamilton JA, Ou-chi N, LeBrasseur NK, Walsh K: Fast/Glycolytic muscle fiber growth reduces

fat mass and improves metabolic parameters in obese mice. Cell Metab. 2008, 7(2):159-72.

13. McBride A, Ghilagaber S, Nikolaev A, Hardie DG: The glycogen-binding domain on the AMPK beta subunit allows the kinase to act as a glycogen sensor. Cell Metab. 2009, 9(1):23-34.

14. Wojtaszewski JF, MacDonald C, Nielsen JN, Hellsten Y, Hardie DG, Kemp BE, Kiens B, Richter EA: Regulation of 5'AMP-activated protein kinase activity and substrate utilization in exercising human skeletal muscle. Am J Physiol Endocrinol Metab 2003, 284(4):E813-22.

15. Creer A, Gallagher P, Slivka D, Jemiolo B, Fink W, Trappe S: Influence of muscle glycogen availability on ERK1/2 and Akt signaling after resistance exercise in human skeletal muscle. J Appl Physiol 2005, 99(3):950-6.

16. Churchley EG, Coffey VG, Pedersen DJ, Shield A, Carey KA, Cameron-Smith D, Hawley JA: Influence of preexercise muscle glycogen content on transcriptional activity of metabolic and myogenic genes in well-trained humans. J Appl Physiol 2007, 102(4):1604-11.

17. Dennis PB, Jaeschke A, Saitoh M, Fowler B, Kozma SC, Thomas G: Mammalian TOR: a homeostatic ATP sensor. Science 2001, 294(5544):1102-5.

18. Camera DM, West DW, Burd NA, Phillips SM, Garnham AP, Hawley JA, Coffey VG: Low muscle glycogen concentration does not suppress the anabolic response to resistance exercise. J Appl Physiol 2012, 113(2):206-14.

19. Lemon PW, Mullin JP: Effect of initial muscle glycogen levels on protein catabolism during exercise. J Appl Physiol 1980, 48(4):624-9.

20. Blomstrand E, Saltin B, Blomstrand E, Saltin B: Effect of muscle glycogen on glucose, lactate and amino acid metabolism during exercise and recovery in human subjects. J Physiol 1999, 514(1):293-302.

21. Ivy JL: Glycogen resynthesis after exercise: effect of carbohydrate intake. Int J Sports Med. 1998, 19(Suppl 2):S142-5.

22. Richter EA, Derave W, Wojtaszewski JF: Glucose, exercise and insulin: emerging concepts. J Physiol 2001, 535(Pt 2):313-22.

23. Derave W, Lund S, Holman GD, Wojtaszewski J, Pedersen O, Richter EA: Contraction-stimulated muscle glucose transport and GLUT-4 surface content are dependent on glycogen content. Am J Physiol 1999, 277(6 Pt 1):E1103-10.

24. Kawanaka K, Nolte LA, Han DH, Hansen PA, Holloszy JO: Mechanisms underlying impaired GLUT-4 translocation in glycogen-supercompensated muscles of exercised rats. Am J Physiol Endocrinol Metab 2000, 279(6):E1311-8.

25. O'Gorman DJ, Del Aguila LF, Williamson DL, Krishnan RK, Kirwan JP: Insulin and exercise differentially regulate PI3-kinase and glycogen synthase in human skeletal muscle. J Appl Physiol 2000, 89(4):1412-9.

26. Berardi JM, Price TB, Noreen EE, Lemon PW: Postexercise muscle glycogen recovery enhanced with a carbohydrate-protein supplement. Med Sci Sports Exerc. 2006, 38(6):1106-13.

27. Ivy JL, Goforth HW Jr, Damon BM, McCauley TR, Parsons EC, Price TB: Early postexercise muscle glycogen recovery is enhanced with a carbohydrate-protein supplement. J Appl Physiol 2002, 93(4):1337-44.

28. Zawadzki KM, Yaspelkis BB 3rd, Ivy JL: Carbohydrate-protein complex increases the rate of muscle glycogen storage after exercise. J Appl Physiol 1992, 72(5):1854-9.

29. Tarnopolsky MA, Bosman M, Macdonald JR, Vandeputte D, Martin J, Roy BD: Postexercise protein-carbohydrate and carbohydrate supplements increase muscle glycogen in men and women. J Appl Physiol 1997, 83(6):1877-83.

30. Jentjens RL, van Loon LJ, Mann CH, Wagenmakers AJ, Jeukendrup AE: Addition of protein and amino acids to carbohydrates does not enhance postexercise muscle glycogen synthesis. J Appl Physiol 2001, 91(2):839-46.

31. Jentjens R, Jeukendrup A: Determinants of post-exercise glycogen synthesis during short-term recovery. Sports Med. 2003, 33(2):117-44.

32. Roy BD, Tarnopolsky MA: Influence of differing macronutrient intakes on muscle glycogen resynthesis after resistance exercise. J Appl Physiol 1998, 84(3):890-6.

33. Parkin JA, Carey MF, Martin IK, Stojanovska L, Febbraio MA: Muscle glycogen storage following prolonged exercise: effect of timing of ingestion of high glycemic index food. Med Sci Sports Exerc. 1997, 29(2):220-4.

34. Fox AK, Kaufman AE, Horowitz JF: Adding fat calories to meals after exercise does not alter glucose tolerance. J Appl Physiol 2004, 97(1):11-6.

35. Biolo G, Tipton KD, Klein S, Wolfe RR: An abundant supply of amino acids enhances the metabolic effect of exercise on muscle protein. Am J Physiol 1997, 273(1 Pt 1):E122-9.

36. Kumar V, Atherton P, Smith K, Rennie MJ: Human muscle protein synthesis and breakdown during and after exercise. J Appl Physiol 2009, 106(6):2026-39.

37. Pitkanen HT, Nykanen T, Knuutinen J, Lahti K, Keinanen O, Alen M, Komi PV, Mero AA: Free amino acid pool and muscle protein balance after resistance exercise. Med Sci Sports Exerc. 2003, 35(5):784-92.

38. Biolo G, Williams BD, Fleming RY, Wolfe RR: Insulin action on muscle protein kinetics and amino acid transport during recovery after resistance exercise. Diabetes 1999, 48(5):949-57.

39. Fluckey JD, Vary TC, Jefferson LS, Farrell PA: Augmented insulin action on rates of protein synthesis after resistance exercise in rats. Am J Physiol 1996, 270(2 Pt 1):E313-9.

40. Denne SC, Liechty EA, Liu YM, Brechtel G, Baron AD: Proteolysis in skeletal muscle and whole body in response to euglycemic hyperinsulinemia in normal adults. Am J Physiol 1991, 261(6 Pt 1):E809-14.

41. Gelfand RA, Barrett EJ: Effect of physiologic hyperinsulinemia on skeletal muscle protein synthesis and breakdown in man. J Clin Invest 1987, 80(1):1-6.

42. Heslin MJ, Newman E, Wolf RF, Pisters PW, Brennan MF: Effect of hyperinsulinemia on whole body and skeletal muscle leucine carbon kinetics in humans. Am J Physiol 1992, 262(6 Pt 1):E911-8.

43. Kettelhut IC, Wing SS, Goldberg AL: Endocrine regulation of protein breakdown in skeletal muscle. Diabetes Metab Rev. 1988, 4(8):751-72.

44. Kim DH, Kim JY, Yu BP, Chung HY: The activation of NF-kappaB through Akt-induced FOXO1 phosphorylation during aging and its modulation by calorie restriction. Biogerontology 2008, 9(1):33-47.
45. Greenhaff PL, Karagounis LG, Peirce N, Simpson EJ, Hazell M, Layfield R, Wackerhage H, Smith K, Atherton P, Selby A, Rennie MJ: Disassociation between the effects of amino acids and insulin on signaling, ubiquitin ligases, and protein turnover in human muscle. Am J Physiol Endocrinol Metab 2008, 295(3):E595-604.
46. Rennie MJ, Bohe J, Smith K, Wackerhage H, Greenhaff P: Branched-chain amino acids as fuels and anabolic signals in human muscle. J Nutr 2006, 136(1 Suppl):264S-8S.
47. Capaldo B, Gastaldelli A, Antoniello S, Auletta M, Pardo F, Ciociaro D, Guida R, Ferrannini E, Sacca L: Splanchnic and leg substrate exchange after ingestion of a natural mixed meal in humans. Diabetes 1999, 48(5):958-66.
48. Power O, Hallihan A, Jakeman P: Human insulinotropic response to oral ingestion of native and hydrolysed whey protein. Amino Acids. 2009, 37(2):333-9.
49. Glynn EL, Fry CS, Drummond MJ, Dreyer HC, Dhanani S, Volpi E, Rasmussen BB: Muscle protein breakdown has a minor role in the protein anabolic response to essential amino acid and carbohydrate intake following resistance exercise. Am J Physiol Regul Integr Comp Physiol 2010, 299(2):R533-40.
50. Tipton KD, Ferrando AA, Phillips SM, Doyle D Jr, Wolfe RR: Postexercise net protein synthesis in human muscle from orally administered amino acids. Am J Physiol 1999, 276(4 Pt 1):E628-34.
51. Miller SL, Tipton KD, Chinkes DL, Wolf SE, Wolfe RR: Independent and combined effects of amino acids and glucose after resistance exercise. Med Sci Sports Exerc. 2003, 35(3):449-55.
52. Koopman R, Beelen M, Stellingwerff T, Pennings B, Saris WH, Kies AK, Kuipers H, van Loon LJ: Coingestion of carbohydrate with protein does not further augment postexercise muscle protein synthesis. Am J Physiol Endocrinol Metab 2007, 293(3):E833-42.
53. Staples AW, Burd NA, West DW, Currie KD, Atherton PJ, Moore DR, Rennie MJ, Macdonald MJ, Baker SK, Phillips SM: Carbohydrate does not augment exercise-induced protein accretion versus protein alone. Med Sci Sports Exerc. 2011, 43(7):1154-61.
54. Borsheim E, Cree MG, Tipton KD, Elliott TA, Aarsland A, Wolfe RR: Effect of carbohydrate intake on net muscle protein synthesis during recovery from resistance exercise. J Appl Physiol 2004, 96(2):674-8.
55. Koopman R, Wagenmakers AJ, Manders RJ, Zorenc AH, Senden JM, Gorselink M, Keizer HA, van Loon LJ: Combined ingestion of protein and free leucine with carbohydrate increases postexercise muscle protein synthesis in vivo in male subjects. Am J Physiol Endocrinol Metab 2005, 288(4):E645-53.

56. Rasmussen BB, Tipton KD, Miller SL, Wolf SE, Wolfe RR: An oral essential amino acid-carbohydrate supplement enhances muscle protein anabolism after resistance exercise. J Appl Physiol 2000, 88(2):386-92.

57. Tang JE, Manolakos JJ, Kujbida GW, Lysecki PJ, Moore DR, Phillips SM: Minimal whey protein with carbohydrate stimulates muscle protein synthesis following resistance exercise in trained young men. Appl Physiol Nutr Metab 2007, 32(6):1132-8.

58. Tipton KD, Elliott TA, Cree MG, Wolf SE, Sanford AP, Wolfe RR: Ingestion of casein and whey proteins result in muscle anabolism after resistance exercise. Med Sci Sports Exerc. 2004, 36(12):2073-81.

59. Tipton KD, Elliott TA, Ferrando AA, Aarsland AA, Wolfe RR: Stimulation of muscle anabolism by resistance exercise and ingestion of leucine plus protein. Appl Physiol Nutr Metab 2009, 34(2):151-61.

60. Phillips SM, Van Loon LJ: Dietary protein for athletes: from requirements to optimum adaptation. J Sports Sci. 2011, 29(Suppl 1):S29-38.

61. Phillips SM: The science of muscle hypertrophy: making dietary protein count. Proc Nutr Soc 2011, 70(1):100-3.

62. Levenhagen DK, Gresham JD, Carlson MG, Maron DJ, Borel MJ, Flakoll PJ: Postexercise nutrient intake timing in humans is critical to recovery of leg glucose and protein homeostasis. Am J Physiol Endocrinol Metab 2001, 280(6):E982-93.

63. Tipton KD, Rasmussen BB, Miller SL, Wolf SE, Owens-Stovall SK, Petrini BE, Wolfe RR: Timing of amino acid-carbohydrate ingestion alters anabolic response of muscle to resistance exercise. Am J Physiol Endocrinol Metab 2001, 281(2):E197-206.

64. Fujita S, Dreyer HC, Drummond MJ, Glynn EL, Volpi E, Rasmussen BB: Essential amino acid and carbohydrate ingestion before resistance exercise does not enhance postexercise muscle protein synthesis. J Appl Physiol 2009, 106(5):1730-9.

65. Tipton KD, Elliott TA, Cree MG, Aarsland AA, Sanford AP, Wolfe RR: Stimulation of net muscle protein synthesis by whey protein ingestion before and after exercise. Am J Physiol Endocrinol Metab 2007, 292(1):E71-6.

66. Coffey VG, Shield A, Canny BJ, Carey KA, Cameron-Smith D, Hawley JA: Interaction of contractile activity and training history on mRNA abundance in skeletal muscle from trained athletes. Am J Physiol Endocrinol Metab 2006, 290(5):E849-55.

67. Timmons JA: Variability in training-induced skeletal muscle adaptation. J Appl Physiol 2011, 110(3):846-53.

68. Adams G, Bamman MM: Characterization and regulation of mechanical loading-induced compensatory muscle hypertrophy. Comprehensive Physiology 2012, 2829:2970.

69. Esmarck B, Andersen JL, Olsen S, Richter EA, Mizuno M, Kjaer M: Timing of postexercise protein intake is important for muscle hypertrophy with resistance training in elderly humans. J Physiol 2001, 535(Pt 1):301-11.

70. Cribb PJ, Hayes A: Effects of supplement timing and resistance exercise on skeletal muscle hypertrophy. Med Sci Sports Exerc. 2006, 38(11):1918-25.

71. Willoughby DS, Stout JR, Wilborn CD: Effects of resistance training and protein plus amino acid supplementation on muscle anabolism, mass, and strength. Amino Acids. 2007, 32(4):467-77.

72. Hulmi JJ, Kovanen V, Selanne H, Kraemer WJ, Hakkinen K, Mero AA: Acute and long-term effects of resistance exercise with or without protein ingestion on muscle hypertrophy and gene expression. Amino Acids. 2009, 37(2):297-308.

73. Verdijk LB, Jonkers RA, Gleeson BG, Beelen M, Meijer K, Savelberg HH, Wodzig WK, Dendale P, van Loon LJ: Protein supplementation before and after exercise does not further augment skeletal muscle hypertrophy after resistance training in elderly men. Am J Clin Nutr 2009, 89(2):608-16.

74. Hoffman JR, Ratamess NA, Tranchina CP, Rashti SL, Kang J, Faigenbaum AD: Effect of protein-supplement timing on strength, power, and body-composition changes in resistance-trained men. Int J Sport Nutr Exerc Metab. 2009, 19(2):172-85.

75. Erskine RM, Fletcher G, Hanson B, Folland JP: Whey protein does not enhance the adaptations to elbow flexor resistance training. Med Sci Sports Exerc. 2012, 44(9):1791-800.

76. Levine JA, Abboud L, Barry M, Reed JE, Sheedy PF, Jensen MD: Measuring leg muscle and fat mass in humans: comparison of CT and dual-energy X-ray absorptiometry. J Appl Physiol 2000, 88(2):452-6.

77. Layman DK: Protein quantity and quality at levels above the RDA improves adult weight loss. J Am Coll Nutr 2004, 23(6 Suppl):631S-6S. PubMed Abstract

78. Norton LE, Layman DK, Bunpo P, Anthony TG, Brana DV, Garlick PJ: The leucine content of a complete meal directs peak activation but not duration of skeletal muscle protein synthesis and mammalian target of rapamycin signaling in rats. J Nutr 2009, 139(6):1103-9.

79. Wilson GJ, Layman DK, Moulton CJ, Norton LE, Anthony TG, Proud CG, Rupassara SI, Garlick PJ: Leucine or carbohydrate supplementation reduces AMPK and eEF2 phosphorylation and extends postprandial muscle protein synthesis in rats. Am J Physiol Endocrinol Metab 2011, 301(6):E1236-42.

80. Atherton PJ, Etheridge T, Watt PW, Wilkinson D, Selby A, Rankin D, Smith K, Rennie MJ: Muscle full effect after oral protein: time-dependent concordance and discordance between human muscle protein synthesis and mTORC1 signaling. Am J Clin Nutr 2010, 92(5):1080-8.

81. Bohe J, Low JF, Wolfe RR, Rennie MJ: Latency and duration of stimulation of human muscle protein synthesis during continuous infusion of amino acids. J Physiol 2001, 532(Pt 2):575-9.

82. Burd NA, Tang JE, Moore DR, Phillips SM: Exercise training and protein metabolism: influences of contraction, protein intake, and sex-based differences. J Appl Physiol 2009, 106(5):1692-701.

83. Breen L, Phillips SM: Interactions between exercise and nutrition to prevent muscle waste during aging. Br J Clin Pharmacol 2012. [Epub ahead of print]

84. Moore DR, Robinson MJ, Fry JL, Tang JE, Glover EI, Wilkinson SB, Prior T, Tarnopolsky MA, Phillips SM: Ingested protein dose response of muscle and albumin protein synthesis after resistance exercise in young men. Am J Clin Nutr 2009, 89(1):161-8.
85. Yang Y, Breen L, Burd NA, Hector AJ, Churchward-Venne TA, Josse AR, Tarnopolsky MA, Phillips SM: Resistance exercise enhances myofibrillar protein synthesis with graded intakes of whey protein in older men. Br J Nutr 2012, 108(10):1780-8.

This chapter was originally published under the Creative Commons Attribution License. Aragon, A. A., and Schoenfeld, B. J. Nutrient Timing Revisited: Is There a Post-Exercise Anabolic Window? Journal of the International Society of Sports Nutrition 2013, 10:5. doi:10.1186/1550-2783-10-5.

CHAPTER 4

NUTRITIONAL THERAPIES FOR MENTAL DISORDERS

SHAHEEN E. LAKHAN and KAREN F. VIEIRA

4.1 INTRODUCTION

Currently, approximately 1 in 4 adult Americans have been diagnosed with a mental disorder, which translates into about 58 million affected people [1]. Though the incidence of mental disorders is higher in America than in other countries, a World Health Organization study of 14 countries reported a worldwide prevalence of mental disorders between 4.3 percent and 26.4 percent [2]. In addition, mental disorders are among the leading causes for disability in the US as well as other countries. Common mental health disorders include mood disorders, anxiety disorders such as post-traumatic stress disorder (PTSD), panic disorders, eating disorders, attention deficit disorder/attention deficit hyperactivity disorder (ADD/ ADHD), and autism. However, the four most common mental disorders that cause disabilities are major depression, bipolar disorder, schizophrenia, and obsessive compulsive disorder (OCD) [3,4].

Typically, most of these disorders are treated with prescription drugs, but many of these prescribed drugs cause unwanted side effects. For example, lithium is usually prescribed for bipolar disorder, but the high-doses of lithium that are normally prescribed causes side effects that include: a dulled personality, reduced emotions, memory loss, tremors, or weight gain [5,6]. These side effects can be so severe and unpleasant that many patients become noncompliant and, in cases of severe drug toxicity, the situation can become life threatening.

Researchers have observed that the prevalence of mental health disorders has increased in developed countries in correlation with the deterioration of the Western diet [7]. Previous research has shown nutritional deficiencies that correlate with some mental disorders [8,9]. The most common nutritional deficiencies seen in mental disorder patients are of omega-3 fatty acids, B vitamins, minerals, and amino acids that are precursors to neurotransmitters [10-16]. Compelling population studies link high fish consumption to a low incidence of mental disorders; this lower incidence rate has proven to be a direct result of omega-3 fatty acid intake [10,17,18]. One to two grams of omega-3 fatty acids taken daily is the generally accepted dose for healthy individuals, but for patients with mental disorders, up to 9.6 g has been shown to be safe and efficacious [19-21]. Western diets are usually also lacking in fruits and vegetables, which further contributes to vitamin and mineral deficiencies.

This article will focus on the nutritional deficiencies that are associated with mental disorders and will outline how dietary supplements can be implemented in the treatment of several disorders (see Table 1 for an overview). The mental disorders and treatments covered in this review do not include the broad and complex range of disorders, but however focuses on the four most common disorders in order to emphasize the alternative or complementary nutritional options that health care providers can recommend to their patients.

4.1.1 MAJOR DEPRESSION

Major depression is a disorder that presents with symptoms such as decreased mood, increased sadness and anxiety, a loss of appetite, and a loss of interest in pleasurable activities, to name a few [22]. If this disorder is not properly treated it can become disabling or fatal. Patients who are suffering from major depression have a high risk for committing suicide so they are usually treated with psychotherapy and/or antidepressants [23]. Depression has for some time now been known to be associated with deficiencies in neurotransmitters such as serotonin, dopamine, noradrenaline, and GABA [22-27]. As reported in several studies, the amino acids tryptophan, tyrosine, phenylalanine, and methionine are often helpful in

TABLE 1: Summary of proposed causes and treatments for common mental health disorders

Mental Disorder	Proposed Cause	Treatment	References	Type of Study
Major Depression	Serotonin deficiency	Tryptophan	[15]	Human pilot clinical trial
			[32]	Double-blind, placebo controlled
	Dopamine/Noradrenaline deficiency	Tyrosine	[30]	Double-blind, placebo controlled
	GABA deficiency	GABA	[36]	Randomized within or between subjects
	Omega-3 deficiency	Omega-3s	[29]	Clinical trial
			[39]	Clinical trial
			[9]	Randomized controlled trial
	Folate/Vitamin B deficiency	Folate/Vitamin B	[13]	Clinical trial
	Magnesium deficiency	Magnesium	[14]	Cases studies
	SAM deficiency	SAM	[37]	Double-blind, placebo controlled
Bipolar Disorder	Excess acetylcholine receptors	Lithium orotate & taurine	[50]	Clinical trial
	Excess vanadium	Vitamin C	[45]	Double-blind, placebo controlled
			[47]	Human pilot clinical trial
	Vitamin B/Folate deficiency	Vitamin B/Folate	[71]	Clinical trial
	L-Tryptophan deficiency	L-Tryptophan	[72]	Clinical trial
	Choline deficiency	Lecithin	[73]	Double-blind, placebo controlled
			[21]	Double-blind, placebo controlled
			[48]	Clinical trial
			[74]	Clinical trial
	Omega-3 deficiency	Omega-3s	[75]	Double-blind, placebo controlled

TABLE 1: *Cont.*

Mental Disorder	Proposed Cause	Treatment	References	Type of Study
Schizophrenia	Impaired serotonin synthesis	Tryptophan	[53]	Open-baseline controlled trial
			[54]	Double-blind, placebo controlled
			[55]	Human pilot open-label trial
	Glycine deficiency	Glycine	[56]	Clinical trial
			[59]	Double-blind, placebo controlled
			[60]	Randomized, placebo controlled
	Omega-3 deficiencies	Omega-3s	[65]	Open-label clinical trial
Obsessive Compulsive Disorder	St. John's wort deficiency	St John's wort	[69]	Randomized, double-blind trial
			[70]	Double-blind, placebo controlled

treating many mood disorders, including depression [28-33]. Tryptophan is a precursor to serotonin and is usually converted to serotonin when taken alone on an empty stomach. Therefore, tryptophan can induce sleep and tranquility and in cases of serotonin deficiencies, restore serotonin levels leading to diminished depression [15,31].

Tyrosine is not an essential amino acid, because it can be made from the amino acid phenylalanine. Tyrosine and sometimes its precursor phenylalanine are converted into dopamine and norepinephrine [34]. Dietary supplements that contain tyrosine and/or phenylalanine lead to alertness and arousal. Methionine combines with ATP to produce S-adenosylmethionine (SAM), which facilitates the production of neurotransmitters in the brain [35-38]. Currently, more studies involving these neurochemicals are needed which exhibit the daily supplemental doses that should be consumed in order to achieve antidepressant effects.

Since the consumption of omega-3 fatty acids from fish and other sources has declined in most populations, the incidence of major depression has increased [10]. Several mechanisms of action may explain how eicosapentaenoic acid (EPA) which the body converts into docosahexaenoic acid (DHA), the two omega-3 fatty acids found in fish oil, elicit antidepressant effects in humans. Most of the proposed mechanisms involve neurotransmitters and, of course, some have more supporting data than others. For example, antidepressant effects may be due to EPA being converted into prostaglandins, leukotrienes, and other chemicals the brain needs. Other theories state that EPA and DHA affect signal transduction in brain cells by activating peroxisomal proliferator-activated receptors (PPARs), inhibiting G-proteins and protein kinase C, as well as calcium, sodium, and potassium ion channels. No matter which mechanism(s) prove to be true, epidemiological data and clinical studies already show that omega-3 fatty acids can effectively treat depression [39]. Consuming omega-3 fatty acid dietary supplements that contain 1.5 to 2 g of EPA per day have been shown to stimulate mood elevation in depressed patients. However, doses of omega-3 higher than 3 g do not present better effects than placebos and may not be suitable for some patients, such as those taking anti-clotting drugs [40].

In addition to omega-3 fatty acids, vitamin B (e.g., folate), and magnesium deficiencies have been linked to depression [9,13,14]. Randomized, controlled trials that involve folate and B12 suggest that patients treated with 0.8 mg of folic acid/day or 0.4 mg of vitamin B12/day will exhibit decreased depression symptoms [9]. In addition, the results of several case studies where patients were treated with 125 to 300 mg of magnesium (as glycinate or taurinate) with each meal and at bedtime led to rapid recovery from major depression in less than seven days for most of the patients [14].

4.1.2 BIPOLAR DISORDER

A patient suffering from major depression may also present symptoms such as recurring episodes of debilitating depression, uncontrollable mania, hypomania, or a mixed state (a manic and depressive episode) which is clinically diagnosed as bipolar disorder [41]. Some biochemical abnormalities in people with bipolar disorder include oversensitivity to acetylcholine, excess vanadium, vitamin B deficiencies, a taurine deficiency, anemia, omega-3 fatty acid deficiencies, and vitamin C deficiency.

Bipolar patients tend to have excess acetylcholine receptors, which is a major cause of depression and mania [42,43]. Bipolar patients also produce elevated levels of vanadium, which causes mania, depression, and melancholy [44,45]. However, vitamin C has been shown to protect the body from the damage caused by excess vanadium. A double-blind, placebo controlled study that involved controlling elevated vanadium levels showed that a single 3 g dose of vitamin C decreases manic symptoms in comparison to placebo [45].

Taurine is an amino acid made in the liver from cysteine that is known to play a role in the brain by eliciting a calming effect. A deficiency of this amino acid may increase a bipolar patient's manic episodes. In addition, eighty percent of bipolar sufferers have some vitamin B deficiencies (often accompanied by anemia) [46]. The combination of essential vitamin supplements with the body's natural supply of lithium reduces depressive and manic symptoms of patients suffering from bipolar disorder [47].

Another well-known factor for mental disorders is that cells within the brain require omega-3 oils in order to be able to transmit signals that enable proper thinking, moods, and emotions. However, omega-3 oils are often present at very low levels in most Americans and bipolar sufferers

[48]. Numerous clinical trials, including double-blind, placebo controlled studies have been performed which show that 1 to 2 grams of omega-3 fatty acids in the form of EPA added to one's daily intake decreases manic/depressive symptoms better than placebo (See Table 1).

Prescription lithium is in the form of lithium carbonate, and doses can be as high as 180 mg. It is these high doses that are responsible for most of lithium's adverse side effects. Some of the more common side effects include a dulled personality, reduced emotions, memory loss, tremors, or weight gain [5,6]. Another form of lithium called lithium orotate, is preferred because the orotate ion crosses the blood-brain barrier more easily than the carbonate ion of lithium carbonate. Therefore, lithium orotate can be used in much lower doses (e.g. 5 mg) with remarkable results and no side effects [49,50]. Clinical trials involving 150 mg daily doses of lithium orotate administered 4 to 5 times a week, showed a reduction of manic and depressive symptoms in bipolar patients [50]. In addition, lithium orotate is available without a prescription, unlike lithium carbonate, which is considered a prescription drug by the Food and Drug Administration (FDA). Studies have also shown that the amino acid-derivative, taurine, as an alternative to lithium, blocks the effects of excess acetylcholine that contributes to bipolar disorder [51].

Numerous studies for bipolar disorder have been published that list specific lifestyle changes as well as amounts of dietary supplements that can be used to treat this disorder. A summary of these results is listed in Table 2.

4.1.3 SCHIZOPHRENIA

Schizophrenia is a mental disorder that disrupts a person's normal perception of reality. Schizophrenic patients usually suffer from hallucinations, paranoia, delusions, and speech/thinking impairments. These symptoms are typically presented during adolescence [52]. Disturbances in amino acid metabolism have been implicated in the pathophysiology of schizophrenia. Specifically, an impaired synthesis of serotonin in the central nervous system has been found in schizophrenic patients [53]. High doses (30 g) of glycine have been shown to reduce the more subtle symptoms

TABLE 2: List of possible causes and treatments for bipolar disorder including specific doses as well as supplementary information

Mental Disorder	Proposed Cause	Treatment	References
Bipolar Disorder	Food allergies	Avoid foods that elicit an allergic response	[76, 77]
	Caffeine	Avoid coffee and other caffeinated beverages	[78]
	Inhibition of lithium from alkalizing agents	Avoid alkalizing agents like bicarbonates	[79]
	Vitamin B6 deficiency	100–200 milligrams/day	[72, 80]
	Vitamin B12 deficiency	300–600 micrograms/day	[71, 81–83]
	Vitamin C deficiency	1–3 grams taken as divided doses	[84–86]
	Folate deficiency	200 micrograms/day	[9, 13, 71, 82, 83, 87, 88]
	Choline deficiency	10–30 grams of phosphatidyl form in divided doses	[73, 89]
	Omega-3 or -6 deficiency	500–1000 milligrams/day	[10, 11, 21, 39, 74, 75, 90–94]
	Phenylalanine deficiency	Initially 500 milligrams/day; can increase to 3–4 grams/day	[95, 96]
	Tryptophan deficiency	50–200 milligrams taken as divided doses	[97–100]
	S-Adenosyl-L-Methionine (SAM) deficiency	800 milligrams	[101–103]
	Melatonin deficiency	3–6 milligrams at 9 pm	[104–106]
	Phosphatidylserine deficiency	100 milligrams with food	[107]

of schizophrenia, such as social withdrawal, emotional flatness, and apathy, which do not respond to most of the existing medications [54-56]. An open-label clinical trial performed in 1996 revealed that 60 g of glycine per day (0.8 g/kg) could be given to schizophrenic patients without producing adverse side effects and that this dose led to a two-fold increase in cerebrospinal fluid (CSF) glycine levels [55]. A second clinical study treated patients with the same dosage divided into 3 doses within 1 week. This form of glycine treatment led to an eight-fold increase in CSF glycine levels [56].

The most consistent correlation found in one study that involved the ecological analysis of schizophrenia and diet concluded that increased consumption of refined sugar results in an overall decreased state of mind for schizophrenic patients, as measured by both the number of days spent in the hospital and poor social functioning [57]. That study also concluded that the dietary predictors of the outcome of schizophrenia and prevalence of depression are similar to those that predict illnesses such as coronary heart disease and diabetes.

A Danish study showed that better prognoses for schizophrenic patients strongly correlate with living in a country where there is a high consumption of omega-3 fatty acids [58]. Eicosapentaenoic acid (EPA), which is found in omega-3 fish oils, has been shown to help depressive patients and can also be used to treat schizophrenia [41,42,59]. Furthermore, studies suggest that supplements such as the commercially available VegEPA capsule, when taken on a daily basis, helps healthy individuals and schizophrenic patients maintain a balanced mood and improves blood circulation [59-65].

The VegEPA capsule contains:

- 280 milligrams of EPA from marine omega-3 fish oil
- 100 milligrams of organic virgin evening primrose omega-6 oil
- 1 milligram of the anti-oxidant vitamin E
- An outer capsule made out of fish gelatine

For schizophrenic patients, docosahexaenoic acid (DHA) supplements inhibit the effects of EPA supplements so it is recommended that the patient only takes the EPA supplement, which the body will convert into the amount DHA it needs [59-65]. Double-blind, placebo controlled studies, randomized, placebo controlled studies, and open-label clinical studies have all shown that approximately 2 g of EPA taken daily in addition to one's existing medication effectively decreases symptoms in schizophrenic patients [59,60,65].

4.1.4 OBSESSIVE-COMPULSIVE DISORDER

Obsessive compulsive disorder (OCD) is an anxiety disorder that causes recurring stressful thoughts or obsessions that are followed by compulsions, which are repeated in an uncontrollable manner as a means of repressing the stressful thought [66]. It is well documented that selective serotonin reuptake inhibitors (SSRIs) help patients with OCD [67]. Therefore, it is clear that nutrients which increase serotonin levels will reduce the symptoms of OCD. As discussed earlier, the amino acid tryptophan is a precursor to serotonin, and tryptophan supplements (which are better than 5-Hydroxytryptophan) will increase serotonin levels and treat OCD [68].

A commercially available supplement called Amoryn has recently proven to help patients suffering from depression, anxiety, and OCD [69,70]. The main ingredient in Amoryn, St. John's wort, has been shown to help OCD patients better deal with their recurring thoughts and compulsions. Two double-blind, placebo-controlled studies were recently performed that compared the affects of a 900 mg daily dose of St. John's wort extract to 20 mg daily doses of Paroxetine (Paxil) or Fluoxetine; which are both SSRIs used to treat OCD. In comparison to patients taking Paxil, those who took the St. John's wort supplement showed a 57% decrease in OCD symptoms and were 47% less likely to exhibit side effects [69]. In comparison to patients taking Fluoxetine, consumption of the St. John's wort extract reduced 48% of OCD patient's symptoms [70]. These results clearly depict how the use nutritional supplements can be effective treatments for mental disorders.

4.2 CONCLUSION

Here we have shown just a few of the many documented nutritional therapies that can be utilized when treating mental disorders. Many of these studies were done in the 1970s and 1980s, but were soon discontinued because they were underfunded. Nutritional therapies have now become a long-forgotten method of treatment, because they were of no interest to pharmaceutical companies that could not patent or own them. Instead, the companies that funded most clinical research spent their dollars investigating synthetic drugs they could patent and sell; these drugs however usually caused adverse side effects.

There is tremendous resistance to using supplements as treatments from clinicians, mostly due to their lack of knowledge on the subject. Others rather use prescription drugs that the drug companies and the FDA researches, monitors and recalls if necessary. However, for some patients, prescription drugs do not have the efficacy of nutritional supplements and they sometimes have far more dangerous side effects. So for clinicians to avoid these supplement therapies because of a lack of knowledge and unwillingness to use treatments not backed by drug companies and the FDA, they are compromising their patients' recovery due to their own laziness or selfishness.

Clinical studies that show the ability of a prescription drug to effectively treat mental disorders will often argue that supplements as treatments, when unmonitored, are more risky than prescription drugs and may ineffectively treat a patient's symptoms. For example one study listed several methods of treatment, none of which include natural compounds, for OCD patients that include: megadoses of SSRIs, intravenous chlomipramine, oral morphine, deep brain stimulation, and functional neurosurgery [67]. Most of these treatments are invasive or unnatural and will inevitably cause severe side effects to the patient, whose symptoms will probably still reoccur over time. Another example of the literature scaring clinicians away from supplement therapies is an article that warns patients about the dangers of consuming high amounts of omega-3 fatty acids. This manuscript involves a patient who was taking approximately 10 times more than the recommended dose of omega-3 supplements [40]. Numerous studies have shown that up 2 grams of EPA (omega-3 fatty acid) taken daily is sufficient for decreasing symptoms of several mental health disorders with no side effects. This publication with a megadose of omega-3 fatty acids stresses the importance of monitoring the consumption of supplements as well as prescribed drugs, preferably through regular consultations with a licensed health care professional.

Proper medical diagnosis and a clear description of all possible treatment options should always be the first plan of action when treating mental disorders. However, the final decision on whether or not to try nutritional supplements as a treatment must be based on the patient preferences. Now with consumers becoming more interested in natural and holistic therapies, nutritional therapies have been well-received, and some studies are again underway in these areas. New well-designed clinical studies are being published daily on the positive effects of nutritional and supplement therapies on all types of disorders and diseases. It will take some time for clinicians to become educated on all the options available, but this is an important task that should not be ignored.

Those with influence in this field should continue to examine natural treatments on the scientific level in order to increase the availability of grant money for this type of research. This will lead to a surge of researchers who will submit proposals for grants enabling laboratories to further investigate the hypothesis that proper nutrition contributes to better mental health.

Psychiatrists treating patients with mental disorders should be aware of available nutritional therapies, appropriate doses, and possible side effects in order to provide alternative and complementary treatments for their patients. This may reduce the number of noncompliant patients suffering from mental disorders that choose not to take their prescribed medications. As with any form of treatment, nutritional therapy should be supervised and doses should be adjusted as necessary to achieve optimal results.

REFERENCES

1.   Kessler RC, Chiu WT, Demler O, Walters EE: Prevalence, severity, and comorbidity of twelve-month DSM-IV disorders in the National Comorbidity Survey Replication (NCS-R). Archives of General Psychiatry 2005, 62(6):617-627.
2.   Demyttenaere K, Bruffaerts R, Posada-Villa J, Gasquet I, Kovess V, Lepine JP, Angermeyer MC, Bernert S, de Girolamo G, Morosini P, Polidori G, Kikkawa T, Kawakami N, Ono Y, Takeshima T, Uda H, Karam EG, Fayyad JA, Karam AN, Mneimneh ZN, Medina-Mora ME, Borges G, Lara C, de Graaf R, Ormel J, Gureje O, Shen Y, Huang Y, Zhang M, Alonso J, Haro JM, Vilagut G, Bromet EJ, Gluzman S, Webb C, Kessler RC, Merikangas KR, Anthony JC, Von Korff MR, Wang PS, Brugha TS, Aguilar-Gaxiola S, Lee S, Heeringa S, Pennell BE, Zaslavsky AM,

Ustun TB, Chatterji S, WHO World Mental Health Survey Consortium: Prevalence, severity, and unmet need for treatment of mental disorders in the World Health Organization World Mental Health Surveys. JAMA 2004, 291(21):2581-2590.

3. Murray CJL, Lopez AD: The Global Burden Of Disease. World Health Organization 1996, 270.

4. American Psychiatric A: Diagnostic and Statistical Manual of Mental Disorders. Fourth edition, text revision Washington DC 2000.

5. Waring WS: Management of lithium toxicity. Toxicol Rev 2006, 25(4):221-230.

6. Vieta E, Rosa AR: Evolving trends in the long-term treatment of bipolar disorder. World J Biol Psychiatry 2007, 8(1):4-11.

7. Young SN: Clinical nutrition: 3. The fuzzy boundary between nutrition and psycopharmacology. CMAJ 2002, 166(2):205-209.

8. Wurtman R, O'Rourke D, Wurtman JJ: Nutrient imbalances in depressive disorders. Possible brain mechanisms. Ann N Y Acad Sci 1989, 575:75-82.

9. Young SN: Folate and depression–a neglected problem. J Psychiatry Neurosci 2007, 32(2):80-82.

10. Hibbeln JR: Fish consumption and major depression. The Lancet 1998, 351(9110):1213.

11. Rudin DO: The major psychoses and neuroses as omega-3 essential fatty acid deficiency syndrome: substrate pellagra. Biol Psychiatry 1981, 16(9):837-850.

12. Rudin DO: The dominant diseases of modernized societies as omega-3 essential fatty acid deficiency syndrome: substrate beriberi. Med Hypotheses 1982, 8(1):17-47.

13. Bell IR, Edman JS, Morrow FD, Marby DW, Mirages S, Perrone G, Kayne HL, Cole JO: B complex vitamin patterns in geriatric and young adult inpatients with major depression. J Am Geriatr Soc 1991, 39(3):252-257.

14. Eby GA, Eby KL: Rapid recovery from major depression using magnesium treatment. Med Hypotheses 2006, 67(2):362-370.

15. Buist R: The therapeutic predictability of tryptophan and tyrosine in the treatment of depression. Int J Clin Nutr Rev 1983, 3:1-3.

16. Chouinard G, Young SN, Annable L: A controlled clinical trial of L-tryptophan in acute mania. Biol Psychiatry 1985, 20(5):546-547.

17. Reis LC, Hibbeln JR: Cultural symbolism of fish and the psychotropic properties of omega-3 fatty acids. Prostaglandins Leukot Essent Fatty Acids 2006, 75(4–5):227-236.

18. Tanskanen A, Hibbeln JR, Hintikka J, Haatainen K, Honkalampi K, Viinamaki H: Fish consumption, depression, and suicidality in a general population. Arch Gen Psychiatry 2001, 58(5):512-513.

19. von Schacky C: A review of omega-3 ethyl esters for cardiovascular prevention and treatment of increased blood triglyceride levels. Vasc Health Risk Manag 2006, 2(3):251-262.

20. Eritsland J: Safety considerations of polyunsaturated fatty acids. Am J Clin Nutr 2000, 71(1 Suppl):197S-201S.

21. Stoll AL, Severus WE, Freeman MP, Rueter S, Zboyan HA, Diamond E, Cress KK, Marangell LB: Omega 3 fatty acids in bipolar disorder: a preliminary double-blind, placebo-controlled trial. Arch Gen Psychiatry 1999, 56(5):407-412.

22. National Institute of Mental Health: Depression. National Institute of Mental Health, National Institutes of Health 2000. US Department of Health and Human Services, Bethesda (MD) [Reprinted September 2002].

23. Rush AJ: The varied clinical presentations of major depressive disorder. The Journal of clinical psychiatry 2007, 68(8 Suppl):4-10.

24. Stockmeier CA: Neurobiology of serotonin in depression and suicide. Ann N Y Acad Sci 1997, 836:220-232.

25. VanPraag HM: Depression, suicide and the metabolism of serotonin in the brain. J Affect Disord 1982, 4(4):275-290.

26. Diehl DJ, Gershon S: The role of dopamine in mood disorders. Compr Psychiatry 1992, 33(2):115-120.

27. Firk C, Markus CR: Serotonin by stress interaction: a susceptibility factor for the development of depression? J Psychopharmacol 2007, in press.

28. Leonard BE: The role of noradrenaline in depression: a review. J Psychopharmacol 1997, 11(4 Suppl):S39-S47.

29. Petty F: GABA and mood disorders: a brief review and hypothesis. J Affect Disord 1995, 34(4):275-281.

30. McLean A, Rubinsztein JS, Robbins TW, Sahakian BJ: The effects of tyrosine depletion in normal healthy volunteers: implications for unipolar depression. Psychopharmacology 2004, 171(3):286-297.

31. Agnoli A, Andreoli V, Casacchia M, Cerbo R: Effect of s-adenosyl-l-methionine (SAMe) upon depressive symptoms. J Psychiatr Res 1976, 13(1):43-54.

32. aan het Rot M, Moskowitz DS, Pinard G, Young SN: Social behaviour and mood in everyday life: the effects of tryptophan in quarrelsome individuals. J Psychiatry Neurosci 2006, 31(4):253-262.

33. Hoes MJ: L-tryptophan in depression. Journal of Orthomolecular Psychiatry 1982, 4:231.

34. Kravitz HM, Sabelli HC, Fawcett J: Dietary supplements of phenylalanine and other amino acid precursors of brain neuroamines in the treatment of depressive disorders. J Am Osteopath Assoc 1984, 84(1 Suppl):119-123.

35. Maurizi CP: The therapeutic potential for tryptophan and melatonin: possible roles in depression, sleep, Alzheimer's disease and abnormal aging. Med Hypotheses 1990, 31(3):233-242.

36. Ruhé HG, Mason NS, Schene AH: Mood is indirectly related to serotonin, norepinephrine and dopamine levels in humans: a meta-analysis of monoamine depletion studies. Mol Psychiatry 2007, 12(4):331-359.

37. DeLeo D: S-adenosylmethionine as an antidepressant: A double blind trial versus placebo. Curr Ther Res 1987, 41(6):865-870.

38. Janicak PG, Lipinski J, Davis JM, Comaty JE, Waternaux C, Cohen B, Altman E, Sharma RP: S-adenosylmethionine in depression. A literature review and preliminary report. Ala J Med Sci 1988, 25(3):306-313.

39. Adams PB, Lawson S, Sanigorski A, Sinclair AJ: Arachidonic acid to eicosapentaenoic acid ratio in blood correlates positively with clinical symptoms of depression. Lipids 1996, 31(Suppl):S157-S161.

40. Grubb BP: Hypervitaminosis A following long-term use of high-dose fish oil supplements. Chest 1990, 97(5):1260.

41. Rihmer Z, Gonda X, Rihmer A: Creativity and mental illness. Psychiatr Hung 2006, 21(4):288-294.
42. Skutsch GM: Manic depression–a disorder of central dopaminergic rhythm. Med Hypotheses 1981, 7(6):737-746.
43. Skutsch GM: Manic depression: a multiple hormone disorder? Biol Psychiatry 1985, 20(6):662-668.
44. Naylor GJ: Vanadium and manic depressive psychosis. Nutr Health 1984, 3:79-85.
45. Naylor GJ, Smith AH: Vanadium: a possible aetiological factor in manic depressive illness. Psychol Med 1981, 11:249-256.
46. Botiglieri T: Folate, vitamin B12, and neuropsychiatric disorders. Nutr Rev 1996, 54:382-390.
47. Hasanah CI, Khan UA, Musalmah M, Razali SM: Reduced red-cell folate in mania. J Affect Disord 1997, 46:95-99.
48. Osher Y, Bersudsky Y, Belmaker RH: Omega-3 eicosapentaenoic acid in bipolar depression: report of a small open-label study. 2005, 66:726-729.
49. Nieper HA: The clinical applications of lithium orotate. A two years study. Agressologie 1973, 14(6):407-411.
50. Sartori HE: Lithium orotate in the treatment of alcoholism and related conditions. Alcohol 1986, 3(2):97-100.
51. O'Donnell T, Rotzinger S, Ulrich M, Hanstock CC, Nakashima TT, Silverstone PH: Effects of chronic lithium and sodium valproate on concentrations of brain amino acids. Eur Neuropsychopharmacol 2003, 13(4):220-227.
52. Castle E, Wessely S, Der G, Murray RM: The incidence of operationally defined schizophrenia in Camberwell 1965–84. British Journal of Psychiatry 1991, 159:790-794.
53. van der Heijden DFFMMA, Tuinier S, Sijben AES, Kahn RS, Verhoeven WMA: Amino acids in schizophrenia: evidence for lower tryptophan availability during treatment with atypical antipsychotics? Journal of Neural Transmission 2005, 112(4):577-585.
54. Javitt DC, Zylberman I, Zukin SR, Heresco-Levy U, Lindenmayer JP: Amelioration of negative symptoms in schizophrenia by glycine. Am J Psychiatry 1994, 151(8):1234-1236.
55. Leiderman E, Zylberman I, Zukin SR, Cooper TB, Javitt DC: Preliminary investigation of high-dose oral glycine on serum levels and negative symptoms in schizophrenia: an open-label trial. Biol Psychiatry 1996, 39(3):213-215.
56. Javitt DC, Silipo G, Cienfuegos A, Shelley AM, Bark N, Park M, Lindenmayer JP, Suckow R, Zukin SR: Adjunctive high-dose glycine in the treatment of schizophrenia. Int J Neuropsychopharmacol 2001, 4(4):385-391.
57. Peet M: International variations in the outcome of schizophrenia and the prevalence of depression in relation to national dietary practices: an ecological analysis. British Journal of Psychiatry 2004, 184:404-408.
58. Christensen O, Christensen E: Fat consumption and schizophrenia. Acta Psychiatr Scand 1988, 78(5):587-591.
59. Peet M: Eicosapentaenoic acid in the treatment of schizophrenia and depression: rationale and preliminary double-blind clinical trial results. Prostaglandins Leukot Essent Fatty Acids 2003, 69(6):477-485.

60. Emsley R, Myburgh C, Oosthuizen P, van Rensburg SJ: Randomized, placebo-controlled study of ethyl-eicosapentaenoic acid as supplemental treatment in schizophrenia. Am J Psychiatry 2002, 159(9):1596-1598.

61. Puri BK, Richardson AJ, Horrobin DF, Easton T, Saeed N, Oatridge A, Hajnal JV, Bydder GM: Eicosapentaenoic acid treatment in schizophrenia associated with symptom remission, normalisation of blood fatty acids, reduced neuronal membrane phospholipid turnover and structural brain changes. Int J Clin Pract 2000, 54(1):57-63.

62. Richardson AJ, Easton T, Gruzelier JH, Puri BK: Laterality changes accompanying symptom remission in schizophrenia following treatment with eicosapentaenoic acid. Int J Psychophysiol 1999, 34(3):333-339.

63. Richardson AJ, Easton T, Puri BK: Red cell and plasma fatty acid changes accompanying symptom remission in a patient with schizophrenia treated with eicosapentaenoic acid. Eur Neuropsychopharmacol 2000, 10(3):189-193.

64. Richardson AJ: The role of omega 3 fatty acids in behaviour, cognition and mood. Scandinavian Journal of Nutrition 2003, 47(2):92-98.

65. Yao JK, Magan S, Sonel AF, Gurklis JA, Sanders R, Reddy RD: Effects of omega-3 fatty acid on platelet serotonin responsivity in patients with schizophrenia. Prostaglandins Leukot Essent Fatty Acids 2004, 71(3):171-176.

66. American Psychiatric A: Quick Reference to the Diagnostic Criteria from DSM-IV-TR. Arlington, VA 2000.

67. Fontenelle LF, Nascimento AL, Mendlowicz MV, Shavitt RG, Versiani M: An update on the pharmacological treatment of obsessive-compulsive disorder. Expert Opin Pharmacother 2007, 8(5):563-583.

68. Yaryura-Tobias JA, Bhagavan HN: L-tryptophan in obsessive-compulsive disorders. Am J Psychiatry 1977, 134(11):1298-1299.

69. Szegedi A, Kohnen R, Dienel A, Kieser M: Acute treatment of moderate to severe depression with hypericum extract WS 5570 (St John's wort): randomised controlled double blind non-inferiority trial versus paroxetine. British Medical Journal 2005, 330(7494):759.

70. Fava M, Alpert J, Nierenberg AA, Mischoulon D, Otto MW, Zajecka J, Murck H, Rosenbaum JF: A double-blind, randomized trial of St. John's wort, fluoxetine, and placebo in major depressive disorder. J Clin Psychopharmacol 2005, 25(5):441-447.

71. Bell IR, Edman JS, Marby DW, Satlin A, Dreier T, Liptzin B, Cole JO: Vitamin B12 and folate status in acute geropsychiatric inpatients: affective and cognitive characteristics of a vitamin nondeficient population. Biol Psychiatr 1990, 27(2):125-137.

72. Green AR, Aronson JK: The pharmacokinetics of oral L-tryptophan: effects of dose and concomitant pyridoxine, allopurinol or nicotinamide administration. Adv Biol Psychiatr 1983, 10:67-81.

73. Cohen BM, Lipinski JF, Altesman RI: Lecithin in the treatment of mania: double-blind, placebo-controlled trials. Am J Psychiatr 1982, 139:1162-1164.

74. Wozniak J, Biederman J, Mick E, Waxmonsky J, Hantsoo L, Best C, Cluette-Brown JE, Laposata M: Omega-3 fatty acid monotherapy for pediatric bipolar disorder: A prospective open-label trial. Eur Neuropsychopharmacol 2007, 17(6–7):440-447.

75. Frangou S, Lewis M, McCrone P: Efficacy of ethyl-eicosapentaenoic acid in bipolar depression: randomised double-blind placebo-controlled study. Br J Psychiatry 2006, 188:46-50.

76. Rix K, Ditchfield J, Freed DL, Goldberg DP, Hillier VF: Food antibodies in acute psychoses. Psychol Med 1985, 15(2):347-354.
77. Davies S, Stewart A: Nutritional Medicine. London, Pan Books 1987, 403.
78. Tondo L, Rudas N: Course of seasonal bipolar disorder influenced by caffeine. J Affective Disorder 1991, 22:249-251.
79. Castrogiovanni P, Pieraccini F: Dietary interferences with lithium therapy. Eur Psychiatr 1996, 11:53-54.
80. Bernstein AL: Vitamin B6 in clinical neurology. Annal NY Acad Sci 1990, 585:250-260.
81. Edwin E, Holten K, Norum FR, Schrumpf A, Skaug OE: Vitamin B12 hypovitaminosis in mental diseases. Acta Med Scand 1965, 177:689-699.
82. Popper CW: Do vitamins or minerals (apart from lithium) have mood-stabilizing effects? J Clin Psychiatry 2001, 62(12):933-944.
83. Haellstroem T: Serum B12 and folate concentrations in mental patients. Acta Psychiatr Scand 1969, 45(1):19-36.
84. Schorah CJ, Morgan DB, Hullin RP: Vitamin C concentrations in patients in a psychiatric hospital. Hum Nutr Clin Nutr 1983, 37C:447-452.
85. Milner G: Ascorbic acid in chronic psychiatric patients: a controlled trial. Br J Psychiatr 1963, 109:294-299.
86. Rimland B: Plasma vitamin C in the prevention and treatment of autism. Autism Res Rev Intl 1998, 12(2):3.
87. Muskiet FAJ, Kemperman RFJ: Folate and long-chain polyunsaturated fatty acids in psychiatric disease. J Nutr Biochem 2006, 17(11):717-727.
88. Taylor MJ, Geddes J: Folic acid as ultimate in disease prevention: Folate also improves mental health. BMJ 2004, 328(7442):768-769.
89. Cohen BM, Miller AL, Lipinski JF, Pope HG: Lecithin in mania: a preliminary report. Am J Psychiatr 1980, 137:242-243.
90. Parker G, Gibson NA, Brotchie H, Heruc G, Rees AM, Hadzi-Pavlovic D: Omega-3 fatty acids and mood disorders. Am J Psychiatry 2006, 163(6):969-978.
91. Hakkarainen R, Partonen T, Haukka J, Virtamo J, Albanes D, Lönnqvist J: Is low dietary intake of omega-3 fatty acids associated with depression? Am J Psychiatry 2004, 161(3):567-569.
92. International medical news group: Depression linked to lower omega-3 fatty acid levels. Family Practice news 2004., 34(8): 54(51)
93. Leaf A: The electrophysiologic basis for the antiarrhythmic and anticonvulsant effects of n-3 polyunsaturated fatty acids: heart and brain. Lipids 2001, 36(Suppl):S107-110.
94. Lieb J: Linoleic acid in the treatment of lithium toxicity and familial tremor. Prostaglandins Med 1980, 4:275-279.
95. Sabelli HC, Fawcett J, Gusovsky F, Javaid JL, Wynn P, Edwards J, Jeffriess H, Kravitz H: Clinical studies on the phenylethylamine hypothesis of affective disorder: urine and blood phenylacetic acid and phenylalanine dietary supplements. J Clin Psychiatry 1986, 47(2):66-70.
96. Simonson M: L-phenylalanine. J Clin Psychiatry 1985, 46(8):355.
97. Bellivier F, Leboyer M, Courtet P, Buresi C, Beaufils B, Samolyk D, Allilaire JF, Feindgold J, Mallet J, Malafosse A: Association between the tryptophan hydroxylase gene and manic-depressive illness. Arch Gen Psychiatr 1998, 55:33-37.

98. Cassidy F, Murry E, Carroll BJ: Tryptophan depletion in recently manic patients treated with lithium. Biol Psychiatr 1998, 43:230-232.
99. Benkelfat C, Seletti B, Palmour RM, Hillel J, Ellenbogen M, Young SN: Tryptophan depletion in stable lithium-treated patients with bipolar disorder in remission. Arch Gen Psychiatry 1995, 52:154-155.
100. Sandyk R: L-tryptophan in neuropsychiatric disorders: a review. Intl J Neurosci 1992, 67:127-144.
101. Carney MWP, Chary TK, Bottiqlieri T, Reynolds EH: The switch mechanism and the bipolar/unipolar dichotomy. Br J Psychiatr 1989, 154:48-51.
102. Tolbert LC, Monti A, Walter-Ryan W, Alacron RD, Bahar B, Keriotis JT, Allison JG, Cates A, Antun F, Smythies JR: Clinical correlations of one-carbon metabolism abnormalities. Prog Neuropsychopharmacol Biol Psychiatr 1988, 12(4):491-502.
103. Carney MWP, Chavy TK, Bottiglieri T, Reynolds LH: Switch and S-adenosylmethionine. Ala J Med Sci 1988, 25(3):316-319.
104. Pacchierotti C, Iapichino S, Bossini L, Pieraccini F, Castrogiovanni P: Melatonin in psychiatric disorders: a review on the melatonin involvement in psychiatry. Front Neuroendocrinol 2001, 22:18-32.
105. Nurnberger JI, Adkins S, Lahiri DK, Mayeda A, Hu K, Lewy A, Miller A, Bowman ES, Miller MJ, Rau L, Smiley C, Davis-Singh D: Melatonin suppression by light in euthymic bipolar and unipolar patients. Arch Gen Psychiatry 2000, 57:572-579.
106. Avery D, Lenz M, Landis C: Guidelines for prescribing melatonin. Ann Med 1998, 30:122-130.
107. Fekkes D, Pepplinkhuizen L, Verheij R, Bruinvels J: Abnormal plasma levels of serine, methionine and taurine in transient, acute, polymorphic psychosis. Psychiatry Res 1994, 51:11-18.

This chapter was originally published under the Creative Commons Attribution License. Lakhan, S. E., and Vierira, K. F. Nutritional Therapies for Mental Disorders. Nutrition Journal 2008, 7:2. doi:10.1186/1475-2891-7-2. http://www.nutritionj.com/content/7/1/2

CHAPTER 5

FRUCTOSE METABOLISM IN HUMANS: WHAT ISOTOPIC TRACER STUDIES TELL US

SAM Z SUN and MARK W EMPIE

5.1 INTRODUCTION

Fructose has been a part of the human diet for many thousands of years, and it is found in highest concentrations in fruits and to a lesser degree in vegetables. Cane, beet, and corn sugars are produced industrially, and their use results in significant quantities of added sugars entering the diet, about half of which is fructose [1]. Cane and beet sugars are comprised of the disaccharide sucrose (glucose bonded to fructose) and are commonly called table sugar or simple sugar. Corn sugars come from corn starch, and mainly consist of high fructose corn syrup 55 (HFCS 55; 55% fructose-41% glucose), HFCS 42 (42% fructose-52% glucose), and corn syrup (glucose and oligoglucose with trace amounts of fructose). During the last several decades, the prevalence of obesity and metabolic syndrome has risen dramatically on a global basis, but more so in the U.S. population. Because the prevalence is chronologically and statistically correlated with the increase of added sugar intakes, particularly HFCS in the U.S. (HFCS is not consumed significantly outside the U.S.), some have proposed the intake of HFCS or fructose as a free monosaccharide may be a cause of various adverse health consequences [2]. Conventional clinical trials and ecological studies have been conducted to assess the hypotheses, but not all results are found to be supportive. Conventional studies often cannot reveal details of interconnecting metabolic pathways when testing fructose or fructose-containing sugars, but they also cannot clearly distinguish a mechanistic cause associated with an observed physiological consequence linked to the sugar consumed. This is because the ordinary diet contains multiple forms of saccharides which are inter-convertible in the body and share many steps of the carbohydrate metabolism pathways.

Over the last decade, a series of controversies have arisen regarding fructose consumption. In 2004, a commentary was written hypothesizing that the "high" fructose content in HFCS was the cause of the obesity rise in America [3]. This was based on the association of the obesity prevalence rise with the replacement of cane and beet sugar by HFCS, even though the fructose content of these two sweetener sources is essentially the same. Later, several dietary studies using calorically high doses of fructose were published to investigate fructose modulation of leptin hormone status, with a suggestion that chronic changes in this hormone level could lead to weight gain [4,5]. However, other studies and evidence based reviews do not always support these findings [6-13]. Recently, Welsh et al. [14] reported that the intake of added sugar has significantly decreased between 1999 and 2008 while the obesity prevalence has continued to rise. The current view is that obesity is a matter of energy balance [15,16]. Next, the fructose moiety in sugars was hypothesized to cause high serum uric acid which could lead to the development of Type-2 diabetes [17]. There is currently no direct proof for a cause and effect relation of urate with diabetes, and NHANES data suggests no relation of serum urate with fructose intake at ordinary dietary consumption levels [18]. Then, another hypothesis has been raised that dietary fructose may potentially lead to Non Alcoholic Fatty Liver Disease (NAFLD) and augmented de-novo triglyceride synthesis, based on an analysis of hormone regulated lipid pathways in the liver [19,20]. It is known that high dietary levels of fructose can increase serum triglycerides. However, all the factors linked to the development of fatty liver disease are not well understood and can include, insulin resistance, inflammation, fat re-deposition, abnormalities in control of reactive oxygen species [21] and uncoupling proteins in mitochondria [22]. NAFLD is currently an important and actively researched field relative to dietary sugar intakes.

Additionally, it is important to understand the practical significance of testing an effect from a single sugar using an unrepresentative dose compared to the true population sugar intake, a question which is currently under debate [23-26]. In many of the intervention studies involved with studying the various hypotheses mentioned above, very high doses of sugars over short term were often applied, the study designs were more similar to toxicological studies, and the studies were only able to draw associative conclusions between applied dose and observed health-related outcomes in the subjects studied. The observed biological changes, although statistically significant by a P-value ruling, were often only fluctuations within normal ranges. These studies rarely measured actual development of disease or the intermediate metabolites characterizing mechanism-based reactions. To begin to prove true effect of a diet component, it is useful to study the component disposal through the common central pathways at the molecular level. These studies are facilitated and detailed by the use of isotope tracer labeled precursors, and this concept is the stimulus for this review.

The questions raised by the above hypotheses reach into the broader metabolome and fluxome. Our understanding of the metabolism of glucose and fructose as separate sugars is founded upon many years of study, and detailed anabolic and catabolic pathways are known [27]. Recently, the extended metabolism of glucose and fructose has been reviewed by Tappy and Le [28]. Glucose and fructose carbons are utilized through the glycolysis, gluconeogenesis, glycogenolysis, tricarboxylic acid (TCA) cycle, lactate production (Cori cycle), pentose phosphate shunt, and lipid synthesis pathways in various physiological compartments to provide substrates for glycogen homeostasis, amino acids, other sugars, fats and energy (e.g. ATP). Glucose and fructose enter the metabolic pathways differently (Figure 1), with glucose being converted to 1,6-diphosphorylated fructose before being cleaved into the three carbon metabolic intermediates, dihydroxy acetone phosphate and glyceraldehyde 3-phosphate. Absorbed fructose is only mono-phosphorylated before being cleaved into glyceraldehyde and dihydroxy acetone phosphate, which is the common intermediate with the glucose pathway. Glucose utilization can be regulated before cleavage, whereas fructose is less regulated. This initial difference has prompted some to hypothesize that, because fructose cleavage by-passes key feedback regulatory steps in the glucose metabolic pathway, this bypass may lead to increases of fatty acid synthesis, which may contribute to causes of obesity [4]. This hypothesis relies on a simplified metabolic pathway analysis and on studies using pure fructose in comparison to pure glucose, a situation which rarely occurs in the American diet [29,30].

In nature, fructose commonly occurs together with glucose, and composition values for some foods have been tabulated by the USDA on its website: http://www.nal.usda.gov/fnic/foodcomp/search/ . The metabolism of food derived sucrose, fruit sugars, honey, and high fructose corn syrup, major sources of fructose and glucose in the diet, are currently under study, and the biological effects resulting from the use of experimentally formulated mixtures of glucose and fructose are relevant to our understanding. The use of mixed sugars are more metabolically predictive of dietary consequences than that from single monosaccharides studied individually, as metabolism of each type of sugar is not independent from the other (discussed below). Metabolic interactions between glucose and fructose significantly impact general sugar metabolism.

Owing to the complexity of fructose and glucose metabolism, conventional feeding study approaches are usually less informative than isotope tracer studies for obtaining a clear picture of mechanisms for utilization of dietary fructose or glucose. It is known that carbon moieties in fructose and glucose can be inter-converted in the liver [31], and thus studying the disposal and metabolic effects of these dietary sugars with respect to one another is most definitively conducted using isotope-labeled sugars as tracers. A number of these isotopic tracer studies exist, and many are found in the literature dated before the year 2000. Although all pathways have not been completely studied for fructose disposal and metabolism under different physiological conditions, a significant number of reports on fructose isotope tracer studies are published. In this work, we have reviewed fructose disposal and metabolism in humans based on isotope tracer studies to better understand from a molecular stand point fructose oxidation, fructose conversion into glucose, fructose conversion into lipids, and fructose conversion into lactate.

5.2 METHOD

Pubmed and Scopus websites were searched using 2 or more key word combinations of fructose, glucose, sucrose, tracer, 13C, 14C, and isotope

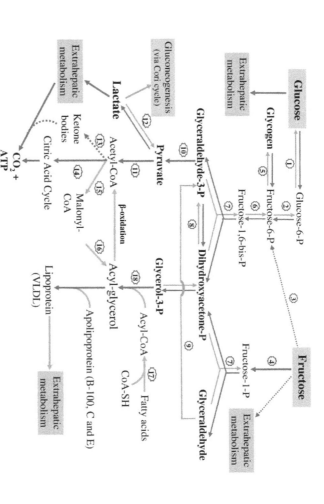

FIGURE 1: Major metabolic pathways and flux of dietary glucose and fructose. P=phosphate. For enzymes numbered in circles: 1=hexokinase/glucokinase or Glucose-6-phosphatase, 2=phosphoglucose isomerase, 3=hexokinase, 4=fructokinase, 5=glycogen synthase or phosphorylase, 6=phosphofructokinase, 7=aldolase, 8=triose phosphate isomerase, 9=triose kinase, 10=several enzymes including pyruvate kinase, 11=pyruvate dehydrogenase complex, 12=lactate dehydrogenase, 13=ketothiolase and other 3 enzymes, 14=enzyme group related to citric acid cycle, 15=acetyl CoA carboxylase, 16=multienzyme complexes, 17=acyl CoA synthase, 18=glycerol-phosphate acyl transferase and triacylglycerol synthase complex. The dashed-line and arrow represents minor pathways or will not occur under a healthy condition or ordinary sugar consumption. The compound names in bold would be major metabolic intermediates or end products of glucose or fructose metabolism.

with limitation of using human studies. When reviewing the metabolic fate of dietary fructose (including oxidation, glucose conversion, glycogen synthesis, lipid conversion and lactate production), the data were obtained from publications that met the following criteria: adult subjects, unbound or bound fructose studied, isotope-tracer used, in English, and with metabolic-related study purposes. In total, 34 papers met the criteria. Other conditions related to study design were not used as exclusion criteria, such as subject fasting status, ways of fructose administration, and sample size. Many of the studies on fructose oxidation were conducted by researchers interested in exercise and athletic performance enhancement. In some studies, fructose ingestions were combined with glucose or with other nutrient intravenous infusion. Dose levels of fructose and administration methods (bolus or several small portions) also varied between studies. As a result, there is significant heterogeneity between studies and the protocol, and quality among the tracer studies cited in this review may not be similar. The expired CO_2 recovery coefficients, correction factors (k factor) for the collection loss of expired CO_2, were not identical in the studies investigating fructose oxidation. This may account for some of the observed variations in oxidation rates of ingested fructose [32].

The following sections will review what tracer studies tell us about the disposal and metabolism of dietary fructose as a single sugar or as a mixture with glucose, either bound or free. In this way, the origins and fate of specific carbons in these sugars may be determined in relation to their partition among metabolic pathways and how the presence of one sugar influences the metabolism of the other. In two sections, studies are included which use non-labeled fructose together with labeled pathway compounds to assess the impact of ingested fructose on metabolites or pool intermediates leading to alterations in relevant end points, such as glucose production or de novo lipogenesis. In this way, fructose carbons themselves are not followed, but broader anabolic responses affected by the fructose load can be measured. This isotopic method makes use of Mass Isotopomer Distribution Analysis (MIDA), a technique reviewed by Hellerstein and Neese (1999) [33]. Although these studies investigate a number of other important physiological parameters, we report here only the results which are directly measured by the tracer itself.

5.2.1 FRUCTOSE ABSORPTION

To place the metabolism of labeled sugars in context, it is helpful to briefly discuss what is known about the uptake of fructose, glucose and sucrose from the gut, interdependencies, and entry into the circulation. These considerations should be taken into account when designing studies. A number of the studies discussed in this section do not use tracer labeled sugars, but are included to provide a comprehensive description. The absorption rate of fructose alone from the small intestine is slower than that of glucose. This is partly due to the differences in the absorption process between the two monosaccharides. Glucose is absorbed from the intestine into the plasma via more than one active glucose co-transporter protein. SGLT1 transports glucose from the intestinal lumen through the apical membrane into the intestinal epithelial cells. Exit from the epithelial cells through the basolateral membrane to the blood is facilitated by GLUT2. Fructose is absorbed at a slower rate from the lower part of duodenum and jejunum both passively and actively by the brush-border membrane transporter 5 (GLUT-5) and transported into blood also by GLUT2 [34,35]. GLUT transporters are primarily made up of 13 multiple homologous proteins (GLUT 1–12 and 14) and they are located throughout the body often exhibiting tissue specificities [36]. The capacity for fructose absorption in humans is not completely clear, but early studies suggested that fructose absorption is quite efficient, though it is less efficient than that of glucose or sucrose [34]. The slower absorption and prolonged contact time with the luminal intestinal wall would be expected to result in the stimulation of regulatory and satiety signals and release of hormones from enteroendocrine cells [37,38].

When fructose is consumed as the sole carbohydrate source, it can be incompletely absorbed, and as a result, produces a hyperosmolar environment in the intestine. A high concentration of solute within the gut lumen draws fluid into the intestine which can produce feelings of malaise, stomachache or diarrhea [39], and results in decreased food intake. However, when glucose is also present, malabsorption is significantly attenuated [40]. Riby et al. [34] compiled data from five studies comparing glucose, fructose and mixtures of the two for degree of absorption, by measurement of breath hydrogen as an indicator of malabsorption. Pure fructose alone produced dose-dependent evidence of malabsorption starting from 12 gram ingestion loads, while glucose and sucrose individually produced no intolerance up to 50 gram ingestion loads. Incremental amounts of added free glucose to a 50 gram fructose load dose-dependently attenuated malabsorption symptoms, and at the equimolar mixture of the two (up to 100 grams total sugars), no malabsorption was observed. Thus, how studies are designed to deliver the various sugars can have an impact on sugar uptake and appearance in the blood.

Sucrose is a valid comparison for glucose-fructose mixtures, as the disaccharide is cleaved by the enzyme sucrase into the mono sugars before being absorbed into the circulation. Comparison of sucrose absorption rates in 32 normal subjects with an equivalent amount of monosaccharide mixture containing glucose and fructose, infused intralumenally to avoid gastric hydrolysis, resulted in the similar absorption rates for each glucose and fructose component of the test [41]. In another study, type-2 diabetic patients were fed sucrose or HFCS with a background diet, resulting in plasma glucose AUC's not being different between sucrose and the HFCS, nor were mean plasma insulin values [42]. It was also shown that mucosal-to-serosal glucose flux was similar between sucrose and glucose + fructose mixture solution, but rates depended on sucrase and sodium-dependent glucose transport in an in vitro study [43]. Other comparison studies in normal men and women [44] and in diabetics [45,46] produced no differences in intestinal uptake between sucrose and honey (a glucose-fructose mixture). Thus, the body appears to handle oral free glucose-fructose mixtures or HFCS similarly as sucrose and that hydrolysis of sucrose does not appear to be rate limiting for uptake.

Once absorbed, glucose is delivered to the liver then to peripheral organs for utilization, and its entrance into muscle and fat cells is insulin dependent. Fructose is primarily delivered to and metabolized in the liver for energy and for two and three carbon precursor production without dependence on insulin. Bolus or divided doses of 50–150 g fructose produce plasma concentrations of 3–11 mg/dl of this sugar [47-52], while glucose can spike upwards of 150 mg/dl and more. Although little dietary fructose appears in the circulation, it can influence plasma glucose concentrations via sugar inter-conversion. In man, studies indicate fructose to glucose conversion may occur to a highly significant degree (reviewed below) and that this conversion occurs via the 3-carbon intermediate pathways. The extent of inter-conversion may be species dependent.

Key points: 1) Fructose is readily absorbed and its absorption is facilitated by the presence of co-ingested glucose. Sucrose, honey, 50:50 glucose-fructose mixtures and HFCS all appear to be similarly absorbed. 2) Fructose itself is retained by the liver, while glucose is mainly released into the circulation and utilized peripherally. And, 3) Plasma levels of fructose are an order of magnitude (10–50 folds) lower than circulating glucose, and fructose elicits only a modest insulin response. This lower glucose and insulin response by the body to fructose intake has been considered desirable for diabetic diets.

5.2.2 FRUCTOSE AND GLUCOSE METABOLIC FLUX

Fructose and glucose metabolic flux is briefly described in Figure 1. The important point of distinction between glucose and fructose metabolism resides in two areas. Absorbed fructose is extracted by, held, and processed in the liver, with little fructose circulating in the blood stream or delivered to peripheral tissues. Absorbed glucose or that produced in the liver from fructose or other precursors is either metabolized in the liver or exported to the blood stream and further to extrahepatic tissues. Most absorbed fructose is cleaved in the liver into glyceraldehyde and dihydroxy acetone phosphate, and these trioses further go to glycerol phosphate and pyruvate metabolic pathways, respectively. With both fructose and glucose, lactate conversion plays an important role in distributing carbohydrate potential energy between gluconeogenesis and acetyl CoA, with entry into the TCA cycle or use in lipid synthesis (Figure 1) [53,54]. Lactate discharge is also a means for fructose carbons to escape the liver and be transported to peripheral tissues. Fructose cleavage to glyceraldehyde can result in the production of glycerol via reduction. It was observed that blood glycerol concentration increased after fructose ingestion in exercise subjects [55,56]. The noted glycerol increases after fructose ingestion are either greater or similar compared with the values after glucose ingestion, and the produced glycerol can be oxidized for energy. However, the metabolic balance between glycerol produced from fructose and central pathway trioses has not been clearly determined.

Given the complexity and interdependencies of energy metabolism and biochemical synthesis arising from sugars, consideration of the flux of carbons among these pathways is critical to understanding the health consequences of consuming these nutrients. Single sugar distribution and fluxes between pathways are not easily studied without isotopic labels. Classically, a limited number of metabolites are characterized in a study and some disposal points can be missed. More recently, a computational technique is being employed utilizing. Nuclear Magnetic Resonance (NMR) or mass spectral analyses of the 13C isotopomer distribution of metabolites, following administration of labeled precursors. These precursors may be uniformly labeled compounds or labeled at specific carbons, depending on the question to be answered. An empirical metabolic flux analysis profile is generated which can be mathematically modeled without being constrained by physical chemistry rigor, as reviewed by Selivanoc and Lee [57,58]. This technique allows one to model metabolite fluxes which may not be well characterized or understood from direct enzymatic or physical chemistry data. A second method is under development to mathematically model general metabolism and interdependencies of pathways using known thermodynamic free energy and kinetic constant parameters for each reaction in the pathway sequences [59], but there is currently insufficient data to apply to fructose metabolism questions.

Each method has advantages and disadvantages, and likely the combination of both is needed for optimal predictive power. In the future, with these tools one should be able to predict outcomes from sugars supply as a function of the organism's energy (ATP) status, oxidation/reduction potential (NADH/NADPH) and nutrient dependent cofactors. Metabolic differences among compartments and their interactions as a whole should be included. Experimentation should account for metabolome interactions, and study results should be interpreted carefully with respect to the experimental conditions employed.

Key points: 1) Fructose is observed to enter all the pathways of disposal as found for glucose glycolysis and the TCA cycle. 2) Three carbon intermediates provide a means for fructose to be released from the liver and to be utilized peripherally, which suggests that physiological effects observed should be integrated with the co-effect and metabolic fluxes arising from all sugars using these pathways.

5.2.3 METABOLIC FATE OF DIETARY FRUCTOSE

The following reviews the data as presented in the papers. As depicted in Figure 1, the interdependence of metabolic pathways of fructose and glucose can influence the flow of metabolites and their temporal appearance as other compounds. Thus, in discussing the disposal of fructose carbons, e.g. through oxidation, one cannot accurately distinguish if the labeled CO_2 arose directly from the fructose itself, or from fructose which had undergone conversion to neoformed glucose, neoformed lactate, or other neoformed compounds. Where the conversions occur, these metabolites can be readily transported out of the liver to other tissues, altering the temporal appearance of metabolites. Further complicating the analysis, some studies used fructose labeled with 13C at different positions of its carbon backbone, uniformly labeled fructose, or 13C naturally enriched fructose. The different labeling may also influence the appearance of the isotope tracer in various metabolites. Using uniformly labeled fructose would limit the potential complications from different labeling positions on isotope tracer appearance in its metabolites. It should also be noted that most of the tracer studies described in the following sections are short-term dietary studies (monitored periods shorter than 8 hours) and may not reflect longer term effects of fructose, such as on de novo lipogenesis, VLDL TG production, or other metabolic specificities.

5.2.4 FRUCTOSE OXIDATION

Multiple studies have been conducted to observe how much fructose and other sugars can be oxidized following ingestion. Table 1 summarizes fructose and other sugar oxidation data reported from tracer studies in humans under different experimental designs. In all, 19 relevant studies were found which met the inclusion criteria of this review. The first 4 studies cited in Table 1 used resting subjects with fructose ingestion levels from 0.5-1.0 g/kg body weight (bw). Within the study monitoring periods, the ingested fructose was oxidized from 30.5% to 59%. The study by Chong and colleagues [48] showed fructose was oxidized faster than glucose (30.5% vs 24.5%). This effect may be due to less regulation of phosphorylation for fructose or to a wider tissue distribution of glucose. Oxidation rates increase as the dose increases but would be attenuated by its rate of absorption when intake amounts are large. Delarue et al. [49] indicated when the fructose administration dose increased from 0.5 to 1 g/kg bw, the oxidation amount of fructose correspondingly increased, such that a similar percent of the given fructose dosage was oxidized (56% and 59%, respectively). However, there is a difference in oxidation rates between normal and diabetic subjects, in that normal subjects could more efficiently oxidize fructose than type-2 diabetics (38.5% vs 31.3% of given dosage) [52].

The other studies in the Table 1 were conducted under conditions of exercise where workloads corresponded to 50-75% of max VO2 uptake. The oxidized amounts of ingested fructose ranged from 37.5% to 62.0%. Except in one study [60] which showed that fructose and glucose had similar oxidation rates (38.8% and 40.5%, respectively), the other studies all observed that glucose was oxidized faster than fructose under the exercise conditions [47,50,55,56,61-65]. A very interesting phenomenon noted is that when fructose and glucose are ingested together (including fructose-containing sucrose), the oxidation rates of the mixed sugars were faster than that of either one of them ingested alone at the same dosage. Adopo et al. reported that, given 100 g fructose, glucose, or fructose + glucose, 73.6% of the mixed sugars were oxidized while the data of fructose and glucose were 43.8% and 48.1% as ingested separately [61]. The series of studies by Jentjens and colleagues [66-69] also reported that fructose plus glucose or sucrose plus glucose consumed together were oxidized faster than glucose alone.

A summary of the sugar oxidation data is shown in Figure 2. The data of obese or diabetic subjects are not included in this figure. In non-exercise subjects, the mean of the oxidized fructose amount was 45.0% ± 10.7 (mean ± SD, range 30.5-59%) of ingested dose within a period of 3-6 hours. Under exercise conditions, this mean was 45.8% ± 7.3 (mean ± SD, range 37.5-62%) within 2-3 hours. When fructose and glucose are ingested in

TABLE 1: Oxidation of Dietary Fructose, Glucose, and Other Sugars in Tracer Studies(1)

Subjects	Exercise	Hours	Sugar dosage (g)	Tracer	Oxidation	Reference
9 M	No	6	0.9 fru/kg bw	13C-fru(L1)	42.9%	[70](2)
9 F	No	6	0.9 fru/kg bw	13C-fru	43%	
8 M+6 F	No	6	0.75 glu/kg bw	13C-glu(L1)	24.5%	[48]
8 M+6 F	No	6	0.75 fru/kg bw	13C-fru(L1)	30.5%	
M+3 F	No	6	0.5 fru/kg bw	13C-fru(L2)	56%	[49](2)
3 M+3 F	No	6	1.0 fru/kg bw	13C-fru	59%	
4 M+4 F	No	3	0.9 fru/kg bw	13C-fru(L3)	38.5%	[52](2)
7 obese F	No	3	0.9 fru/kg bw	13C-fru	34.9%	
8 type-2 (4 M)	No	3	0.9 fru/kg bw	13C-fru	31.3%	
10 M	yes	2	0.6 malt/min	13C-glu(L4)	81.7% (0.49 g/ min)	[82]
10 M	yes	2	0.3 fru+0.6 malt/min	13C-fru(L4)	62.0% (0.18 g/ min)	
10 M	yes	2	0.5 fru+0.6 malt/min	13C-fru	54.0% (0.27 g/ min)	
10 M	yes	2	0.7 fru+0.6 malt/min	13C-fru	52.0% (0.36 g/ min)	
6 M	Yes	2	100 galactose	13C-galactose(L1)	23.7%	[60]
6 M	Yes	2	100 fru	13C-fru(L1)	38.8%	
6 M	Yes	2	100 glu	13C-glu(L1)	40.5%	
6 M	yes	2	100 fru	13C-fru(L1)	43.8%	[47]
6 M	yes	2	100 glu	13C-glu(L1)	48.1%	
6 M	yes	2	100 fru+120 sucr	13C-fru	42.0% (42 g)	

TABLE 1: *Cont.*

Subjects	Exercise	Hours	Sugar dosage (g)	Tracer	Oxidation	Reference
6 M	yes	2	100 glu + 120 sucr	13C-glu	50.2% (50.2 g)	
18 M	yes	2	1.33 fru/kg bw	13C-fru(L1)	36.7% (35.7 g)	[62]
18 M	yes	2	1.33 glu/kg bw	13C-glu(L1)	57.2% (56.1 g)	
6 M	yes	2	100 fru	13C-fru(L1)	45.8%	[61](2,3)
6 M	yes	2	100 glu	13C-glu(L1)	58.3%	
6 M	yes	2	50 fru + 50 glu	13C-fru + 13C-glu	73.6%	
6 M	yes	3	150 fru	13C-fru(L2)	38.0%	[50]
6 M	yes	3	150 glu	13C-glu(L2)	54.0%	
5 M	yes	2	1.33 fru/kg bw	13C-fru(L4)	51.0% (49 g)	[65]
5 M	yes	2	1.33 glu/kg bw	13C-glu(L4)	60.4% (58 g)	
5 M3	yes	2	1.33 fru/kg bw	13C-fru	37.5% (36 g)	
5 M3	yes	2	1.33 glu/kg bw	13C-glu	58.3% (56 g)	
6 M	yes	2	1.33 fru/kg bw	13C-fru(L4)	54.0% (53 g)	[64](3)
6 M	yes	2	1.33 glu/kg bw	13C-glu(L4)	72.0% (70 g)	
6 M	yes	2	100 fru	13C-fru(L4)	54%	[55]
6 M	yes	2	100 glu	13C-glu(L4)	67%	
7 M	yes	3	140 fru	13C-fru(L4)	56%	[63](3)
7 M	yes	3	140 glu	13C-glu(L4)	75%	
10 M	yes	2	1.0 fru/kg bw	13C-fru(L2)	43.0% (30 g)	[56]

TABLE 1: *Cont.*

Subjects	Exercise	Hours	Sugar dosage (g)	Tracer	Oxidation	Reference
10 M	yes	2	1.0 glu/kg bw	13C-glu(L2)	37.1% (26 g)	
8 M	yes	2	0.5 fru + 1.0 glu/ min	13C- fru + 13C- glu(L2)	72.7% (1.09 g/ min)	[68](3,4)
8 M	yes	2	1.5 glu/min	13C-glu	50.7% (0.76 g/ min)	
8 M	yes	2	1.2 sucr/min	13C-sucr(L1)	78.3% (0.94 g/ min)	[67](4)
8 M	yes	2	1.2 glu/min	14C-glu(L1)	58.3% (0.70 g/ min)	
8 M	yes	2	0.6 sucr + 0.6 glu/min	13C-su- cr + 14C-glu	70.8% (0.85 g/ min)	
9 M	yes	2.5	1.8 glu/min	13C-glu(L2)	53.3% (0.96 g/ min)	[69](4)
9 M	yes	2.5	0.6 sucr + 1.2 glu/min	13C- sucr + 13C- glu(L2)	62.2% (1.12 g/ min)	
9 M	yes	2.5	0.6 malt + 1.2 glu/min	13C- malt + 13C- glu(L2)	52.2% (0.94 g/ min)	
8 M	yes	2	1.8 glu /min	14C-glu(L1)	38.7% (0.75 g/ min)	[66](4)
8 M	yes	2	0.6 fru + 1.2 glu/ min	13C- fru + 14C- glu(L1)	64.4% (1.16 g/ min)	

(1): hours = study monitoring hours, fru = fructose, glu = glucose, sucr = sucrose, malt = maltose; bw = body weight; M = male, F = female; type-2 = type-2 diabetes. In the reference column, (2) = with glucose infusion. (3) = non-fasting subjects. (4) = oxidation data of the 2nd hour or the last 1.5 hours. In the column of "Tracer", superscripted L1 = labeled uniformly, L2 = naturally enriched, L3 = labeled at position 1, and L4 = 13C position not indicated.

combination, either as fructose plus glucose, as sucrose, or as sucrose plus one of the 2 mono-sugars, the mean oxidized amount of the mixed sugars increased to 66.0%±8.2 (mean±SD, range 52.2-73.6%). The oxidation data of glucose alone is 58.7%±12.9 (mean±SD, range 37.1-81.0%).

Key points: 1) A significant amount of ingested fructose is oxidized by the body to produce energy. 2) Under resting conditions, fructose may be preferentially or similarly utilized to produce energy as glucose, and under exercise, glucose appeared to be more preferentially used to produce energy by the body. 3) When fructose and glucose are ingested together, the mixed sugars will be oxidized significantly faster than either one of the sugars ingested alone. And, 4) Fructose metabolism could be very different between normal and obese/diabetic subjects. A potential consideration with these oxidation studies is that with shorter time frames of measurement or incomplete oxidation and with only partial labeling, position of the isotope label can influence the rate of appearance of the isotope in the exhaled carbon dioxide (CO_2). Temporal isotope appearance in CO_2 can be altered if some of the fructose carbons are not completely oxidized in the time frame of measurement due to diversion to non-oxidative pathways.

5.2.5 FRUCTOSE-GLUCOSE CONVERSION

The disposal pathway for fructose is not solely by direct oxidation, as some absorbed fructose will be converted to glucose. A number of studies have determined the extent of the conversion, which can only be clearly done using tracers. Table 2 tabulates the data from various studies with different experimental conditions. Tran and colleagues [70] studied the conversion of fructose to glucose as compared between men and women. After a 3 times ingestion of a fructose-containing beverage (3x0.3 g/kg bw), 37.4% of the fructose was converted to glucose in men during 6 hours. This value is significantly higher than the conversion rate of 28.9% observed in women. Similarly, using an equal fructose dosage, Paquot et al. [52] noted the conversion percent from fructose to glucose was 36.4% in 8 normal subjects (4M+4F), which is comparable with Tran's data. However, the conversion proportion appeared to be lower in obese and diabetic subjects (29.5% and 30.2%, respectively). In a dosing study monitored over a period

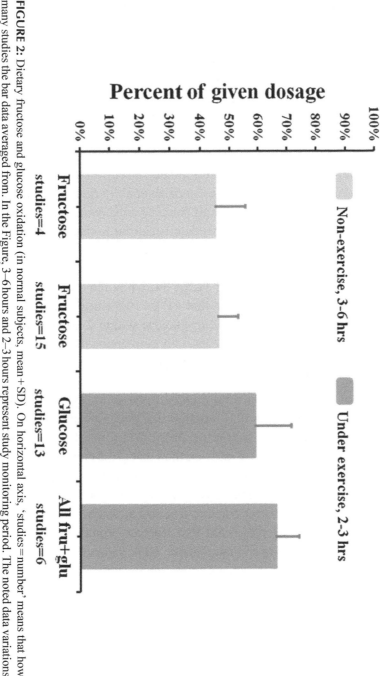

FIGURE 2: Dietary fructose and glucose oxidation (in normal subjects, mean + SD). On horizontal axis, 'studies = number' means that how many studies the bar data averaged from. In the Figure, 3–6 hours and 2–3 hours represent study monitoring period. The noted data variations between studies could be due to the differences of sugar dosages, tracer labeling forms, sugar administration methods, subject characteristics, and/or measurement errors. Also, the produced CO_2 from labeled sugar oxidation can arise directly from sugar molecules themselves, or other compounds converted from the sugars, such as glucose, lactate, or fatty acid from fructose.

of 6 hours, using 0.5 and 1 gram/kg bw, the conversion from fructose to glucose was reported to be 54% and 50.7% of given dosages, respectively [49]. Surmely et al. [71] infused fructose at 3 mg/kg bw per minute for the first 3 hours, followed by doubling the infusion dosage for the next 3 hours. It was noted that the subsequent higher infusion dose level somewhat slowed the fructose conversion percentage, 22% and 28% for high and low dose levels respectively. Under exercise conditions, Lecoultre et al. [51] reported that 29% of ingested fructose (96 g) is converted to glucose when a steady state of carbohydrate flux was reached (1.7-2 hrs from the beginning of study). With repeated administration to achieve a high dose level and under exercise, Jandrian et al. [50] reported that 55-60% of circulating glucose comes from fructose conversion during the latter half of the monitoring period. This data is similar to that observed by Delarue et al. [49] who reported the amount of glucose synthesized from fructose was 57% of overall glucose appearance in the circulation after an ingestion of fructose at dosage of 1 g/kg bw, while the subjects were not under exercise. These data suggest that 41% ±10.5 (mean ± SD, range 29-54%) of fructose can be converted to glucose within 2–6 hours after ingestion in normal non-exercise subjects. This conversion may be lower in women compared to men, and obese and diabetic subjects may also have lower conversion capability.

For the conversion from dietary fructose to glycogen, data are very limited. Nilsson et al. [72] reported that a significantly higher amount of glycogen was determined in the liver (274.6 mmol glycosyl unit per kg wet tissue) after fructose infusion than that (76.2 mmol glycosyl unit) after glucose infusion; and no difference of glycogen increase in muscle was noted (23.0 and 24.4 mmol glycosyl units per kg after fructose and glucose infusions, respectively). Another fructose infusion study (non-exercise) by Dirlewanger et al. [73] noted that fructose stimulates total glucose output, glucose cycling and intrahepatic UDP galactose turnover, which was used as a marker for increased glycogen synthesis. Blom and colleagues [74] reported that dietary fructose could be about half as efficient as glucose or sucrose to replenish muscle glycogen after exercise. In that study, healthy young subjects exhaustively exercised on bicycle ergometers, and ingested 0.7 g/kg bw of fructose, glucose, or sucrose divided in 3 doses. The rates of glycogen synthesis in muscle corresponding to each sugar

TABLE 2: Conversion from Fructose to Glucose, Tracer Studies in Adults(1)

Subjects	Exercise	Hours	Fru dosages	Tracer	Blood glu (mmol/L)	Fru to glu conversion	Reference
9M	No	6	3x0.3 g/kg bw	13C-fru(L1)	5.94%(2)	37.4%	
9F	No	6	3x0.3 g/kg bw	13C-fru	4.87%(2)	28.9%	[70]
4M+4F	No	3	3x0.3 g/kg bw	13C-fru(L3)	5.2	36.4%(3)	
7 obese F	No	3	3x0.3 g/kg bw	13C-fru	5.3	29.5%(3)	
8 type-2 (4 M)	No	3	3x0.3 g/kg bw	13C-fru	7.7	30.2%(3)	[52]
3M+3F	No	6	0.5 g/kg bw	13C-fru(L2)	4.56	54.0%	
3M+3F	No	6	1.0 g/kg bw	13C-fru	4.66	50.7%	[49]
3M+3F	No	0-3	3 mg/kg/min4	13C-fru(L1)	NA	28.0%(4)	
7M	No	4-6	6 mg/kg/min4	13C-fru	NA	22.0%(4)	[71]
7M	Yes	2	96 g fru+144 g glu	13C-fru(L1)	6.2	29%(5)	[51]
6M	Yes	3	6x25 g	13C-fru(L2)	NA	55-60%(6)	[50]

For superscript numbers: (1), except Jandrain's study [50], the subjects in the other studies were under glucose infusion; Hours = study monitoring hours, fru = fructose, glu = glucose; bw = body weight; M = male, F = female; type-2 = type-2 diabetes; NA = not available. (2), increases from baseline. (3), data were calculated based on reported parameters. (4), fructose administrated by infusion. (5), under steady state of carbohydrate flux. (6), percent of circulating glucose in the 2nd half of study hours. In the column of "Tracer", superscripted L1 = labeled uniformly, L2 = naturally enriched, and L3 = labeled at position 1.

treatment were observed as 0.32, 0.58, and 0.62 mmol/kg per hour, respectively. The data indicate that energy status plays a role in how the body handles fructose distribution and conversion to glycogen. A more recent study reported that a part of dietary fructose was converted to glycogen based on surge of blood 13C-glucose concentration following a glucagon administration after 4 hours of 13C-labeled fructose intake (0.72 g/kg-bw) [75].

Although it was reported that a significant amount of fructose in the circulation could be used to produce glycogen in liver via first conversion to glucose, no isotope tracer studies were found to directly quantitate 13C-carbons from dietary fructose incorporated into glycogen in humans. Considering that most of absorbed fructose is extracted and metabolized in liver, the data from the fructose infusion studies noted above may not be representative for orally administered fructose.

Lastly, a number of studies using labeled glucose have examined how dietary fructose loads affect glucose production and disposal [76-78]. In these three studies, 6,6-deuterium labeled glucose was infused as a glucose metabolism tracer into male subjects after a 4–7 day fructose feeding, with fructose representing >25% energy in the diet. Results indicated that hepatic glucose production in normal subjects did not change [76], or had no effect on whole-body insulin-mediated glucose disposal [78]. Using a 2-step hyperinsulinemic euglycemic clamp, healthy offspring of Type 2 diabetics fed a high fructose diet exhibited higher fasting hepatic glucose levels compared to controls [77].

Key points: 1) Fructose is converted to glucose to variable extents, depending on exercise condition, gender, and health status. This interconversion occurs at the triose phosphate intersection of the glucose-fructose pathways. 2) A portion of fructose is incorporated into glycogen after conversion to glucose, but the extent is not known. 3) Fructose feeding has an effect on hepatic glucose production and whole body glucose disposal. And, 4) Fructose may be processed differently in obese population or population with higher diabetes risk.

5.2.6 FRUCTOSE-LACTATE CONVERSION

Another significant and perhaps underappreciated metabolic pathway of dietary fructose is its conversion to lactate. Earlier tracer studies observed that blood lactate concentration was increased after fructose or fructose + glucose ingestion compared to that after glucose ingestion alone [56,66,68,79,80]. It was also observed that sucrose ingestion also caused a higher blood lactate response than did glucose [67,81]. However, no detailed data were reported to clarify how much of the ingested fructose was converted into the lactate in these studies.

Recently, Lecoultre et al. [51] conducted a tracer study in 7 men while under exercise. Within 100 minutes, 96 g fructose with 144 g glucose were co-ingested. The lactate conversion from 13C-labeled fructose was calculated using the parameters between 100 and 120 minutes when steady state of carbohydrate flux was assumed. As a result, 28% of fructose ingested was converted to lactate (35 micromol/kg-bw/min). Most of the converted lactate (25/28 or 89.3%) from fructose was oxidized mainly by working skeletal muscle (31 micromol/kg-bw/min). The non-oxidative fructose disposal was 0.52 grams per minute accounting for about 40% of the fructose ingested. The rate of appearance of glucose from fructose conversion was 19.8 micromol/kg-bw/min or 29% of the fructose dose. The authors also indicated that the increased lactate production and oxidation would be an essential explanation of faster oxidation of fructose + glucose co-ingested than glucose ingested alone.

In the tracer study by Rowland and colleagues [82], blood lactate concentration changes were compared in 10 men under exercise using oral test solutions of 13C-labeled glucose + 14C-labeled fructose (at 0.6 g/min glucose + 0, 0.3, 0.5, or 0.7 g/min fructose). During the 2-hour study period and compared to glucose alone, plasma lactate amount increased 31% and 24% under the glucose + fructose ingestions at 0.6 + 0.5 g/min and 0.6 + 0.7 g/min, respectively. However, the study did not indicate conversion percentage from the labeled fructose or glucose dosages.

Key points: 1) Clearly, a significant amount of fructose can be converted to lactate, but quantitative metabolic data of dietary fructose to lactate conversion is very limited. The effects of fructose dose, administration method, physical activity, and subject characteristics on fructose-lactate metabolism remain to be further studied. 2) Labeling patterns of isotope tracer in fructose will have an influence on the measured isotope appearance in lactate, if the studied sugar is not uniformly labeled.

5.2.7 FRUCTOSE-LIPIDS CONVERSION

A significant number of clinical studies have been performed to investigate the influence of fructose intake on blood triglyceride (TG) concentrations. However, tracer studies aimed at revealing metabolic conversion from labeled fructose carbons to TG are extremely limited. In contrast to the conversion from fructose to glucose, the metabolic pathway from fructose to TG conversion can be much more complicated due to the complex distribution and diversity of blood lipid compositions in the body. De novo lipogenesis from sugars can occur in the liver and end up as packaged VLDL TG and/or as intrahepatocellular lipids. There are currently no convenient methods to quantitate overall DNL and intrahepatic lipid deposition. The fractional contribution of sugars to de novo lipogenesis and VLDL TG are commonly determined using tracer enrichment data of blood samples. The time periods of liver de novo lipogenesis from sugars and the factors influencing it are not completely understood, and are impacted by the concentrations and tracer characteristics of the various substrates drawn from lipid precursor pools. De novo lipogenesis may also occur in adipose tissue or muscles, but there are no adequate methods available to quantitate it. A more expansive discussion of de novo lipogenesis and methodological considerations is an appropriate subject for a separate review.

Perhaps because of these difficulties, only two tracer studies were found that investigated conversion of labeled dietary fructose carbons into plasma lipids. Chong et al. [48] studied the effect of fructose on postprandial lipidemia in fourteen adults (8 men) who were orally administered 13C-labeled fructose or 13C-labeled glucose at a dose of 0.75 g/kg bw, together with an 2 H-labeled oil mix (85% palm oil and 15% sunflower oil) at 0.5 g/kg bw. Blood lipid changes were monitored in a 6-hour period. It was observed that plasma TG concentration rose more significantly after fructose ingestion (from baseline 1240 μmol/L (≈110 mg/dl) to its plateau of 2350 μmol/L (≈208 mg/dl)) than that after glucose ingestion (from baseline 1240 μmol/L to its plateau of 1700 μmol/L(≈150 mg/dl)). However, the concentration increases of 13C-enriched TG-fatty acids and TG-glycerol from the labeled fructose in the Sf 20–400 lipid fraction (including VLDL) were very small within the monitoring period. The plateau value of 13C-palmitate concentration was about 0.022 μmol/L (≈0.002 mg/dl), 13C-myristate was about 0.0015 μmol/L (≈0.0001 mg/dl), and 13C-TG-glycerol was about 1.4 μmol/L (≈0.124 mg/dl), suggesting that fructose carbons were not substantially transferred into plasma TG molecules during the time period monitored. The authors indicated that the lipogenic potential of fructose seems to be small, since the results showed that only 0.05% and 0.15% of fructose were converted to de novo fatty acids and TG-glycerol at 4 hour, respectively. The reported data should be viewed in the context of the 4-hour time period and whether further conversion would be observed at extended times was not illustrated. It was observed by Vedala and colleagues [83], using labeled fatty acids, acetate and glycerol as precursors, that a meaningful portion of de novo synthesized triglyceride would appear in blood at later times, and rates of this delayed secretion were significantly different among normal, hypertriglyceridemic, and diabetic subjects. However, this study did not specifically measure fructose conversion using labeled sugars.

In another study, Tran et al. [70] reported that 13C-labeled fructose consumption at 3x0.3 g/kg body weight caused a small but significant increase of 13C-enrichment in VLDL palmitate in 8 men compared with that found in 9 women (no increase) during a 6-hour monitoring period. However, compared to baselines, plasma TG and non-esterified fatty acid concentrations decreased 5.3% and 32.9% in men and 3.3% and 24.4% in women, respectively. The data indicate that the conversion from fructose to fatty acid occurred, however, no blood lipid concentrations increased. Although the authors reported that 42.9% and 43% of the ingested fructose was oxidized and 37.4% and 28.9% was converted into glucose in men and women during the 6-hour monitoring period, conversion rate or percent from fructose into fatty acid or triglyceride was not reported. This study also noted that men processed dietary fructose differently than women, and the given fructose lowered postprandial plasma lipids. It was discussed that although fructose is a potent lipogenic substrate, the observed fat synthesis arising from fructose carbons appeared to be quantitatively minor compared with other pathways of fructose disposal, but it may nevertheless have a significant impact on plasma and tissue lipids. In this same study, respiratory quotient (RQ) measurements found differences between genders, with male subjects increasing their RQ by 3% and females maintaining theirs. This data suggest that the increase of blood TG frequently observed in men compared to women after high dose fructose ingestion could be due to fat sparing during energy utilization.

There are several studies which used labeled acetate, administered by intravenous infusion as a precursor of lipid synthesis, to assess the fructose stimulation of de novo lipogenesis (DNL). This technique uses the approach of Mass Isotopomer Distribution Analysis (MIDA) to estimate the infused subunit (acetate) appearance in newly synthesized fatty acids and further predict the effect of dietary fructose on fractional DNL. The advantages and limitations of the method were well reviewed by Hellerstein in 1996 [84]. Parks et al. [85] investigated the influence of fructose-containing drinks on blood lipid changes using infused 13C-acetate. Six healthy subjects were randomly administrated 86 grams (mean) of glucose, glucose + fructose (50:50) or glucose + fructose (25:75) in drinks by a crossover-designed trial. Four hours after fructose ingestion, a standard lunch was consumed. Compared to glucose, more palmitate synthesis in triglyceride-rich lipoprotein (TRL) TG was noted after fructose-containing drinks, but not after the lunch. No significant differences were observed for TRL-TG concentrations between glucose and fructose-containing drink arms after baseline correction. Plasma TG concentration was decreased after glucose preload and stayed constant after fructose-containing drink preloads. Following the lunch, TG concentrations increased for all treatments. The authors reported that the after lunch TG-AUC data from fructose-containing drink treatments were significantly larger than that of glucose drink treatment. However, this AUC data was calculated over the entire study time period. Due to the negative TG rise during the glucose preload phase, the difference between the glucose and fructose arms was accentuated.

Similarly, in Stanhope and colleagues' study [86], 13C-labeled acetate infusion was used to measure fractional DNL in a 10-week intervention involving 18 overweight or obese subjects consuming either glucose (n = 8) or fructose beverages (n = 10) delivering 25% of daily energy. The percent changes of fractional hepatic DNL were not significantly different from baseline following 9-week glucose consumption for both fasting and postprandial measurements. In the fructose beverage group, the percent changes of fractional hepatic DNL were also not significantly different between baseline and following 9-weeks for fasting data, but were significantly increased for postprandial data (2-7% during 11-hour monitoring). The actual amount of the DNL was not reported.

Faeh et al. conducted a shorter term crossover study using a 6-day intervention [78]. Seven men were fed hypercaloric (+800-1000 kcal/d) diets, with the additional 25% of energy provided through a fructose solution. Fractional hepatic DNL was measured via 13C-labeled acetate infusion. The % changes from baseline for plasma TG and hepatic DNL were found to be significantly increased for the hypercaloric fructose diet compared to isocaloric control diets. The authors noted that the results could not truly differentiate the effects of the high-fructose intake per se and that of the total carbohydrate energy overfeeding. This study also found that fish oil added to the diets containing fructose attenuated this hyperlipidemic response somewhat [78].

Clearly, the 3 studies discussed above [78,85,86] assess effects of dietary fructose with or without over energy intakes on the utilization of acetate in the circulation, which is designed to feed directly into lipid synthesis. In humans, acetate concentrations in blood are fairly low. As indicated in the Human Metablome Database [87], normal blood concentrations of acetate are 41.9 ± 15.1 (SD) µmol/L in adults aged 18 years and over. For earlier data, Richards et al. [88] reported in 1976 that the normal value of blood acetate was 25 ± 2 µmol/L. Beyond alcohol consumption, common dietary intakes have no or limited influence on blood acetate concentration [89,90]. In the studies of Parks, Stanhope, and Faeh, acetate was constantly infused at 0.5-0.55 g/hr (about 7000 µmol/hr) for 25, 26 and 9.5 hours, respectively. Although the data of blood acetate concentration were not reported in those studies, it would be important to determine whether the acetate infusion significantly raised blood acetate concentrations such that this could have a meaningful impact on metabolic response to the fructose challenge. The coexistence of the infused acetate and intermediate metabolites of fructose, including regulatory elements of citrate, malate, and lactate, could prime the pathway of DNL. As detailed above, dietary fructose (up to 25% of daily energy and 3 g/day-kg in these studies) can metabolically be converted into lactate and further result in blood lactate concentration increases. Beynen and colleagues' hepatic cell study [91] indicated that lactate and acetate both stimulate fatty acid synthesis, and lactate can induce activation of acetyl-CoA carboxylase, a key enzyme for fatty acid synthesis. Thus, the meaning of stimulation of de novo lipogenesis observed from use of infused intermediate metabolite tracer and its interaction with the sugars studied should be considered carefully, along with the study methodologies being validated for the dosage of infused tracer. Additionally, how the infused acetate can truly represent intrahepatocellular acetyl-CoA pool is another key point to be clarified.

Key points: 1) The above tracer studies indicate the complex relationship between dietary fructose and lipid synthesis. The observed increases in plasma TG and DNL in these studies can arise from both increased lipid synthesis and decreased lipid clearance, and the relative contributions were not addressed in any detail. And, 2) The intake levels, health status, and gender of subjects are all important factors influencing sugar-lipid relationships. The influence of fructose consumption on plasma lipids and de novo lipogenesis remains controversial and understudied.

5.2.8 INFLUENCE OF EXOGENOUS SUGARS ON UTILIZATION OF ENDOGENOUS ENERGY SOURCES

After sugar ingestion, body utilization of energy sources will change. As exogenous carbohydrate is used as a fuel source, the oxidation rates of stored energy, typically, endogenous carbohydrates and fat, will decrease. The extent of the decrease is usually driven by ingested sugar type, intake amount, and status of body energy need (such as vigorous exercise or screen watching). Under exercise, glucose is more likely to be preferentially oxidized than fructose, and this scenario will go in the opposite direction under a resting state. Although data are limited related to detailed shifting of energy sources under different conditions, some studies using subjects under exercise may provide a basic concept of energy source shifting after sugar ingestion. Jentjens and colleagues conducted a series of studies [66-69,80] using exercise subjects under somewhat comparable conditions, and reported some data related to the energy source shifting. The subjects were given drinks containing glucose, sucrose, glucose + fructose, or glucose + sucrose at dosage 0 (control), 1.2, 1.5, 1.8, or 2.4 g/min and under exercise workloads around 50% VO2 max uptake. For controls (0 gram of sugar intake), the oxidation rates of fat and endogenous carbohydrate were between 0.77-0.95 g/min and 1.43-1.85 g/min, respectively. Compared to the control, glucose-containing drinks decreased fat oxidation rates by 21.6-41.7% (calculated based on reported data) and endogenous carbohydrate oxidation rates by 8.5-31.5%, except that one of the 5 studies noted endogenous carbohydrate oxidation rates increased (3.8% for medium and 14.1% for high glucose intake). For fructose-containing drink arms, either for glucose + fructose, sucrose, or glucose + sucrose, the fat oxidation rates were lowered by 19.5-47.4% and endogenous carbohydrate oxidation rates were lowered by 13.0-31.6%. These percent decreases appeared to be positively correlated to the sugar intake levels and the ratios of fructose in the mixed sugar drinks. The other two studies [60,79] with similar settings as Jentjens and colleagues' work also reported comparable data of decreasing fat and endogenous carbohydrate oxidation after sugar preloads.

Key points: 1) Together with other sugar inter-conversion data and the RQ data of Tran et al. [70], the shifting of energy sources after sugar ingestion may indicate that the utilization of exogenous and endogenous energy is closely regulated according to the energy balance of body. 2) Beyond specific health and physiological conditions, physical activity, over energy consumption, dietary macronutrient composition, and other lifestyle factors would also play critical roles in the body's utilization of dietary sugars. In view of these factors, how energy is quantitatively balanced with fructose loading is an area yet to be delineated.

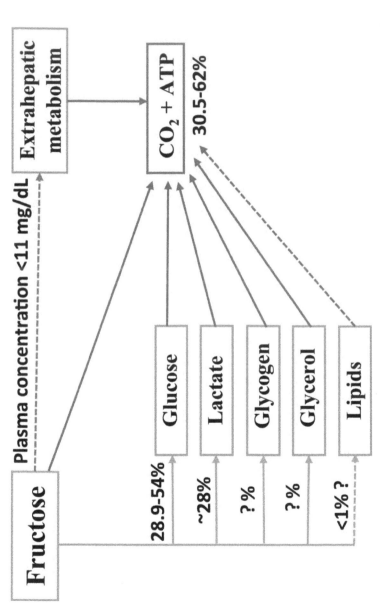

FIGURE 3: Metabolic fate of dietary fructose carbons. The data are obtained within study periods less than or equal to 6 hours. After 50–150 gm fructose ingestion, the peak of fructose concentration in plasma would be between 3–11 mg/dL. The percent data above arrow lines or under box are the estimated amounts of ingested fructose doses via the pathway, and the question mark represents that the data remain to be further confirmed. The dash-line represents presumably minor pathways.

5.3 SUMMARY

Figure 3 summarizes the major metabolic fates of dietary fructose based on the data obtained from the reviewed isotope tracer studies. The mean oxidation rate of dietary fructose was 45.0% (ranged 30.5-59%) of ingested doses in normal subjects within a period of 3–6 hours. With exercise conditions, the mean oxidation rate of fructose came to 45.8% (ranged 37.5-62%) within 2–3 hours. When fructose was ingested together with glucose, the mean oxidation rate of the mixed sugars increased to 66.0% (ranged 52.2-73.6%) under similar exercise conditions. Secondly, the mean conversion rate from fructose to glucose was 41% (ranged 29-54%) of ingested dose in 3–6 hours after ingestion in normal non-exercise subjects. This value may be higher in subjects under exercise. The conversion amount from fructose to glycogen remains to be further clarified. Thirdly, at short time periods ($\leq$ 6 hours), it appeared that only a small percent of fructose carbons enter the pathway of liponeogenesis after fructose ingestion. The hyperlipidemic effect of dietary fructose observed in both tracer and non-tracer studies may involve other metabolic mechanisms and this could relate to energy source shifting and lipid sparing. Lastly, fructose can be catabolized into lactate and cause an increase of blood lactate concentrations. Approximately a quarter of ingested fructose could be converted into lactate within a few of hours and this is a means to release fructose-derived carbons from the liver for extrahepatic utilization. Even though the reviewed tracer studies may not be fully representative of real-life diets and the obtained data are limited, this review provides a basic outline how fructose is utilized after it is consumed by humans.

REFERENCES

1. USDA-ERS: Food Availability: Spreadsheets-Added sugar and sweeteners. 2010. http://wwwersusdagov/Data/FoodConsumption/FoodAvailSpreadsheetshtm#sweets
2. Johnson RJ, Segal MS, Sautin Y, Nakagawa T, Feig DI, Kang DH, Gersch MS, Benner S, Sanchez-Lozada LG: Potential role of sugar (fructose) in the epidemic of hypertension, obesity and the metabolic syndrome, diabetes, kidney disease, and cardiovascular disease. Am J Clin Nutr 2007, 86:899-906.
3. Bray GA, Nielsen SJ, Popkin BM: Consumption of high-fructose corn syrup in beverages may play a role in the epidemic of obesity. Am J Clin Nutr 2004, 79:537-543.

4. Teff KL, Elliott SS, Tschop M, Kieffer TJ, Rader D, Heiman M, Townsend RR, Keim NL, D'Alessio D, Havel PJ: Dietary fructose reduces circulating insulin and leptin, attenuates postprandial suppression of ghrelin, and increases triglycerides in women. J Clin Endocrinol Metab 2004, 89:2963-2972.
5. Teff KL, Grudziak J, Townsend RR, Dunn TN, Grant RW, Adams SH, Keim NL, Cummings BP, Stanhope KL, Havel PJ: Endocrine and metabolic effects of consuming fructose- and glucose-sweetened beverages with meals in obese men and women: influence of insulin resistance on plasma triglyceride responses. J Clin Endocrinol Metab 2009, 94:1562-1569. |
6. Melanson KJ, Zukley L, Lowndes J, Nguyen V, Angelopoulos TJ, Rippe JM: Effects of high-fructose corn syrup and sucrose consumption on circulating glucose, insulin, leptin, and ghrelin and on appetite in normal-weight women. Nutrition 2007, 23:103-112.
7. Sun SZ, Empie MW: Lack of findings for the association between obesity risk and usual sugar-sweetened beverage consumption in adults–a primary analysis of databases of CSFII-1989–1991, CSFII-1994–1998, NHANES III, and combined NHANES 1999–2002. Food Chem Toxicol 2007, 45:1523-1536.
8. Dolan LC, Potter SM, Burdock GA: Evidence-based review on the effect of normal dietary consumption of fructose on development of hyperlipidemia and obesity in healthy, normal weight individuals. Crit Rev Food Sci Nutr 2010, 50:53-84.
9. Dolan LC, Potter SM, Burdock GA: Evidence-based review on the effect of normal dietary consumption of fructose on blood lipids and body weight of overweight and obese individuals. Crit Rev Food Sci Nutr 2010, 50:889-918.
10. Livesey G: Fructose ingestion: dose-dependent responses in health research. J Nutr 2009, 139:1246S-1252S.
11. Livesey G, Taylor R: Fructose consumption and consequences for glycation, plasma triacylglycerol, and body weight: meta-analyses and meta-regression models of intervention studies. Am J Clin Nutr 2008, 88:1419-1437.
12. Sievenpiper JL, Carleton AJ, Chatha S, Jiang HY, de Souza RJ, Beyene J, Kendall CW, Jenkins DJ: Heterogeneous effects of fructose on blood lipids in individuals with type 2 diabetes: systematic review and meta-analysis of experimental trials in humans. Diabetes Care 2009, 32:1930-1937. |
13. Sievenpiper JL, de Souza RJ, Mirrahimi A, Yu ME, Carleton AJ, Beyene J, Chiavaroli L, Di Buono M, Jenkins AL, Leiter LA, et al.: Effect of Fructose on Body Weight in Controlled Feeding Trials: A Systematic Review and Meta-analysis. Ann Intern Med 2012, 156:291-304.
14. Welsh JA, Sharma AJ, Grellinger L, Vos MB: Consumption of added sugars is decreasing in the United States. Am J Clin Nutr 2011, 94:726-734. |
15. Hall KD, Heymsfield SB, Kemnitz JW, Klein S, Schoeller DA, Speakman JR: Energy balance and its components: implications for body weight regulation. Am J Clin Nutr 2012, 95:989-994.
16. Swinburn B, Sacks G, Ravussin E: Increased food energy supply is more than sufficient to explain the US epidemic of obesity. Am J Clin Nutr 2009, 90:1453-1456.
17. Johnson RJ, Perez-Pozo SE, Sautin YY, Manitius J, Sanchez-Lozada LG, Feig DI, Shafiu M, Segal M, Glassock RJ, Shimada M, et al.: Hypothesis: could excessive fructose intake and uric acid cause type 2 diabetes? Endocr Rev 2009, 30:96-116. |

18. Sun SZ, Flickinger BD, Williamson-Hughes PS, Empie MW: Lack of association between dietary fructose and hyperuricemia risk in adults. Nutr Metab (Lond) 2010, 7:16.

19. Lim JS, Mietus-Snyder M, Valente A, Schwarz JM, Lustig RH: The role of fructose in the pathogenesis of NAFLD and the metabolic syndrome. Nat Rev Gastroenterol Hepatol 2010, 7:251-264.

20. Lustig RH: Fructose: metabolic, hedonic, and societal parallels with ethanol. J Am Diet Assoc 2010, 110:1307-1321.

21. Gambino R, Musso G, Cassader M: Redox balance in the pathogenesis of nonalcoholic Fatty liver disease: mechanisms and therapeutic opportunities. Antioxid Redox Signal 2011, 15:1325-1365.

22. Cortez-Pinto H, Machado MV: Uncoupling proteins and non-alcoholic fatty liver disease. J Hepatol 2009, 50:857-860.

23. Sievenpiper JL, de Souza RJ, Kendall CW, Jenkins DJ: Is fructose a story of mice but not men? J Am Diet Assoc 2011, 111:219-220. author reply 220–212

24. Lustig RH: Author's Response to Letter to the editor. J Am Diet Assoc 2011, 111:220-222.

25. Livesey G: More on mice and men: fructose could put brakes on a vicious cycle leading to obesity in humans. J Am Diet Assoc 2011, 111:986-990. author reply 990–983

26. Lustig RH: Author's Response to Letters to The Editor. J Am Diet Assoc 2011, 111:990-993.

27. Mayes PA: Intermediary metabolism of fructose. Am J Clin Nutr 1993, 58:754S-765S.

28. Tappy L, Le KA: Metabolic effects of fructose and the worldwide increase in obesity. Physiol Rev 2010, 90:23-46.

29. Sun SZ, Anderson GH, Flickinger BD, Williamson-Hughes PS, Empie MW: Fructose and non-fructose sugar intakes in the US population and their associations with indicators of metabolic syndrome. Food Chem Toxicol 2011, 49:2875-2882.

30. Marriott BP, Cole N, Lee E: National estimates of dietary fructose intake increased from 1977 to 2004 in the United States. J Nutr 2009, 139:1228S-1235S.

31. Chandramouli V, Kumaran K, Ekberg K, Wahren J, Landau BR: Quantitation of the pathways followed in the conversion of fructose to glucose in liver. Metabolism 1993, 42:1420-1423.

32. Folch N, Peronnet F, Pean M, Massicotte D, Lavoie C: Labeled CO(2) production and oxidative vs nonoxidative disposal of labeled carbohydrate administered at rest. Metabolism 2005, 54:1428-1434.

33. Hellerstein MK, Neese RA: Mass isotopomer distribution analysis at eight years: theoretical, analytic, and experimental considerations. Am J Physiol 1999, 276:E1146-E1170.

34. Riby JE, Fujisawa T, Kretchmer N: Fructose absorption. Am J Clin Nutr 1993, 58:748S-753S.

35. Ferraris RP: Dietary and developmental regulation of intestinal sugar transport. Biochem J 2001, 360:265-276. |

36. Scheepers A, Joost HG, Schurmann A: The glucose transporter families SGLT and GLUT: molecular basis of normal and aberrant function. JPEN J Parenter Enteral Nutr 2004, 28:364-371.
37. Read N, French S, Cunningham K: The role of the gut in regulating food intake in man. Nutr Rev 1994, 52:1-10.
38. Lavin JH, Wittert GA, Andrews J, Yeap B, Wishart JM, Morris HA, Morley JE, Horowitz M, Read NW: Interaction of insulin, glucagon-like peptide 1, gastric inhibitory polypeptide, and appetite in response to intraduodenal carbohydrate. Am J Clin Nutr 1998, 68:591-598.
39. Ravich WJ, Bayless TM: Carbohydrate absorption and malabsorption. Clin Gastroenterol 1983, 12:335-356.
40. Latulippe ME, Skoog SM: Fructose Malabsorption and Intolerance: Effects of Fructose with and without Simultaneous Glucose Ingestion. Crit Rev Food Sci Nutr 2011, 51:583-592. |
41. Gray GM, Ingelfinger FJ: Intestinal absorption of sucrose in man: interrelation of hydrolysis and monosaccharide product absorption. J Clin Invest 1966, 45:388-398.
42. Akgun S, Ertel NH: The effects of sucrose, fructose, and high-fructose corn syrup meals on plasma glucose and insulin in non-insulin-dependent diabetic subjects. Diabetes Care 1985, 8:279-283.
43. Heitlinger LA, Li BU, Murray RD, McClung HJ, Sloan HR, DeVore DR, Powers P: Glucose flux from dietary disaccharides: all sugars are not absorbed at equal rates. Am J Physiol 1991, 261:G818-G822.
44. Macdonald I, Turner LJ: Serum-fructose levels after sucrose or its constituent monosaccharides. Lancet 1968, 1:841-843.
45. Bornet F, Haardt MJ, Costagliola D, Blayo A, Slama G: Sucrose or honey at breakfast have no additional acute hyperglycaemic effect over an isoglucidic amount of bread in type 2 diabetic patients. Diabetologia 1985, 28:213-217.
46. Samanta A, Burden AC, Jones GR: Plasma glucose responses to glucose, sucrose, and honey in patients with diabetes mellitus: an analysis of glycaemic and peak incremental indices. Diabet Med 1985, 2:371-373.
47. Burelle Y, Peronnet F, Massicotte D, Brisson GR, Hillaire-Marcel C: Oxidation of 13C-glucose and 13C-fructose ingested as a preexercise meal: effect of carbohydrate ingestion during exercise. Int J Sport Nutr 1997, 7:117-127.
48. Chong MF, Fielding BA, Frayn KN: Mechanisms for the acute effect of fructose on postprandial lipemia. Am J Clin Nutr 2007, 85:1511-1520.
49. Delarue J, Normand S, Pachiaudi C, Beylot M, Lamisse F, Riou JP: The contribution of naturally labelled 13C fructose to glucose appearance in humans. Diabetologia 1993, 36:338-345.
50. Jandrain BJ, Pallikarakis N, Normand S, Pirnay F, Lacroix M, Mosora F, Pachiaudi C, Gautier JF, Scheen AJ, Riou JP, et al.: Fructose utilization during exercise in men: rapid conversion of ingested fructose to circulating glucose. J Appl Physiol 1993, 74:2146-2154.
51. Lecoultre V, Benoit R, Carrel G, Schutz Y, Millet GP, Tappy L, Schneiter P: Fructose and glucose co-ingestion during prolonged exercise increases lactate and glucose

fluxes and oxidation compared with an equimolar intake of glucose. Am J Clin Nutr 2010, 92:1071-1079.

52. Paquot N, Schneiter P, Jequier E, Gaillard R, Lefebvre PJ, Scheen A, Tappy L: Effects of ingested fructose and infused glucagon on endogenous glucose production in obese NIDDM patients, obese non-diabetic subjects, and healthy subjects. Diabetologia 1996, 39:580-586.

53. Miller BF, Fattor JA, Jacobs KA, Horning MA, Navazio F, Lindinger MI, Brooks GA: Lactate and glucose interactions during rest and exercise in men: effect of exogenous lactate infusion. J Physiol 2002, 544:963-975. |

54. Roef MJ, de Meer K, Kalhan SC, Straver H, Berger R, Reijngoud DJ: Gluconeogenesis in humans with induced hyperlactatemia during low-intensity exercise. Am J Physiol Endocrinol Metab 2003, 284:E1162-E1171.

55. Guezennec CY, Satabin P, Duforez F, Merino D, Peronnet F, Koziet J: Oxidation of corn starch, glucose, and fructose ingested before exercise. Med Sci Sports Exerc 1989, 21:45-50.

56. Decombaz J, Sartori D, Arnaud MJ, Thelin AL, Schurch P, Howald H: Oxidation and metabolic effects of fructose or glucose ingested before exercise. Int J Sports Med 1985, 6:282-286.

57. Selivanov VA, Sukhomlin T, Centelles JJ, Lee PW, Cascante M: Integration of enzyme kinetic models and isotopomer distribution analysis for studies of in situ cell operation. BMC Neurosci 2006, 7(Suppl 1):S7. ||

58. Lee WN, Go VL: Nutrient-gene interaction: tracer-based metabolomics. J Nutr 2005, 135:3027S-3032S.

59. Vinnakota KC, Wu F, Kushmerick MJ, Beard DA: Multiple ion binding equilibria, reaction kinetics, and thermodynamics in dynamic models of biochemical pathways. Methods Enzymol 2009, 454:29-68. |

60. Burelle Y, Lamoureux MC, Peronnet F, Massicotte D, Lavoie C: Comparison of exogenous glucose, fructose and galactose oxidation during exercise using 13C-labelling. Br J Nutr 2006, 96:56-61.

61. Adopo E, Peronnet F, Massicotte D, Brisson GR, Hillaire-Marcel C: Respective oxidation of exogenous glucose and fructose given in the same drink during exercise.

62. J Appl Physiol 1994, 76:1014-1019.

63. Massicotte D, Peronnet F, Adopo E, Brisson GR, Hillaire-Marcel C: Effect of metabolic rate on the oxidation of ingested glucose and fructose during exercise. Int J Sports Med 1994, 15:177-180.

64. Massicotte D, Peronnet F, Allah C, Hillaire-Marcel C, Ledoux M, Brisson G: Metabolic response to [13C]glucose and [13C]fructose ingestion during exercise. J Appl Physiol 1986, 61:1180-1184.

65. Massicotte D, Peronnet F, Brisson G, Bakkouch K, Hillaire-Marcel C: Oxidation of a glucose polymer during exercise: comparison with glucose and fructose. J Appl Physiol 1989, 66:179-183.

66. Massicotte D, Peronnet F, Brisson G, Boivin L, Hillaire-Marcel C: Oxidation of exogenous carbohydrate during prolonged exercise in fed and fasted conditions. Int J Sports Med 1990, 11:253-258.

67. Jentjens RL, Moseley L, Waring RH, Harding LK, Jeukendrup AE: Oxidation of combined ingestion of glucose and fructose during exercise. J Appl Physiol 2004, 96:1277-1284.

68. Jentjens RL, Shaw C, Birtles T, Waring RH, Harding LK, Jeukendrup AE: Oxidation of combined ingestion of glucose and sucrose during exercise. Metabolism 2005, 54:610-618.

69. Jentjens RL, Underwood K, Achten J, Currell K, Mann CH, Jeukendrup AE: Exogenous carbohydrate oxidation rates are elevated after combined ingestion of glucose and fructose during exercise in the heat. J Appl Physiol 2006, 100:807-816.

70. Jentjens RL, Venables MC, Jeukendrup AE: Oxidation of exogenous glucose, sucrose, and maltose during prolonged cycling exercise. J Appl Physiol 2004, 96:1285-1291.

71. Tran C, Jacot-Descombes D, Lecoultre V, Fielding BA, Carrel G, Le KA, Schneiter P, Bortolotti M, Frayn KN, Tappy L: Sex differences in lipid and glucose kinetics after ingestion of an acute oral fructose load. Br J Nutr 2010, 104:1139-1147.

72. Surmely JF, Paquot N, Schneiter P, Jequier E, Temler E, Tappy L: Non oxidative fructose disposal is not inhibited by lipids in humans. Diabetes Metab 1999, 25:233-240.

73. Nilsson LH, Hultman E: Liver and muscle glycogen in man after glucose and fructose infusion. Scand J Clin Lab Invest 1974, 33:5-10.

74. Dirlewanger M, Schneiter P, Jequier E, Tappy L: Effects of fructose on hepatic glucose metabolism in humans. Am J Physiol Endocrinol Metab 2000, 279:E907-E911.

75. Blom PC, Hostmark AT, Vaage O, Kardel KR, Maehlum S: Effect of different post-exercise sugar diets on the rate of muscle glycogen synthesis. Med Sci Sports Exerc 1987, 19:491-496.

76. Coss-Bu JA, Sunehag AL, Haymond MW: Contribution of galactose and fructose to glucose homeostasis. Metabolism 2009, 58:1050-1058. |

77. Sobrecases H, Le KA, Bortolotti M, Schneiter P, Ith M, Kreis R, Boesch C, Tappy L: Effects of short-term overfeeding with fructose, fat and fructose plus fat on plasma and hepatic lipids in healthy men. Diabetes Metab 2010, 36:244-246.

78. Le KA, Ith M, Kreis R, Faeh D, Bortolotti M, Tran C, Boesch C, Tappy L: Fructose overconsumption causes dyslipidemia and ectopic lipid deposition in healthy subjects with and without a family history of type 2 diabetes. Am J Clin Nutr 2009, 89:1760-1765.

79. Faeh D, Minehira K, Schwarz JM, Periasamy R, Park S, Tappy L: Effect of fructose overfeeding and fish oil administration on hepatic de novo lipogenesis and insulin sensitivity in healthy men. Diabetes 2005, 54:1907-1913.

80. Hulston CJ, Wallis GA, Jeukendrup AE: Exogenous CHO oxidation with glucose plus fructose intake during exercise. Med Sci Sports Exerc 2009, 41:357-363.

81. Jentjens RL, Jeukendrup AE: High rates of exogenous carbohydrate oxidation from a mixture of glucose and fructose ingested during prolonged cycling exercise. Br J Nutr 2005, 93:485-492.

82. Daly ME, Vale C, Walker M, Littlefield A, George K, Alberti M, Mathers J: Acute fuel selection in response to high-sucrose and high-starch meals in healthy men. Am J Clin Nutr 2000, 71:1516-1524.

83. Rowlands DS, Thorburn MS, Thorp RM, Broadbent S, Shi X: Effect of graded fructose coingestion with maltodextrin on exogenous 14C-fructose and 13C-glucose oxidation efficiency and high-intensity cycling performance. J Appl Physiol 2008, 104:1709-1719.

84. Vedala A, Wang W, Neese RA, Christiansen MP, Hellerstein MK: Delayed secretory pathway contributions to VLDL-triglycerides from plasma NEFA, diet, and de novo lipogenesis in humans. J Lipid Res 2006, 47:2562-2574.

85. Hellerstein MK, Schwarz JM, Neese RA: Regulation of hepatic de novo lipogenesis in humans. Annu Rev Nutr 1996, 16:523-557.

86. Parks EJ, Skokan LE, Timlin MT, Dingfelder CS: Dietary sugars stimulate fatty acid synthesis in adults. J Nutr 2008, 138:1039-1046. |

87. Stanhope KL, Schwarz JM, Keim NL, Griffen SC, Bremer AA, Graham JL, Hatcher B, Cox CL, Dyachenko A, Zhang W, et al.: Consuming fructose-sweetened, not glucose-sweetened, beverages increases visceral adiposity and lipids and decreases insulin sensitivity in overweight/obese humans. J Clin Invest 2009, 119:1322-1334. |

88. Human Metablome Database - Acetic acid. 2011. http://wwwhmdbca/metabolites/HMDB00042

89. Richards RH, Dowling JA, Vreman HJ, Feldman C, Weiner MW: Acetate levels in human plasma. Proc Clin Dial Transplant Forum 1976, 6:73-79.

90. Lundquist F: Production and Utilization of Free Acetate in Man. Nature 1962, 193:579-580.

91. Lundquist F, Tygstrup N, Winkler K, Mellemgaard K, Munck-Petersen S: Ethanol metabolism and production of free acetate in the human liver. J Clin Invest 1962, 41:955-961. |

92. Beynen AC, Buechler KF, Van der Molen AJ, Geelen MJ: The effects of lactate and acetate on fatty acid and cholesterol biosynthesis by isolated rat hepatocytes. Int J Biochem 1982, 14:165-169.

This chapter was originally published under the Creative Commons Attribution License. Sun, S. Z., and Emple, M. W. Fructose Metabolism in Humans: What Isotopic Tracer Studies Tell Us. Nutrition & Metabolism 2012, 9:89. doi:10.1186/1743-7075-9-89.

CHAPTER 6

VITAMIN C IN HUMAN HEALTH AND DISEASE IS STILL A MYSTERY? AN OVERVIEW

K. AKHILENDER NAIDU

6.1 HISTORICAL PERSPECTIVE

The sea voyager/sailors developed a peculiar disease called scurvy when they were on sea. This was found to be due to eating non-perishable items and lack of fresh fruits and vegetables in their diet. A British naval Physician, Lind [1] documented that there was some substance in citrus fruits that can cure scurvy. He developed a method to concentrate and preserve citrus juice for use by sailors. British Navy was given a daily ration of lime or lemon juice to overcome ascorbic acid deficiency. Ascorbic acid was first isolated from natural sources and structurally characterized by Szent-Gyorgyi, Waugh and King [2,3]. This vitamin was first synthesized by Haworth and Hirst [4]. Currently ascorbic acid is the most widely used vitamin supplement through out the world.

6.2 SOURCES OF ASCORBIC ACID

Ascorbic acid is widely distributed in fresh fruits and vegetables. It is present in fruits like orange, lemons, grapefruit, watermelon, papaya, strawberries, cantaloupe, mango, pineapple, raspberries and cherries. It is also found in green leafy vegetables, tomatoes, broccoli, green and red peppers, cauliflower and cabbage.

Most of the plants and animals synthesize ascorbic acid from D-glucose or D-galactose. A majority of animals produce relatively high levels of ascorbic acid from glucose in liver (Fig 1).

However, guinea pigs, fruit eating bats, apes and humans can not synthesize ascorbic acid due to the absence of the enzyme L-gulonolactone oxidase. Hence, in humans ascorbic acid has to be supplemented through food and or as tablets.

Ascorbic acid is a labile molecule, it may be lost from foods during cooking/processing even though it has the ability to preserve foods by virtue of its reducing property. Synthetic ascorbic acid is available in a wide variety of supplements viz., tablets, capsules, chewable tablets, crystalline powder, effervescent tablets and liquid form. Buffered ascorbic acid and esterfied form of ascorbic acid as ascorbyl palmitate is also available commercially. Both natural and synthetic ascorbic acid are chemically identical and there are no known differences in their biological activities or bio-availability.

6.3 CHEMISTRY OF ASCORBIC ACID

L-ascorbic acid ($C_6H_8O_6$) is the trivial name of Vitamin C. The chemical name is 2-oxo-L-threo-hexono-1,4-lactone-2,3-enediol. L-ascorbic and dehydroascorbic acid are the major dietary forms of vitamin C [5]. Ascorbyl palmitate is used in commercial antioxidant preparations. All commercial forms of ascorbic acid except ascorbyl palmitate are soluble in water. L-ascorbic acid and its fatty acid esters are used as food additives, antioxidants, browning inhibitors, reducing agents, flavor stabilizers, dough modifiers and color stabilizers. Ascorbyl palmitate has been used for its greater lipid solubility in antioxidant preparations. In foods, pH influences the stability of ascorbic acid. It exhibits maximal stability between pH 4 and 6 [5]. Cooking losses of ascorbic acid depend on degree of heating, surface area exposed to water, oxygen, pH and presence of transition metals.

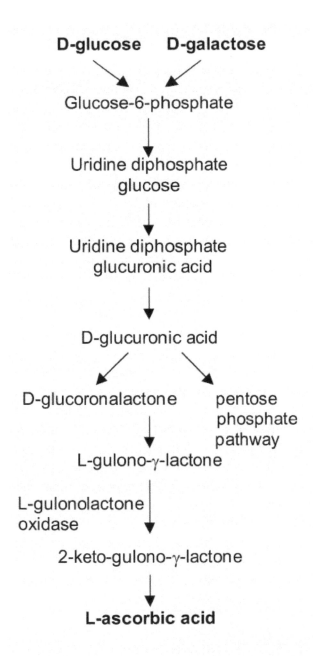

FIGURE 1: Biosynthesis of L-Ascorbic acid in animals

6.4 CATABOLISM OF ASCORBIC ACID

Ascorbic acid present in foods is readily available and easily absorbed by active transport in the intestine [6]. Most of it (80–90%) will be absorbed when the in take is up to 100 mg/day, whereas at higher levels of intake (500 mg/day) the efficiency of absorption of ascorbic acid rapidly declines. Ascorbic acid is sensitive to air, light, heat and easily destroyed by prolonged storage and over processing of food.

Ascorbic acid being a water soluble compound is easily absorbed but it is not stored in the body. The average adult has a body pool of 1.2–2.0 g of ascorbic acid that may be maintained with 75 mg/d of ascorbic acid. About 140 mg/d of ascorbic acid will saturate the total body pool of vitamin C [7]. The average half life of ascorbic acid in adult human is about 10–20 days, with a turn over of 1 mg/kg body and a body pool of 22 mg/kg at plasma ascorbate concentration of 50 μmol/ L [8,9]. Hence ascorbic acid has to regularly supplemented through diet or tablets to maintain ascorbic acid pool in the body.

The major metabolites of ascorbic acid in human are dehydroascorbic acid, 2,3-diketogulonic acid and oxalic acid (Fig 2). The main route of elimination of ascorbic acid and its metabolites is through urine. It is excreted unchanged when high doses of ascorbic acid are consumed. Ascorbic acid is generally non-toxic but at high doses (2–6 g/day) it can cause gastrointestinal disturbances or diarrhea [10,11]. The side effects are generally not serious and can be easily reversed by reducing intake of ascorbic acid. Furthermore, there is no consistent and compelling data on serious health effects of vitamin C in humans [11].

A deficiency of ascorbic acid leads to scurvy. It is characterized by spongy swollen bleeding gums, dry skin, open sores on the skin, fatigue, impaired wound healing and depression [13]. Scurvy is of rare occurrence nowadays due to adequate intake ascorbic acid through fresh vegetables and fruits and or supplementation as tablets.

6.5 DIETARY RECOMMENDATIONS OF ASCORBIC ACID

The new average daily intake level that is sufficient to meet the nutritional requirement of ascorbic acid or recommended dietary allowances (RDA)

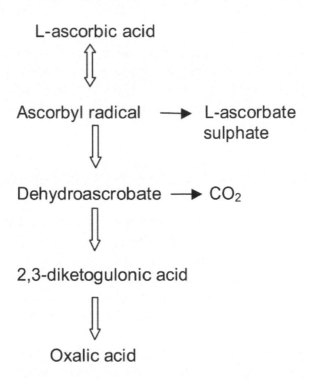

FIGURE 2: Catabolism of Ascorbic acid

for adults (>19 yr) are 90 mg/day for men and 75 mg/day for women [14]. Consumption of 100 mg/day of ascorbic acid is found to be sufficient to saturate the body pools (neutrophils, leukocytes and other tissues) in healthy individuals. Based on clinical and epidemiological studies it has been suggested that a dietary intake of 100 mg/day of ascorbic acid is associated with reduced incidence of mortality from heart diseases, stroke and cancer [15]. However, stress, smoking, alcoholism, fever, viral infections cause a rapid decline in blood levels of ascorbic acid.

Smoking is known to increase the metabolic turnover of ascorbic acid due to its oxidation by free radicals and reactive oxygen species generated by cigarette smoking [16]. It has been suggested that a daily intake of at least 140 mg/day is required for smokers to maintain a total body pool similar to that of non-smokers consuming 100 mg/day [17]. Based on latest literature reports, it has been recommended that the RDA for ascorbic acid should be 100–120 mg/day to maintain cellular saturation and optimum risk reduction of heart disease, stroke and cancer in healthy individuals [18]. There is no scientific evidence to show that even very large doses of vitamin C are toxic or exert serious adverse health effects [11,19]. Furthermore, the panel on dietary antioxidants and related compounds suggested that in vivo data do not clearly show a relationship between excess vitamin C intake and kidney stone formation, pro-oxidant effects, excess iron absorption [20].

6.6 PHYSIOLOGICAL FUNCTIONS OF ASCORBIC ACID

The physiological functions of ascorbic acid are largely dependent on the oxido-reduction properties of this vitamin. L-ascorbic acid is a co-factor for hydroxylases and monooxygenase enzymes involved in the synthesis of collagen, carnitine and neurotransmitters [21]. Ascorbic acid accelerates hydroxylation reactions by maintaining the active center of metal ions in a reduced state for optimal activity of enzymes hydroxylase and oxygenase.

Ascorbic acid plays an important role in the maintenance of collagen which represents about one third of the total body protein. It constitutes the principal protein of skin, bones, teeth, cartilage, tendons, blood vessels, heart valves, inter vertebral discs, cornea and eye lens. Ascorbic acid is essential to maintain the enzyme prolyl and lysyl hydroxylase in an active form. The hydroxylation of proline and lysine is carried out by the enzyme prolyl hydroxylase using ascorbic acid as co-factor. Ascorbic acid deficiency results in reduced hydroxylation of proline and lysine, thus affecting collagen synthesis.

Ascorbic acid is essential for the synthesis of muscle carnitine (β-hydroxy butyric acid). [22]. Carnitine is required for transport and transfer of fatty acids into mitochondria where it can be used for energy production. Ascorbic acid acts as co-factor for hydroxylations involved in carnitine synthesis. Further, ascorbic acid acts as co-factor for the enzyme dopamine-β-hydroxylase, which catalyzes the conversion of neurotransmitter dopamine to norepinephrine. Thus ascorbic acid is essential for synthesis of catecholamines. In addition, ascorbic acid catalyzes other enzymatic reactions involving amidation necessary for maximal activity of hormones oxytocin, vasopressin, cholecystokinin and alpha-melanotripin [23].

Ascorbic acid is also necessary for the transformation of cholesterol to bile acids as it modulates the microsomal 7 α-hydroxylation, the rate limiting reaction of cholesterol catabolism in liver. In ascorbic acid deficiency, this reaction becomes slowed down thus, resulting in an accumulation of cholesterol in liver, hypercholesterolemia, formation of cholesterol gall stones etc [24].

6.7 ASCORBIC ACID AND IRON

Ascorbic acid is known to enhance the availability and absorption of iron from non-heme iron sources [25]. Ascorbic acid supplementation is found to facilitate the dietary absorption of iron. The reduction of iron by ascorbic acid has been suggested to increase dietary absorption of non-heme iron [26]. It is well known that in the presence of redox-active iron, ascorbic acid acts as a pro-oxidant in vitro and might contribute to the formation of hydroxyl radical, which eventually may lead to lipid, DNA or protein oxidation [27]. Thus, ascorbic acid supplementation in individuals with high iron and or bleomycin-detectable iron (BDI) in some preterm infants could be deleterious because it may cause oxidative damage to biomolecules [28-31]. However, no pro-oxidant effect was observed on ascorbic acid supplementation on DNA damage in presence or absence of iron [32].

6.8 ASCORBIC ACID IN HEALTH AND DISEASE

6.8.1 ASCORBIC ACID AND COMMON COLD

The most widely known health beneficial effect of ascorbic acid is for the prevention or relief of common cold. Pauling [33] suggested that ingestion of 1–2 g of ascorbic acid effectively prevents/ ameliorate common cold. The role of oral vitamin C in the prevention and treatment of colds remains controversial despite many controlled trials. Several clinical trails with varying doses of ascorbic acid showed that ascorbic acid does not have significant prophylactic effect, but reduced the severity and duration of symptoms of cold during the period of infection. Randomized and non-randomized trials on vitamin C to prevent or treat the common cold showed that consumption of ascorbic acid as high as 1.0 g/day for several winter months, had no consistent beneficial effect on the incidence of common cold. For both preventive and therapeutic trials, there was a consistent beneficial but generally modest therapeutic effect on duration of cold symptoms. There was no clear indication of the relative benefits of different regimes of vitamin C doses. However, in trials that tested vitamin C after cold symptoms occurred, there was some evidence of greater benefits with large dose than with lower doses [34].

There has been a long-standing debate concerning the role of ascorbic acid in boosting immunity during cold infections. Ascorbic acid has been shown to stimulate immune system by enhancing T-cell proliferation in response to infection. These cells are capable of lysing infected targets by producing large quantities of cytokines and by helping B cells to synthesize immunoglobulins to control inflammatory reactions. Further, it has been shown that ascorbic acid blocks pathways that lead to apoptosis of T-cells and thus stimulate or maintain T cell proliferation to attack the infection. This mechanism has been proposed for the enhanced immune response observed after administration of vitamin C during cold infections [35].

6.8.2 ASCORBIC ACID AND WOUND HEALING

Ascorbic acid plays a critical role in wound repair and healing/regeneration process as it stimulates collagen synthesis. Adequate supplies of ascorbic acid are necessary for normal healing process especially for post-operative patients. It has been suggested that there will be rapid utilization of ascorbic acid for the synthesis of collagen at the site of wound/ burns during post-operative period [36]. Hence, administration of 500 mg to 1.0 g/day of ascorbic acid are recommended to accelerate the healing process [8].

6.8.3 ASCORBIC ACID AND ATHEROSCLEROSIS

Lipid peroxidation and oxidative modification of low density lipoproteins (LDL) are implicated in development of atherosclerosis [37]. Vitamin C protects against oxidation of isolated LDL by different types of oxidative stress, including metal ion dependent and independent processes [38]. Addition of iron to plasma devoid of ascorbic acid resulted in lipid peroxidation, whereas endogenous and exogenous ascorbic acid was found to inhibit the lipid oxidation in iron-over loaded human plasma [39]. Similarly, when ascorbic acid was added to human serum supplemented with Cu^{2+}, antioxidant activity rather than pro-oxidant effects were observed [40].

Ascorbic acid is known to prevent the oxidation of LDL primarily by scavenging the free radicals and other reactive oxygen species in the aqueous milieu [41]. In addition, in vitro studies have shown that physiological concentrations of ascorbic acid strongly inhibit LDL oxidation by vascular endothelial cells [42]. Adhesion of leukocytes to the endothelium is an important step in initiating atherosclerosis. In vivo studies have demonstrated that ascorbic acid inhibits leukocyte-endothelial cell interactions induced by cigarette smoke [43,44] or oxidized LDL [45]. Further, lipophilic derivatives of ascorbic acid showed protective effect on lipid-peroxide induced endothelial injury [46].

A number of studies have been carried out in humans to determine the protective effect of ascorbic acid supplementation (500–100 mg/day) on in vivo and ex vivo lipid peroxidation in healthy individuals and smoker. The findings are inconclusive as ascorbic acid supplementation showed a reduction or no change in lipid peroxidation products [10,47-50]. In this context, it is important to note that during ex vivo LDL oxidation studies, water soluble ascorbic acid is removed during initial LDL isolation step itself. Therefore, no change in ex vivo would be expected [15]. Overall, both in vitro and in vivo experiments showed that ascorbic acid protects isolated LDL and plasma lipid peroxidation induced by various radical or oxidant generating systems. However, a recent report demonstrated that large doses of exogenous iron (200 mg) and ascorbic acid (75 mg) promoted the release of iron from iron binding proteins and also enhanced in vitro lipid peroxidation in serum of guinea pigs. This finding supports the hypothesis that high intake of iron along with ascorbic acid could increase in vivo lipid peroxidation of LDL and therefore could increase risk of atherosclerosis [51]. However, Chen et al., [52] demonstrated that ascorbic acts as an antioxidant towards lipids even in presence of iron over load in in vivo systems.

Numerous studies have looked at the association between ascorbic acid intake and the risk of developing cardiovascular disease (CHD). A large prospective epidemiological study in Finnish men and women suggested that high intake of ascorbic acid was associated with a reduced risk of death from CHD in women and not in men [53]. Similarly, another study showed that high intake of ascorbic acid in American men and women appeared to benefit only women [54,55]. A third American cohort study suggested that cardiovascular mortality was reduced in both sexes by vitamin C [56]. In the UK, a study showed that the risk of stroke in those with highest intake of vitamin C was only half that of subjects with the lowest intake and no evidence suggestive of lower rate of CHD in those with high vitamin C intake [57]. However, a recent meta analysis on the role of ascorbic acid and antioxidant vitamins showed no evidence of significant benefit in prevention of CHD [58]. Thus, no conclusive evidence is available on the possible protective effect of ascorbic acid supplementation on cardiovascular disease.

6.8.4 ASCORBIC ACID AND CANCER

Nobel laureate Pauling and Cameron advocated use of high doses of ascorbic acid (> 10 g/day) to cure and prevent cold infections and in the treatment of cancer [34,59]. The benefits included were increased sense of well being/ much improved quality of life, prolongation of survival times in terminal patients and complete regression in some cases [60-62]. However, clinical studies on cancer patients carried out at Mayo Clinic showed no significant differences between vitamin C and placebo groups in regard to survival time [63]. Cameron and Pauling [23] believed that ascorbic acid combats cancer by promoting collagen synthesis and thus prevents tumors from invading other tissues. However, researchers now believe that ascorbic acid prevents cancer by neutralizing free radicals before they can damage DNA and initiate tumor growth and or may act as a pro-oxidant helping body's own free radicals to destroy tumors in their early stages [64-66].

Extensive animal, clinical and epidemiological studies were carried out on the role of ascorbic acid in the prevention of different types of cancers. A mixture of ascorbic acid and cupric sulfate significantly inhibited human mammary tumor growth in mice, while administered orally [67]. Ascorbic acid decreased the incidence of kidney tumors by estradiol or diethylstilbesterol in hamsters due to decrease in the formation of genotoxic metabolites viz., diethylstilbesterol-4'-4"-qunione [68]. Ascorbic acid and its derivatives were shown to be cytotoxic and inhibited the growth of a number of malignant and non-malignant cell lines in vitro and in vivo [69-72]. Ascorbic acid has been reported to be cytotoxic to some human tumor cells viz., neuorblastoma [73], osteosarcoma and retinoblastoma [74]. A number of ascorbic acid isomers/ derivatives were synthesized and tested on tumor cell lines. Roomi et al., 1998 [75] demonstrated that substitution at 2- or 6- and both at 2,6-positions in ascorbic acid have marked cytotoxicity on malignant cells. Ascorbate-6-palmitate and ascorbate-6-stearate, the fatty acid esters of ascorbic acid were found to be more potent inhibitors of growth of murine leukemia cells compared to ascorbate 2-phosphate, ascorbate 6-phosphate and or ascorbate 6-sulfate respectively [75].

Among ascorbic acid derivatives, fatty acid esters of ascorbic acid viz., ascorbyl palmitate and ascorbyl stearate have attracted considerable interest as anticancer compounds in view of their lipophilic nature as they can easily cross cell membranes and blood brain barrier [76]. Ascorbic acid and ascorbyl esters have been shown to inhibit the proliferation of mouse glioma and human brain tumor cells viz., glioma (U-373) and glioblastoma (T98G) cells and renal carcinoma cells [77-79]. Ascorbyl stearate was found to be more potent than sodium ascorbate in inhibiting proliferation of human glioblastoma cells [80]. Ascorbyl-6-O-palmitate and ascorbyl-2-O-phosphate-6-O-palmitate also showed anti-metastatic effect by inhibiting invasion of human fibrosarcoma HT-1080 cells through matrigel and pulmonary metastasis of mouse melanoma model systems [81].

Numerous reports are available in literature on cytotoxic and anti-carcinogenic effect of ascorbic acid and its derivatives in different tumor model systems. However, the molecular mechanisms underlying the anti-carcinogenic potential of ascorbic acid are not completely elucidated.

Recently, Naidu et al [80] demonstrated that ascorbyl stearate inhibited cell proliferation by interfering with cell cycle, reversed the phenotype and induced apoptosis by modulation of insulin-like growth factor 1-receptor expression in human brain tumor glioblastoma (T98G) cells. They also studied the effect of ascorbyl stearate on cell proliferation, cell cycle, apoptosis and signal transduction in a panel of human ovarian and pancreatic cancer cells. Treatment with ascorbyl stearate resulted in concentration-dependent inhibition of cell proliferation and also clonogenicity of ovarian/ pancreatic cancer cells [82,83]. The anti-proliferative effect was found to be due to the arrest of cells in S/G2-M phase of cell cycle, with increased fraction of apoptotic cells. The cell cycle perturbations were found to be associated with ascorbyl stearate induced reduction in the expression and phosphorylation of IGF-I receptor, while the expression of EGFR and PDGFR remained unchanged. These changes were also associated with activated ERK1/2 but late reduction in AKT phosphorylation. Overexpression of IGF-I receptor in OVCAR-3 cells had no protective effect, however ectopic expression of a constitutively active AKT2 did offer protection from the cytotoxic effects of ascorbyl stearate. In conclusion, ascorbyl stearate-induced anti-proliferative and apoptotic effects in ovarian cancer were found to be mediated through cell cycle arrest and modulation of the IGF-IR and PI3K/AKT2 survival pathways [83].

A plethora of epidemiological studies were carried out to find out the association of ascorbic acid with various types of cancers including breast, esophageal, lung, gastric, pancreatic, colorectal, prostate, cervical and ovarian cancer etc. The results were found to be inconclusive in most types of cancers except gastric cancer [84]. One of the most consistent epidemiological findings on vitamin C has been an association with high intake of ascorbic acid or vitamin C rich foods and reduced risk of stomach cancer. Considerable biochemical and physiological evidence suggests that ascorbic acid functions as a free radical scavenger and inhibit the formation of potentially carcinogenic N-nitroso compounds from nitrates, nitrite in stomach and thus offer protection against stomach cancer [85-87].

Low intake of ascorbic acid and other vitamins was associated with an increased risk of cervical cancer in two of three studies reported [88-91]. This relationship needs further study because the results suggest that other nutrients including vitamin E, carotenoids, retinoic acid either individually or in synergy with ascorbic acid may impart a protective effect against various cancers. Current evidences suggest that vitamin C alone may not be sufficient as an intervention in the treatment of most active cancers, as it appears to be preventive than curative. However, vitamin C supplementation has shown to improve the quality of life and extend longevity in cancer patients, hence it could be considered as an adjuvant in cancer therapy.

Dehydroascorbic acid, the oxidized form of ascorbic acid was shown to cross the blood brain barrier by means of facilitative transport and was suggested to offer neuroprotection against cerebral ischemia by augmenting antioxidant levels of brain [92].

6.9 CONTROVERSIES ON HEALTH BENEFITS OF ASCORBIC ACID

6.9.1 DOES ASCORBIC ACID ACTS AS ANTIOXIDANT OR PRO-OXIDANT?

Vitamin C is an important dietary antioxidant, it significantly decreases the adverse effect of reactive species such as reactive oxygen and nitrogen species that can cause oxidative damage to macromolecules such as lipids, DNA and proteins which are implicated in chronic diseases including cardiovascular disease, stroke, cancer, neurodegenerative diseases and cataractogenesis [93].

As shown in Table 2, ascorbic acid is a potent water soluble antioxidant capable of scavenging/ neutralizing an array of reactive oxygen species viz., hydroxyl, alkoxyl, peroxyl, superoxide anion, hydroperoxyl radicals and reactive nitrogen radicals such as nitrogen dioxide, nitroxide, peroxynitrite at very low concentrations [15]. In addition ascorbic acid can regenerate other antioxidants such as α-tocopheroxyl, urate and β-carotene radical cation from their radical species [94]. Thus, ascorbic acid acts as co-antioxidant for α-tocopherol by converting α-tocopheroxyl radical to α-tocopherol and helps to prevent the α-tocopheroxyl radical mediated peroxidation reactions [95].

$$AH^- + Fe^{3+} \rightarrow A\bullet^- + Fe^{2+} + H^+$$
$$AH^- + Cu^{2+} \rightarrow A\bullet^- + Cu^+ + H^+$$
$$H_2O_2 + Fe^{2+} \rightarrow HO\bullet + Fe^{3+} + {}^-OH$$
$$H_2O_2 + Cu^+ \rightarrow HO\bullet + Cu^{2+} + {}^-OH$$
$$LOOH + Fe^{2+} \rightarrow LO\bullet + Fe^{3+} + {}^-OH$$
$$LOOH + Cu^+ \rightarrow LO + Cu^+ + {}^-OH$$
$$HO\bullet, LO\bullet \rightarrow Lipid\ peroxidation$$

Adapted from Carr and Frei [15]

These radical species are highly reactive and can trigger lipid peroxidation reactions. Thus the question arises whether vitamin C acts as a pro-oxidant in in vivo conditions? The answer appears to be "no" as though these reactions occur readily in vitro, its relevance in in vivo has been a matter of debate concerning ready availability of catalytically active free metal ions in vivo [94]. In biological systems, iron is not freely available, but it is bound to proteins like transferrin, hemoglobin and ferretin. Mobilization of iron from these biomolecules may be required before it can catalyze lipid peroxidation. Further, the concentration of free metal ions in in vivo is thought to be very low as iron and other metals are sequestered by various metal binding proteins [94]. Another factor that may affect pro-oxidant vs antioxidant property of ascorbic acid is its concentration. The in vitro data suggest that at low concentrations ascorbic acid act as a pro-oxidant, but as an antioxidant at higher levels [96]. Moreover, a recent report demonstrated that large doses of exogenous iron (200 mg) and ascorbic acid (75 mg) promote the release of iron from iron binding proteins and also enhance in vitro lipid peroxidation in serum of guinea pigs. This finding supports the hypothesis that high intake of iron along with ascorbic acid could increase in vivo lipid peroxidation of LDL and therefore could increase risk of atherosclerosis [52]. However, another study demonstrated that in iron-overloaded plasma, ascorbic acid acts as an antioxidant and prevent oxidative damage to lipids in vivo [97].

6.9.2 IS ASCORBIC ACID HARMFUL TO CANCER PATIENTS?

Agus et al [98] have reported that the tumor cells contain large amounts of ascorbic acid, although the role of ascorbic acid in tumors is not yet known. They have established that vitamin C enters through the facilitative glucose transporters (GLUTs) in the form of dehydroascorbic acid, which is then reduced intracellularly and retained as ascorbic acid. It is speculated that high levels of ascorbic acid in cancer cells may interfere with chemotherapy or radiation therapy since these therapies induce cell death by oxidative mechanism. Thus, ascorbic acid supplementation might make cancer treatment less effective because, ascorbic acid being a strong antioxidant may scavenge or neutralize the oxidative stress induced by chemotherapy in cancer patients. However, more studies are needed to understand the role of ascorbic acid in tumors cells and the speculative contraindication of ascorbic acid for cancer chemotherapy.

6.9.3 DOES ASCORBIC ACID CAUSE CANCER ?

Recently, it has been reported that lipid hydroperoxide can react with ascorbic acid to form products that could potentially damage DNA, suggesting that it may form genotoxic metabolites from lipid hydroperoxides implicating that ascorbic acid may enhance mutagenesis and risk of cancer. Lee et al [99], demonstrated that ascorbic acid induces decomposition of lipid hydroperoxide (13-(S)-hydroperoxy-(Z,E)-9,11-octadecadienoic acid;(13-HPODE) in presence of transition metals to DNA-reactive bifunctional electro-philes namely 4-oxo-2-nonenal, 4,5-epoxy-2(E)-decenal and 4-hydroxy-2-nonenal. 4-oxo-2-nonenal being a genotoxin can react with DNA bases to form mutations [100] or apoptosis [101].

Thus, the above process can give rise to substantial amounts of DNA damage in vivo. However there are many questions, which need to be considered before we accept the hypothesis that ascorbic acid can cause cancer by producing genotoxic metabolites from lipids. The hydroperoxides formed through lipid peroxidation reaction are rapidly reduced to aldehydes by a number of enzymes. Further, ascorbic acid being a strong antioxidant effectively inhibits the formation of lipid peroxides as ascorbic acid forms the first line of antioxidant defense mechanism in human plasma. The formation of lipid hydroperoxides occur only after ascorbic acid has been exhausted. Hence, interaction of ascorbic acid and hydroperoxide may not arise in human plasma. Recently, high intracellular vitamin C was reported to prevent oxidation-induced mutations in human cells [102]. Thus, the physiological relevance of these results is yet to be established in in vivo experiments.

6.10 CONCLUSION

Ascorbic acid is one of the important and essential vitamins for human health. It is needed for many physiological functions in human biology. Fresh fruits, vegetables and also synthetic tablets supplement the ascorbic acid requirement of the body. However, stress, smoking, infections and burns deplete the ascorbic acid reserves in the body and demands higher doses of ascorbic acid supplementation. Based on available biochemical, clinical and epidemiological studies, the current RDA for ascorbic acid is suggested to be 100–120 mg/day to achieve cellular saturation and optimum risk reduction of heart diseases, stroke and cancer in healthy individuals. In view of its antioxidant property, ascorbic acid and its derivatives are widely used as preservatives in food industry. Many health benefits have been attributed to ascorbic acid namely antioxidant, anti-atherogenic and anti-carcinogenic activity. Lately some of these beneficial effects of ascorbic acid are contradicted. The relation between ascorbic acid and cancer is still a debatable as the molecular mechanism underlying anti-carcinogenic activity of ascorbic acid is not clearly elucidated. Regarding the pro-oxidant activity of vitamin C in presence of iron, there is compelling evidence for antioxidant protection of lipids by ascorbic acid both with and without iron co-supplementation in animals and humans. Current evidences also suggest that ascorbic acid protects against atherogenesis by inhibiting LDL oxidation. The data on vitamin C and DNA damage are conflicting and inconsistent. However, more mechanistic and human in vivo studies are warranted to establish the beneficial claims on ascorbic acid. Thus, though ascorbic acid was discovered in 17th century, the role of this important vitamin in human health and disease still remains a mystery in view of many beneficial claims and contradictions.

TABLE 1: Ascorbic acid content in selected foods

Fruits	mg/100 g edible portion
Banana	8–16
Apple	3–30
Mango	10–15
Pineapple	15–25
Cherry	15–30
Papaya	39
Orange	30–50
Grape fruit	30–70
Lemon	40–50
Strawberry	40–70
Currant black	150–200
Rose hips	250–800
Vegetables	
Onion	10–15
Tomato	10–20
Egg plant	15–20
Radish	25
Spinach	35–40
Cabbage	30–70
Cauliflower	50–70
Broccoli	80–90
Coriander	90
Brussels sprout	100–120
Pepper	150–200
Parsley	200–300

Adapted from Johnson et al [12]

TABLE 2: Reactive species scavenged by ascorbic acid

Chemical species	Reaction rate (M-1s-1)
Reactive oxygen species	
Hydroxyl radical	1.1×1010
Alkoxyl radical	1.6×109
Peroxy radical	1.2×106
Superoxide anion/ hydroperoxy radical	1.0×105
Reactive nitrogen species	
Dinitrogen trioxide/dinitrogen tetroxide	1.2×109
Peroxynitrite/peroxynitrous acid	235
Antioxidant derived radicals	
Alpha-tocopherol radical	2×105
Urate radical	1×106
Thiyl/sulphenyl radical	6×108

Adapted from Carr and Frei [15]

REFERENCES

1. Lind J: A treatise of scurvy. Printed by Sands, Murray and Cochran for Kincaid, A and Donaldson, A. Edinburgh 1753.
2. Svirbely JL, Szent-Gyorgyi A: The chemical nature of vitamin C. Biochem J 1932, 26865-870.
3. Waugh WA, King CG: Isolation and identification of vitamin C. J Biol Chem 1932, 97:325-331.
4. Haworth WN, Hirst EL: Synthesis of ascorbic acid. J Soc Chem Ind (London) 1933, 52:645-647.
5. Moser U, Bendich A: Vitamin C. In Handbook of Vitamins. Edited by Machlin LJ. Marcel Dekker, New York; 1990:Ch5.
6. Sauberlich HE: Bioavailability of vitamins. Prog Food Nutr Sci 1985, 9:1-33.
7. Sauberlich HE: Ascorbic acid. In Present knowledge in Nutrition. Edited by Brown ML. Nutrition Foundation, Washington DC; 1990.
8. Hellman L, Burns JJ: Metabolism of L-ascorbic acid-1-C14 in man. J Biol Chem 1958, 230:923-930.
9. Kallner A, Horing D, Hartman D: Kinteics of ascorbic acid in humans. In Ascorbic acid: Chemistry, metabolism and uses. Edited by Seib PA, Tolbert BM. Advances in Chemistry Series No.200, American Chemical Society, Washington, DC; 1982:385-400.
10. Anderson D, Phillips BJ, Yu T, Edwards AJ, Ayesh R, Butterworth KR: The effect of vitamin C supplementation on biomarkers of oxygen radical generated damage in

human volunteers with low or high cholesterol levels. Environ Mol Mutagens 1997, 30:161-174.

11. Johnson CS: Biomarkers for establishing a tolerable upper intake level for vitamin C. Nutr Rev 1999, 57:71-77.

12. Johnson CS, Steinberg FM, Rucker RB: Ascorbic acid. In Hand book of Vitamins. Edited by Rucker RB, Sultie JW, McCormick, DB, Machlin LJ. Marcel Dekker Inc, New York; 1998:529-585.

13. Olson RE: Water soluble vitamins. In Principles of Pharmacology. Edited by Munson PL, Mueller RA, Bresse GR. Chapman and Hall, New York; 1999:Ch 59.

14. Frei B, Traber M: The new US dietary reference for vitamins C and E. Redox Rep 2001, 6:5-9.

15. Carr AC, Frei B: Does vitamin C act as pro-oxidant under physiological conditions? FASEB J 1999, 13:1007-1024.

16. Frei B, Forte TM, Ames BN, Cross CE: Gas-phase oxidants of cigarette smoke induce lipid peroxidation and changes in lipoprotein properties in human blood plasma: protective effects of ascorbic acid. Biochem J 1981, 277:133-138.

17. Kallner A, Hartmann D, Hornig D: On the requirement of ascorbic acid in man: steady-state turnover and body pool in smokers. Am J Clin Nutr 1981, 34:1347-1355.

18. Carr AC, Frei B: Toward new recommended dietary allowance for vitamin C based on antioxidant and health effects in humans. Am J Clin Nutr 1999, 69:1086-1107.

19. Bendich A: Vitamin C safety in humans. In Vitamin C in Health and Disease. Edited by Packer L, Fuchs J. Marcel Dekker Inc. New York; 1997:369-379.

20. Food and Nutrition Board: Dietary reference intakes for vitamin C, vitamin E, selenium and carotenoids. National Academy Press, Washington, DC 2000.

21. Levin M: New concepts in the biology and biochemistry of ascorbic acid. New Engl J Med 1986, 31:892-902.

22. Hulse JD, Ellis SR, Henderson LM: Carnitine biosynthesis-beta hydroxylation of trimethyllysine by an α-keto glutarate dependent mitochondrial dioxygenase. J Biol Chem 1978, 253:1654-1659.

23. Cameron E, Pauling L: Ascorbic acid and the glycosaminoglycans. Oncology 1973, 27:181-192.

24. Ginter E, Bobek P, Jurcovicova M: Role of ascorbic acid in lipid metabolism. In Ascorbic acid, chemistry, metabolism and uses. Edited by Seith PA, Toblert, BM. American Chemical Society, Washington, DC; 1982:381-393.   Hallberg L: Bioavailability of dietary iron in man. Annu Rev Nutr 1981, 1:123-127.

25. Bendich A, Cohen M: Ascorbic acid safety: analysis factors affecting iron absorption. Toxicol Lett 1990, 51:189-190.

26. Samuni A, Aronovitch J, Godinger D, Chevion M, Czapski G: On the cytotoxicity of vitamin C and metal ions: A site specific Fenton mechanism. Eur J Biochem 1983, 137:119-124.

27. Minetti M, Forte T, Soriani M, Quaresima V, Menditto A, Ferrari M: Iron Induced ascorbate oxidation in plasma as monitored by ascorbate free radical formation: No spin trapping evidence for the hydroxyl radical in iron-over loaded plasmas. Biochem J 1992, 282:459-465.

28. Berger TM, Mumby S, Gutteridge JMC: Ferrous ion detected in iron-overloaded cord blood plasma from preterm and term babies: Implication for oxidation stress. Free Rad Res 1995, 22:555-559.

29. Halliwell B: Vitamin C: Antioxidant or pro-oxidant in vivo ? Free Rad Res 1996, 25:439-454.

30. Herbert V, Shaw S, Jayatileke E: Vitamin C driven free radicals generation from iron. J Nutr 1996, 126:1213-1220.

31. Proteggente AR, Rehman A, Halliwell B, Rice-Evans CA: Potential problems of ascorbic acid and iron supplementation: Pro-oxidant effect in vivo ? Biochem Biophys Res Commun 2000, 277:535-540.

32. Pauling L: Vitamin C and common cold. Freeman, San Francisco, CA 1970.

33. Douglas RM, Chalker EB, Treacy B: Vitamin C for preventing and treating the common cold. Cochrane Database Syst Rev 2000, 2:CD000980.

34. Campbell JD, Cole M, Bunditrutavorn B, Vell AT: Ascorbic acid is a potent inhibitor of various forms of T cell apoptosis. Cell Immunol 1999, 194:1-5.

35. Shukla SP: Level of ascorbic acid and its oxidation in the liver of Scorpion. Palamnaeus bengalensis. Experentia 1969, 25:602-604.

36. Steinbrecher UP, Zhang H, Lougheed M: Role of oxidative modified LDL in atherosclerosis. Free Rad Biol Med 1990, 9:155-168.

37. Frei B: Vitamin C as an antiatherogen: mechanism of action. In In Vitamin C in Health and disease. Edited by Packer L, Fuchs J. Marcel and Dekker, Inc., New York; 1997:163-182.

38. Berger TM, Polidori MC, Dabhag A, Evans PJ, Halliwell B, Marrow JD, Roberts LJ, Frei B: Antioxidant activity of viamin C in iron-over loaded human plasma. J Biol Chem 1992, 272:15656-15660.

39. Dasgupta A, Zdunek T: In vitro lipid peroxidation of human serum catalyzed by copper ion: antioxidant rather than pro-oxidant role of ascorbate. Life Sci 1992, 50:2875-2882.

40. Frei B, England L, Ames BN: Ascorbate is an outstanding antioxidant in human blood plasma. Proc Natl Acad Sci USA 1989, 86:6377-6381.

41. Martin A, Frei B: Both intracellular and extracellular vitamin C inhibit atherogenic modification of LDL by human vascular endothelial cells. Atheroscler Thromb Vasc Biol 1997, 17:1583-1590.

42. Lehr HA, Frei B, Arfors KE: Vitamin C prevents cigarette smoke-induced leukocyte aggregation and adhesion to endothelium in vivo. Proc Natl Acd Sci USA 1994, 91:7688-7692.

43. Lehr HA, Weyrich AS, Saetzler RK, Jurek A, Arfors KE, Zimmerman GA, Prescott SM, McIntyre TM: Vitamin C blocks inflammatory platelet-activating factor mimetics created by cigarette smoking. J Clin Invest 1997, 99:2358-2364.

44. Lehr HA, Frei B, Olofsson AM, Carew TE, Arfors KE: Protection from oxidized LDL induced leukocyte adhesion to microvascular and macrovascular endothelium in vivo by vitamin C but not by vitamin E. Circulation 1995, 91:1552-1532.

45. Kaneko T, Kaji K, Mastuo M: Protective effect of lipophilic derivatives of ascorbic acid on lipid peroxide-induced endothelial injury. Arch Biochem Biophys 1993, 304:176-180.

46. Fuller CJ, Grundy SM, Norkus EP, Jialal I: Effect ascorbate supplementation on low density lipoprotein oxidation in smokers. Atherosclerosis 1996, 119:139-150.

47. Nyyssonen K, Poulsen HE, Hayn M, Agerbo P, Porkkalo Sarataho E, Kaikkonen J, Salonen R, Salonen JT: Effect of supplementation of smoking men with plain or slow release ascorbic acid on lipoprotein oxidation. Eur J Clin Nutr 1997, 51:154-163.

48. Samman S, Brown AJ, Beltran C, Singh S: The effect of ascrobic acid on plasma lipids and oxidisability of LDL in male smokers. Eur J Clin Nutr 1997, 51:472-477.

49. Wen Y, Cooke T, Feely J: The effect of pharmacological supplemen-tation with vitamin C on low density lipoprotein oxidation. Br J Clin Pharma 1997, 44:94-97.

50. Kapsokefalou M, Miller DD: Iron loading and large doses of intravenous ascorbic acid promote lipid peroxidation in whole serum in guinea pigs. Br J Nutr 85:681-687.

51. Chen K, Suh J, Carr AC, Marrow JD, Zeind J, Frei B: Vitamin C suppresses lipid damage in vivo even in the presence of iron over-load. Am J Physiol Endocrinol Metab 2000, 279:E1406-1212.

52. Knekt P, Reunanen A, Jarvinen R, Seppanen R, Heliovaara M, Aromaa A: Antioxidant vitamin intake and coronary mortality in a longitudinal population study. Am J Epidemiol 1994, 139:1180-1189.

53. Manson JE, Stampfer MJ, Willett WC, et al.: A prospective study of vitamin C and incidence of coronary heart disease in women. Circulation 1982, 85:865-875.

54. Rimm EB, Stampfer MJ, Ascherio A, Giovanno E, Colditz GA, Willettt WC: Vitamin E consumption and risk of coronary heart disease in men. N Engl J Med 1993, 328:1450-1456.

55. Enstrom JE, Kanim LE, Klein MA: Vitamin C intake and mortality among a sample of the United States population. Epidemiology 1992, 3:194-202.

56. Gale CR, Martyn CN, Winter PD, Cooper C: Vitamin C and risk of death from stroke and coronary heart disease in cohort of elderly people.

57. Br Med J 1995, 310:1563-1566.

58. Ness A, Egger M, Davey-Smith G: Role of antioxidant vitamins in prevention of cardiovascular disease. Br Med J 1999, 319:577-579.

59. Cameron E, Pauling L: In: Cancer and Vitamin C. W.W.Norton &; Company, Inc, New York; 1979:132.

60. Cameron E, Pauling L: Supplemental ascorbate in the supportive treatment of cancer: Prolongation of survival times in terminal human cancer. Proc Natl Acad Sci USA 1976, 73:3685-3689.

61. Cameron E, Pauling L: Supplemental ascorbate in the supportive treatment of cancer: Reevaluation of prolongation of survival times in terminal human cancer. Proc Natl Acad Sci USA 1978, 75:4538-4542.

62. Murata A, Morsige F, Yamaguchi H: Prolongation of survival times of terminal cancer patients by administration of large doses of ascorbate. Int J Vit Nutr Res Suppl 1982, 23:103-113.

63. Moertel CG, Fleming TR, Creagan ET, Rubin J, O'Connell MJ, Ames MM: High dose vitamin C versus placebo in the treatment of patients with advanced cancer who have had no prior chemothrerapy : A randomized double blind comparison. N Engl J Med 1985, 312:137-141.

64. Block G: Vitamin C and cancer prevention: the epidemiological evidence. Am J Clin Nutr 1991, 53:270S-282S.
65. Frei B: Reactive oxygen species and antioxidant vitamins: Mechanism of action. Am J Med 1994, 97:5S-13S.
66. Uddin S, Ahmad S: Antioxidant protection against cancer and other human diseases. Comprehen Therap 1995, 21:41-45.
67. Tsao CS: Inhibiting effect of ascorbic acid on growth of human mammary tumor xenografts. Am J Clin Nutr 1991, 54:1274S-1280S.
68. Liehr JG: Vitamin C reduces the incidence and severity of renal tumors induced by estradiol or diethylstibesterol. Am J Clin Nutr 1991, 54:1256S-1260S.
69. Park CH, Kimler BF: Growth modulation of human leukemic, pre-leukemic and myeloma progenitor cell by L-ascorbic acid. Am J Clin Nutr 1991, 54:1241S-1246S.
70. Eckert-Maksic M, Kovacek I, Maksic ZB, Osmak M, Paveli K: Effect of ascorbic acid and its derivatives on different tumors in vivo and in vitro. In Molecules in Natural Science and Medicine. An Encomium for Linus Pauling. Edited by Maksic ZB, Eckert-Maksic M. Ellis Horwood, New York; 1991:509-524.
71. Murakami K, Muto N, Fukasawa GK, Yamamoto I: Comparison of ascorbic acid and ascorbic acid 2-O-L-glucosidase on the cytotoxicity and bioavailability to low density culture of fibroblast. Biochem Pharmacol 1992, 44:2191-2197.
72. Roomi MW, House D, Eckert_Maksic M, Maksic ZB, Tsao CS: Growth suppression of malignant leukemia cell line in vitro by ascorbic acid (Vitamin C) and its derivatives. Cancer Lett 1998, 122:93-99.
73. Pavelic K: L-ascorbic acid induced DNA strand breaks and cross links in human neuroblastoma cell. Brain Res 1985, 342:369-373.
74. Medina MA, de Veas RG, Schweigerer L: Ascorbic acid is cytotoxic for peidoatric tumor cells cultured in vitro. Biochem Mol Biol Inter 1994, 34:871-874.
75. Roomi MW, House D, Tsao CS: Cytotoxic effect of substitution at 2-, 6-, and 2,6-positions in ascorbic acid on malignant cell line. Cancer Biochem Biophys 1998, 16:295-300.
76. Banks WA, Kastin AJ: Peptides and blood brain barrier: lipophilicity as predictor of permeability. Brain Res Bull 1985, 15:287-292.
77. Naidu AK, Wiranowska M, Kori SH, Prockop LD, Kulkarni AP: Inhibition of human glioma cell proliferation and glutathione-S-transferase by ascorbyl esters and interferon. Anticancer Res 1993, 13:1469-1471.
78. Naidu AK, Wiranowska M, Kori SH, Prockop LD, Kulkarni AP: Inhibition of cell proliferation and glutathione-S-transferase by ascorbyl esters and interferon in mouse glioma. J Neuro-Oncol 1993, 16:1-10.
79. Makino Y, Sakagami H, Takeda M: Induction of cell death by ascorbic acid derivatives in human renal carcinoma and glioblastoma cell lines. Anticancer Res 1999, 19:3125-3132.
80. Naidu KA, Tang JL, Naidu KA, Prockop LD, Nicosia SV, Coppola D: Antiproliferative and apoptotic effect of ascorbyl stearate in human glioblastoma multiforme cell: Modulation of insulin-like growth factor-I receptor (IGF-IR) expression. J Neuro-Oncol 2001, 54:15-22.
81. Liu JW, Nago N, Kageyama K, Miwa N: Anti-metastatic effect of an autooxidation-resistant and lipophilic ascorbic acid derivative through inhibition of tumor invasion. Anticancer Res 2000, 20:113-118.

82. Naidu AK, Karl RC, Naidu KA, Coppola D: The antiproliferative and pro-apoptotic effect of Ascorbyl Stearate in Human pancreatic cancer cells : Association with decreased expression of insulin-like growth factor receptor-1. Digest Dis Sci 2003, 48:230-237.

83. Naidu AK, Naidu KA, Sun M, Dan HC, Nicosia SV, Cheng JQ, Coppola D: Ascorbyl stearate inhibits proliferation and induces apoptosis of human ovarian carcinoma cells by targeting PI3k/akt pathway. Communicated to J Biol Chem 2003.

84. Sauberlich HE: Vitamin C and Cancer. In Nutrition and disease update cancer. Edited by Carroll KK, Kritchevsky D. AOCS Press, Champaign, Ilinois; 1994:111-157.

85. Schorah CJ, Sobala M, Collis N, Primrose JN: Gastric juice ascorbic acid: effects of disease and implications for gastric carcinogenesis. Am J Clin Nutr 1991, 53:287S-293S.

86. Sobala GM, Pignaetelli B, Schorah CJ, Bartsch H, Sanderson M, Dixon MF, Shires S, King RFG, Axon ATR: Levels of nitrite, nitrate, N-nitroso compounds, ascorbic acid in gastric juice of patients with and without precancerous conditions of the stomach. Carcinogenesis 1991, 12:193-198.

87. Drake IM, Davies MJ, Mapstone NP, Dixon MF, Schorah CJ, White KL, Chamers DM, Axon AT: Ascorbic acid may protect against human gastric cancer by scavenging mucosal oxygen radicals. Carcinogenesis 1996, 17:559-562.

88. Brock KE, Berry G, Mock PA, MacLennan R, Truswell AS, Brinton LA: Nutrients in diet and plasma and risk in situ cervical cancer. J Natl Cancer Inst 1988, 80:580-585.

89. Verreault R, Chu J, Mandelson M, Shy K: A case study of diet and invasive cancer. Int J Cancer 1989, 43:1050-1054.

90. Potischman N, Brinton LA: Nutrition and cervical neoplasia. Cancer Causes Control 1996, 7:113-126.

91. Rock CL, Michael CW, Reynolds RK, Ruffin MT: Prevention of cervix cancer. Crit Rev Oncol Hematol 2000, 33:169-183.

92. Huang J, Agus DB, Winfree CJ, Kiss S, Mack WJ, McTaggart RA, Choudhri TF, Kim LJ, Mocco J, Pinsky DJ, Fox WD, Israel RJ, Boyd TA, Golde DW, Connolly ES Jr: Dehydroascorbic acid, a blood-brain barrier transportable form of vitamin C, mediates potent cerebroprotection in experimental stroke. Proc Natl Acad Sci U S A 2001, 98:11720-11724.

93. Halliwell B, Gutteridge JMC: Free radicals in Biology and Medicine. Oxford University Press, Oxford 1999.

94. Halliwell B, Gutteridge JMC: Oxygen free radicals and iron in relation to biology and medicine: some problem and concepts. Arch Biochem Biophys 1986, 246:501-514.

95. Neuzil J, Thomas SR, Stocker R: Requirement for promotion or inhibition by α-tocopheroxyl radical induced plasma lipoprotein lipid peroxidation. Free Rad Biol Med 1997, 22:57-71.

96. Buettner GR, Jurkiewicz BA: Catalytic metals, ascorbate and free radicals: combinations to avoid. Rad Res 1996, 145:532-541.

97. Berger TM, Poldori MC, Dabbagh A, Evans PJ, Halliwell B, Morrow JD, Roberts II J, Frei B: Antioxidant activity of vitamin C in iron-overloaded human plasma. J Biol Chem 1997, 279:15636-15660.

98. Agus DB, Vera JC, Golde DW: Stromal cell oxidation: A mechanism by which tumors obtain vitamin C. Cancer Res 1999, 59:4555-4558.
99. Lee SH, Oe T, Bliar IA: Vitamin C induced decomposition of lipid hydroperoxides to endogenous genotoxins. Science 2001, 292:2083-2086.
100. Marnett LJ: Oxyradicals and DNA damage. Carcinogenesis 2000, 21:361-370.
101. Johnson TM, Yu ZX, Ferrans VJ, Lowenstein T, Finkel T: Reactive oxygen species are downstream mediators of p53 dependent apotposis. Proc Natl Acd Sci USA 1996, 93:11848-11852.
102. Lutsenko EA, Carcamo JM, Golde DW: Vitamin C prevents DNA mutation induced by oxidative stress. J Biol Chem 2002, 277:16895-16899.

This chapter was originally published under the Creative Commons Attribution License. Naidu, K. A. Vitamin C in Human Health and Disease Is Still a Mystery? An Overview. Nutrition Journal 2003, 2:7. doi:10.1186/1475-2891-2-7. http://www.nutritionj.com/content/2/1/7/

DAILY EGG CONSUMPTION IN HYPERLIPIDEMIC ADULTS: EFFECTS ON ENDOTHELIAL FUNCTION AND CARDIOVASCULAR RISK

VALENTINE NJIKE, ZUBAIDA FARIDI, SUPARNA DUTTA, ANJELICA L GONZALEZ-SIMON, and DAVID L KATZ

7.1 BACKGROUND

As of the early 1970's, a reduction in consumption of eggs, a concentrated source of cholesterol (one yolk provides ~215 mg of cholesterol), had been widely recommended in an effort to lower blood cholesterol and reduce the risk of heart disease[1]. In 1973, the American Heart Association (AHA) guidelines specifically advocated exclusion of eggs from the diet, accompanying the advised cholesterol restriction[2]. More recent AHA guidelines no longer advise for or against egg or egg yolk consumption, admitting that there is a lack of scientific evidence for selecting a target level for dietary cholesterol[3]. This is partially due to individual differences in serum cholesterol responses to dietary cholesterol. The recommended intake of daily dietary cholesterol continues to be 300 mg/day or less for healthy adults and less than 200 mg/day for persons with elevated cholesterol or heart disease[3]. Given the widespread nature of this recommendation, there is surprisingly little evidence that egg consumption increases blood cholesterol levels, thereby increasing cardiovascular risk [4].

Data from recent studies show that consumption of one or two eggs per day, when part of a low fat diet, does not adversely affect the lipid profile[5,6]. In fact, the preclusion of eggs from the diet may represent a potential reduction in overall dietary quality. As an inexpensive functional food with an exceptional nutritional profile [7,8], eggs are an excellent natural source of folate, riboflavin, selenium, choline, vitamin B12, and fat-soluble vitamins A, D, E, and K. Eggs also provide high-quality, bioavailable protein [9,10] with little total fat. Compared to other animal protein sources, eggs contain proportionately less saturated fat, which has generally been recognized as a strong dietary determinant of elevated low-density lipoprotein (LDL) levels and increased risk of coronary heart disease (CHD) [11] although this topic is not without controversy [12].

As a dietary substitute for eggs, egg substitute is comprised of 99% egg whites and provide 12 key vitamins and nutrients, including riboflavin, B12, folate, and pantothenic acid, while excluding the cholesterol contribution of the egg yolk [13]. Although nutritionally similar to eggs, egg substitute contains emulsifiers, stabilizers, and artificial color and are on average three times as expensive as regular eggs.

Endothelial function refers to arterial vasomotor responses mediated predominantly by the release of nitric oxide (vasodilating), and endothelin (vasoconstricting) from the vascular endothelium [14,15], and plays an important role in the pathogenesis of atherosclerosis, hypertension, cardiovascular disease, and diabetes [15-17]. Endothelial dysfunction correlates strongly with both coronary disease and its risk factors [18,19] and reverses in response to risk modification efforts [15,20-27]. Endothelial dysfunction has increasingly been viewed as an indicator of coronary risk[19], and its amelioration as an indicator of risk reduction [20,23].

The relationship of egg consumption to coronary outcomes depends not only on the cholesterol content of eggs themselves, but on the composition of the total diet. It is a common misconception that dietary cholesterol increases serum cholesterol which increases CHD risk [28,29]; however, research has failed to provide substantial evidence of this assumed relationship [30]. In our previous trial, daily ingestion of eggs did not produce adverse effects on cardiac risk, as indicated by endothelial function and lipid profile, in healthy adults [31]. To the best of our knowledge, no study has ever compared the effects of egg versus egg substitute consumption on cardiovascular risk. Therefore, we performed a randomized cross-over trial to assess the effects of egg or egg substitute consumption on endothelial function and lipid panel in hyperlipidemic adults.

7.2 SUBJECTS AND METHODS

7.2.1 SUBJECTS

Forty adults (16 men and 24 women) with diagnosed hyperlipidemia were recruited from Southwestern Connecticut; largely through mass media print advertisements and posters. Eligible subjects were 35 years of age or older if they were male or post-menopausal and not currently using hormone replacement therapy if they were female. Additionally, eligible subjects were non-smokers, and hyperlipidemic as defined by serum total cholesterol >240 mg/dL, and/or LDL cholesterol >160 mg/dL, and/or a total cholesterol/HDL ratio >5.7. All ethnic and minority groups were equally eligible. Individuals with a current eating disorder, a restricted diet, diagnosed coronary disease, diabetes, or sleep apnea were excluded from the study. Additional exclusion criteria included the regular use of lipid-lowering medication, insulin or glucose sensitizing medication, vasoactive medication or nutriceuticals, high dose vitamin E or C, and fiber supplements.

Individuals who responded to recruitment efforts (n = 172) were prescreened using a semi-structured telephone interview. Those who met initial screening criteria (n = 40) underwent a clinical screening evaluation (weight, height, body mass index (BMI), and blood pressure measurements) performed by a clinical research specialist, along with laboratory testing (fasting total cholesterol, HDL, LDL, and triglycerides levels) (Figure 1).

All participants provided informed consent and were compensated monetarily for their time. The study protocol was approved by the Institutional Review Board (IRB) of Griffin Hospital (Derby, CT).

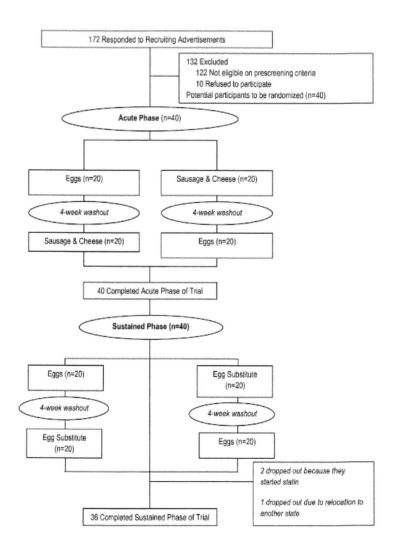

FIGURE 1: Flow of Participants through the Trial.

7.2.2 STUDY DESIGN

This study was a randomized, single-blind crossover trial with investigators blinded to treatment assignments. The trial consisted of an acute and a sustained phase. In the acute phase, 40 participants were randomly assigned to consume one of the two sequence permutations of a single dose of breakfast of three medium hardboiled eggs and a sausage/cheese breakfast sandwich (Table 1). In the sustained phase, participants were randomly assigned to one of the two sequence permutations of two medium hardboiled eggs and 1/2 cup of egg substitute breakfast daily for six weeks. Randomization was conducted by the data manager using a SAS (SAS version 9.1; SAS Institute, Cary, NC) algorithm. Each treatment assignment was separated by a four-week washout period. The study participants fasted overnight before undergoing endothelial function assessment. Due to the obvious dietary makeup of each treatment assignment, it was not possible to blind participants to their assignment; however, the ultrasonographer was strictly blinded to participants' treatment assignment.

7.2.3 OUTCOME MEASURES

7.2.3.1 ENDOTHELIAL FUNCTION ASSESSMENT

The brachial artery reactivity studies (BARS) methodology employed is comparable to those of other leading labs[16,32-36], and is described in "Guidelines for ultrasound assessment of endothelial-dependent flow-mediated vasodilation of the brachial artery"[16]. Participants were required to lie at rest in the quiet, temperature-controlled, softly lit room for at least 15 minutes before scanning was initiated. The baseline diameter of the brachial artery was measured from two-dimensional ultrasound images using a high frequency, 10-15 MHz, vascular ultrasound transducer (Sonos 4500; Phillips Medical Systems, Andover, MA). Arterial flow-velocity was measured by means of a pulsed Doppler signal at a 70° angle to the vessel, with the range gate in the center of the artery. Flow was determined by multiplying the arterial cross-sectional area (πr^2) by the Doppler flow velocity. The timing of each image frame with respect to the cardiac cycle was determined with simultaneous ECG gating during image acquisition via the high-quality mainframe ultrasound system. The arterial diameter was measured at a fixed distance from an anatomical marker, such as a bifurcation, with ultrasonic calipers. Measurements were taken from the anterior to the posterior "m" line in diastole. The brachial artery was imaged at a location 3-7 cm above the antecubital fossa in the longitudinal plane. A segment with clear anterior and posterior intimal interfaces between the lumen and vessel wall was selected for continuous 2D gray scale imaging. The transmit (focus) zone was set to the depth of the near wall because of difficulty in differentiating the near from the far wall "m" line (the interface between media and adventitia)[16,33]. Images were acquired on videotape and magnetic optical disk for evaluation and analysis subsequent to the examination. Diameter was obtained from m-line to m-line, over a consistent segment of vessel at least 10-15 mm in length.

TABLE 1: Composition of the Breakfasts in the Acute Phase

Nutrition Content	Sausage & Cheese	3 Medium Eggs
Calories (kcal)	310	189
Calories from fat (Kcal)	260	118
Fat (gm)	29	13
Saturated fat (gm)	12	4
Sodium (mg)	720	185
Protein (g)	13	17
Calories from protein (kcal)	52	66
Carbohydrate (g)	0	1

To create a flow stimulus in the brachial artery, a sphygmomanometer cuff was placed on the upper arm proximal to the transducer. A baseline blood flow and diameter were acquired. Arterial occlusion was created by cuff inflation to 50 mm Hg above the systolic blood pressure. The cuff remains inflated for 5 minutes. This causes ischemia and consequent dilation of downstream resistance vessels via auto-regulatory mechanisms[16]. Cuff deflation induces a brief high-flow state through the brachial artery (reactive hyperemia) to accommodate the dilated resistance vessels. The resulting increase in shear stress causes the normal brachial artery to dilate.

A pulsed Doppler signal was obtained within 15 seconds of cuff release to assess hyperemic velocity, and a longitudinal image of the artery was recorded continuously from 20 seconds to 2 minutes after cuff deflation. All images were coordinated with a continuous ECG monitor and obtained at end-diastole. The resultant coefficient of intra observer reliability was 0.9.

Flow-mediated dilation (FMD) was measured as the percent change in brachial artery diameter from pre-cuff inflation to 60-seconds post-cuff release. In addition to brachial diameter at 60 seconds post-cuff release, flow after cuff deflation within the first 15 seconds was used as an indicator of stimulus strength, hyperemic flow being the stimulus for endothelial reactivity. To account for potential variability in stimulus strength, FMD was divided by flow at 15 seconds[16]. post-cuff deflation to create a stimulus-adjusted response measure[31,37,38].

7.2.3.2 LIPID PROFILE

Serum was drawn for lipid assessments about twenty minutes prior to endothelial function assessment. The lipid profile was determined as follows: Total cholesterol (Tchol), triglycerides (TRIG), and high-density lipoprotein (HDL) were obtained by direct measurements. Very-low-density lipoprotein (VLDL) and low-density-lipoprotcin (LDL) were obtained by calculation: VLDL = TRIG/5; and LDL = Tchol - (VLDL + HDL) [39].

7.2.3.3 BODY WEIGHT

Body weight was measured for all study participants at the beginning and end of the sustained phase. Body weight was measured to the nearest 0.5 pound using a balance-type medical scale. Participants were measured in the morning, unclothed with the exception of undergarments.

7.2.3.4 BLOOD PRESSURE

Blood pressure was determined with the use of the Datascope Accutorr Plus automatic digital blood pressure device (Datascope Corp, Mahwah,

NJ) with the participant supine after a 5-min period of rest. Both systolic and diastolic pressures were calculated as the mean value of 2 readings 5 minutes apart. All measurements were obtained by one investigator.

7.2.4 STATISTICAL ANALYSIS

Statistical analysis was conducted using SAS software (Version 9.1, SAS Institute, Cary, NC). A two-tailed p-value of ≤ 0.05 was considered statistically significant. Two-way repeated measures ANOVA, with treatment and time as the main effects, were performed to compare treatment-specific outcome measures responses, accounting for time differences. Within-treatment effects for outcome measures were assessed using paired t-tests. The combined effect of independent variables (age, blood pressure, LDL, BMI and treatment sequence) and treatment assignment on all outcome measures was assessed with generalized linear modeling. All analyses of endpoints were based on the intention-to-treat principle.

Sample size was predicated on 80% power to detect a minimal difference of 3.5% change in FMD between the egg and egg substitute treatments at six weeks. A two-tailed alpha level of 0.05 was set with an allowance for 10% attrition and noncompliance.

7.3 RESULTS

Forty hyperlipidemic participants participated in this study. Sixty percent of the participants were female. Participants ranged in age from 35 to 77 years, with a mean age of 60 years (Table 2). Four participants dropped out of the study after the acute phase. One participant dropped out because the participant was unwilling to consume eggs or egg substitute daily for six weeks, another dropped out because of relocation to another state, and two dropped out because they started using lipid lowering medication (statin).

TABLE 2: Demographic and Baseline Characteristics

Variable	Values	Range
Gender		
Female	24 (60%)	
Male	16 (40%)	
Race		
White	39 (97.5%)	
African American	1 (2.5%)	
Age (years)	59.9 ± 9.6	35 to 77
Brachial Artery Diameter (mm)	4.0 ± 0.8	2.8 to 5.4
Systolic Blood Pressure (mmHg)	131.4 ± 15.7	94 to 173
Diastolic Blood Pressure (mmHg)	73.0 ± 12.6	45 to 105
Framingham 10-years Risk (%)	6.6 ± 5.8	1 to 30
Body Mass Index (kg/m2)	28.7 ± 4.7	20.4 to 38.4
Weight (kg)	76.3 ± 21.8	45 to 105

Values are mean ± SD except otherwise stated

7.3.1 ACUTE PHASE

After a single dose of eggs, endothelial function did not change from baseline as compared to sausage and cheese ($p = 0.99$). Accounting for the strength of the stimulus that determines vasodilatations (SARM), our findings on endothelial function persisted (Table 3.)

7.3.2 SUSTAINED PHASE

Daily consumption of egg substitute for six weeks improved endothelial function relative to egg consumption ($p < 0.01$). These findings persisted controlling for the variation of the strength of the stimulus that causes the vasodilatation (Table 4.)

TABLE 3: Acute Phase: Mean change in Outcome Measures after Treatment Assignment

Variable	Egg (n = 40)	Sausage & Cheese (n = 40)	p-value
Baseline Brachial Artery Diameter (mm)	4.0 ± 0.8	4.0 ± 0.7	0.79*
Flow Mediated Dilatation (%)			
Baseline	5.9 ± 4.6	5.2 ± 3.6	0.45*
Post-prandial	6.3 ± 5.3	5.6 ± 4.5	
Change	0.4 ± 1.9 (P = 0.22)	0.4 ± 2.4 (P = 0.34)	0.99
Adjusted change†	0.3 ± 2.1 (P = 0.46)	0.4 ± 2.1 (P = 0.31)	0.84
Stimulus adjusted response measure			
Baseline	0.10 ± 0.12	0.06 ± 0.06	0.05*
Post-prandial	0.08 ± 0.09	0.07 ± 0.06	
Change	-0.02 ± 0.11 (P = 0.31)	0.00 ± 0.08 (P = 0.89)	0.35

*Values are mean ± SD; p-value obtained from ANOVA for repeated measurements except otherwise stated; Change = Post-prandial - Baseline; *p-value obtain from Student's t-test; p-values in parenthesis indicate within-group p-values; † obtained from generalized linear models, controlling for age, blood pressure, LDL and BMI*

Daily consumption of egg substitute for six weeks significantly lowered total cholesterol as compared to egg consumption ($p < 0.01$) and also lowered LDL as compared to egg consumption ($p = 0.01$). However, daily consumption of egg substitute for six weeks did not significantly lowered total cholesterol to HDL ratio as compared to egg consumption ($p = 0.38$).

Daily consumption of egg or egg substitute for six weeks did not show significant increase in BMI as compared to egg consumption ($p = 0.56$)

7.4 DISCUSSION

Our findings in this study expand on existing evidence that short-term egg consumption does not adversely affect endothelial function, in a population not previously examined: hyperlipidemic adults. Moreover, we observed that consuming eggs daily did not unfavorably influence serum cholesterol or other measures of the lipid profile. While the subjects demonstrated impaired endothelial function at baseline (i.e. relative to healthy

TABLE 4: Sustained Phase: Mean Change in Outcome Measures after Six Weeks of Treatment

Variable	Egg (n = 36)	Egg substitute (n = 36)	p-value
Endothelial Function			
Flow Mediated Dilatation (%)			
Baseline	5.6 ± 3.9	5.8 ± 3.9	0.78
6 Weeks	5.3 ± 4.1	6.9 ± 4.0	
Change	-0.1 ± 1.5 (P = 0.80)	1.0 ± 1.2 (P < 0.01)	<0.01
Adjusted change†	-0.2 ± 1.3 (P = 0.35)	0.9 ± 1.4 (P < 0.01)	< 0.01
Stimulus adjusted response measure			
Baseline	0.08 ± 0.10	0.06 ± 0.06	0.39
6 Weeks	0.08 ± 0.11	0.09 ± 0.09	
Change	0.01 ± 0.05 (P = 0.54)	0.03 ± 0.06 (P < 0.01)	0.07
Lipid Panel			
Total Cholesterol (mg/dL)			
Baseline	244 ± 24	244 ± 24	1.00
6 Weeks	239 ± 27	227 ± 27	
Change	-5 ± 21 (P = 0.10)	-18 ± 18 (P < 0.01)	< 0.01
Low Density Lipoprotein (mg/dL)			
Baseline	168 ± 17	168 ± 17	
6 Weeks	165 ± 24	154 ± 24	
Change	-2 ± 19 (P = 0.30)	-14 ± 20 (P < 0.01)	0.01
High Density Lipoprotein (mg/dL)			
Baseline	52 ± 15	52 ± 15	1.00
6 Weeks	51 ± 14	50 ± 13	
Change	-1 ± 11 (P = 0.53)	-2 ± 10 (P = 0.15)	0.63
Triglycerides (mg/dL)			
Baseline	132 ± 52	132 ± 52	1.00
6 Weeks	118 ± 47	116 ± 50	
Change	-14 ± 37 (P = 0.02)	-18 ± 43 (P = 0.03)	0.83
Total Cholesterol to High Density Lipoprotein Ratio			
Baseline	5.0 ± 1.3	5.0 ± 1.3	1.00
6 Weeks	5.0 ± 1.2	4.8 ± 1.3	
Change	-0.06 ± 0.66 (P = 0.54)	-0.21 ± 0.82 (P = 0.11)	0.38
Body composition			

TABLE 4: *Cont.*

Variable	Egg (n = 36)	Egg substitute (n = 36)	p-value
Weight (kg)			
Baseline	81 ± 19	81 ± 19	1.00
6 Weeks	82 ± 18	82 ± 18	
Change	0.4 ± 2.3 (P = 0.33)	0.7 ± 2.4 (P = 0.08)	0.52
Body Mass Index (kg/m2)			
Baseline	29.2 ± 4.5	29.2 ± 4.5	1.00
6 Weeks	29.3 ± 4.3	29.5 ± 4.5	
Change	0.2 ± 0.8 (P = 0.13)	0.4 ± 0.9 (P = 0.04)	0.56

Values are mean ± SD; p-value obtained from ANOVA for repeated measurements except otherwise stated; p-values in parenthesis indicate within-group p-values; *p-value obtain from student ttest; Change = 6 Weeks - Baseline; † obtained from generalized linear models, controlling for age, blood pressure, LDL and BMI

endothelial function), the acute induction of endothelial dysfunction by the test meal high in saturated fat was not observed. Egg substitute, which is made from 99% egg whites, is lower in calories relative to whole eggs, lacks cholesterol and fat, and is fortified with vitamins, also lowered cholesterol and triglyceride levels. In addition, egg substitute led to a decrease in LDL and significantly improved endothelial function, as compared to sustained egg consumption. To the best of our knowledge, this is the first study to provide evidence that egg or egg substitute consumption does not adversely affect endothelial function in hyperlipidemic adults.

The acute phase findings were consistent with those of the sustained phase. Single, acute doses of egg did not adversely affect endothelial function. The sausage and cheese breakfast sandwich, designed to demonstrate acute dysfunction in the endothelium, surprisingly also did not adversely affect endothelial function. This finding is consistent with the prior study of egg ingestion in healthy adults conducted at our lab[31], but is at odds with the reported literature. This may be explained by differences in gastric transit times for different types of food [31]. Future studies should evaluate postprandial brachial artery dilation at different time points to assess the effect of time. Several studies have examined FMD serially over time following meal ingestion[40,41].

While rich in cholesterol, eggs are also nutritious. Data from NHANES III reveal that egg consumption is an important nutritional contribution to the average American diet, providing a relatively inexpensive source of amino acids and essential fatty acids[9]. Eggs provide arginine, a precursor to nitric oxide, which in turn plays a central role in endothelial function[42]. Endothelial function is an arterial vasomotor response mediated predominantly by the release of nitric oxide (vasodilating), and endothelin (vasoconstricting) from the vascular endothelium[26]. This system plays an important role in the pathogenesis of atherosclerosis, cardiovascular disease, and other chronic diseases[19,43].

The relative importance of dietary cholesterol to cardiovascular risk, and the association between dietary and serum cholesterol, are both subject to ongoing debate[44-46]. The association between dietary cholesterol and coronary events and mortality is generally positive but rather weak, and derived largely from ecological and prospective cohort studies with variable follow-up[47-50]. In most of these studies, it is difficult to determine the effects of cholesterol independent of dietary fat. One large prospective cohort followed U.S. male physicians for over 20 years. The study monitored egg consumption and documented new cases of heart failure during follow-up. Results failed to find a correlation between occasional egg consumption and heart failure, although an increased risk of heart failure was related to participants who reported consuming more than one egg per day[49]. Another study found no impact of egg intake on cardiovascular risk, specifically stroke, ischemic stroke, and coronary artery disease. However, Nakamura et al demonstrated significant correlation with serum cholesterol concentrations in women consuming more than two eggs per day [51]. In a recent study by Djoussé et al, egg consumption was associated with increased risk of diabetes [52]. Djoussé et al also linked egg consumption to increased mortality and even more so in diabetics [53]. Overall, scientific studies of the relationship between egg consumption, cardiovascular disease and mortality[30,54], have thus been somewhat inconsistent.

Little, if any, epidemiological evidence exists supporting a direct link between egg consumption and cardiovascular disease or mortality risk. Previous studies have shown weak positive associations between intake of dietary cholesterol and serum cholesterol, while others failed to find any association[30]. Hu and colleagues analyzed data from two large co-horts, the Health Professional Follow-Up Study and the Nurses' Health Study, to assess the effect of egg consumption on cardiovascular events and deaths[47]. After a mean of 8 years of follow-up, no overall significant association was observed between egg consumption and risk of CHD in both males and females. Hu et al [47] also reported that the relative risk of CHD was the same whether the participants consumed less than, or more than, one egg per week.

While our study provides valuable data regarding egg ingestion in hy-perlipidemic adults, it is not without limitations. The study's small sample size almost entirely derived from one community in Southwestern Con-necticut may limit the generalization these of results. Furthermore, the duration of egg consumption during this study limits the ability to pre-dict the long-term effects. Variables potentially confounding the correla-tion between nutrient intakes and endothelial responses included physical activity, vasoactive medication use, and genetic factors. Unmeasured or inaccurately measured dietary intake data could also have confounded re-sults. Three-day food diaries, used to track dietary intake, did not indicate any significant unintended changes in overall dietary pattern, although changes in diet or behavior that were not captured may have impacted findings. Adjustment for potential confounders was managed through ap-plication of strict eligibility criteria, randomization, and crossover design. Also of note, the intended provocation of endothelial dysfunction by a single meal high in saturated fat failed to demonstrate a deleterious effect. Finally, endothelial function was measured only one time after treatment assignments and not monitored for a prolonged time.

7.5 CONCLUSIONS

In light of the persistent uncertainties and lack of observational evidence regarding the effects of egg consumption on serum cholesterol and cardiac risk, the application of this methodology and technology in further studies is appropriate, and very much needed. Short of a randomized controlled trial of egg consumption and cardiovascular events, endothelial function testing offers one of the best available means to evaluate the role of egg ingestion on cardiac risk. To date, the evidence generally mitigates against an association between moderate egg consumption and increased cardiac risk. Further testing in at-risk samples, including individuals with established coronary disease, is now justified to clarify the place of eggs in a judicious and heart-healthy diet.

REFERENCES

1. Report of Inter-Society Commission for Heart Disease Resources: Prevention of Cardiovascular Disease. Primary prevention of the atherosclerotic diseases. Circulation 1970, 42(6):A55-95.
2. American Heart Association: Diet and Coronary Heart Disease. Dallas, TX: American Heart Association; 1973.
3. Krauss RM, Eckel RH, Howard B, Appel LJ, Daniels SR, Deckelbaum RJ, Erdman JWJ, Kris-Ehterton P, Goldberg IJ, Kotchen TA, Lichtenstein AH, Mitch WE, Mullis R, Robinson K, Wylie-Rosett J, St Jeor S, Suttie J, Tribble DL, Bazzarre TL: AHA Dietary Guidelines: Revision 2000: A statement for healthcare professionals from the Nutrition Committee of the American Heart Association. Circulation 2000, 102(18):2284-2299.
4. Harman NL, Leeds AR, Griffin BA: Increased dietary cholesterol does not increase plasma low density lipoprotein when accompanied by an energy-restricted diet and weight loss. Eur J Nutr 2008, 47(6):287-293.
5. Biostatistical fact sheet -- risk factors; high blood cholesterol and other lipids [http://www.americanheart.org/downloadable/heart/1017696825613chollip.pdf] webcite http://www.americanheart.org/downloadable/heart/1017696825613chollip.pdf webcite. Accessed December 27, 2002
6. American Heart Association: An eating plan for healthy adults. The new 2000 food guidelines. Our American Heart Association diet. Dallas, TX 2000.
7. USDA: Nutrient Database for Standard Reference. [http://www.nal.usda.gov/fnic/foodcomp/search/] 2009.
8. Herron KL, Fernandez ML: Are the current dietary guidelines regarding egg consumption appropriate? J Nutr 2004, 134(1):187-190.
9. Song WO, Kerver JM: Nutritional contribution of eggs to American diets. J Am Coll Nutr 2000, 19(5 Suppl):556S-562S.
10. Bowes , Church : Bowes and Church's Food Values of Proteins Commonly Used. 16th edition. J.B. Lippincott Company; 1994.
11. Mente A, de Koning L, Shannon HS, Anand SS: A systematic review of the evidence supporting a causal link between dietary factors and coronary heart disease. Arch Intern Med 2009, 169(7):659-669.
12. Siri-Tarino PW, Sun Q, Hu FB, Krauss RM: Meta-analysis of prospective cohort studies evaluating the association of saturated fat with cardiovascular disease. Am J Clin Nutr 91(3):535-546.

13. Vorster HH, Beynen AC, Berger GM, Venter CS: Dietary cholesterol--the role of eggs in the prudent diet. S Afr Med J 1995, 85(4):253-256.
14. Luscher TF, Barton M: Biology of the endothelium. Clin Cardiol 1997, 20(11 Suppl 2):II-3-10.
15. Vogel RA: Measurement of endothelial function by brachial artery flow-mediated vasodilation. Am J Cardiol 2001, 88(suppl):31E-34E.
16. Corretti MC, Anderson TJ, Benjamin EJ, Celermajer D, Charbonneau F, Creager MA, Deanfield J, Drexler H, Gerhard-Herman M, Herrington D, Vallance P, Vita J, Vogel R: Guidelines for the ultrasound assessment of endothelial-dependent flow-mediated vasodilation of the brachial artery. A report of the International Brachial Artery Reactivity Task Force. J Am Coll Cardiol 2002, 16(2):257-265.
17. Nash DT: Insulin resistance, ADMA levels, and cardiovascular disease. JAMA 2002, 287(11):1451-1452.
18. McVeigh GE, Brennan GM, Johnston GD, McDermott BJ, McGrath LT, Henry WR, Andrews JW, Hayes JR: Impaired endothelium-dependent and independent vasodilation in patients with type 2 (non-insulin-dependent) diabetes mellitus. Diabetologia 1992, 35(8):771-776.
19. Neunteufl T, Katzenschlager R, Hassan A, Klaar U, Schwarzacher S, Glogar D, et al.: Systemic endothelial dysfunction is related to the extent and severity of coronary artery disease. Atherosclerosis 1997, 129:111-118.
20. O'Driscoll G, Green D, Taylor RR: Simvastatin, an HMG-coenzyme A reductase inhibitor, improves endothelial function within 1 month. Circulation 1997, 95(5):1126-1131.
21. O'Driscoll G, Green D, Rankin J, Stanton K, Taylor R: Improvement in endothelial function by angiotensin converting enzyme inhibition in insulin-dependent diabetes mellitus. J Clin Invest 1997, 100(3):678-684.
22. Treasure CB, Klein JL, Weintraub WS, Talley JD, Stillabower ME, Kosinski AS, Zhang J, Boccuzzi SJ, Cedarholm JC, Alexander RW: Beneficial effects of cholesterol-lowering therapy on the coronary endothelium in patients with coronary artery disease. N Engl J Med 1995, 332(8):481-487.
23. Anderson TJ, Gerhard MD, Meredith IT, Charbonneau F, Delagrange D, Creager MA, Selwyn AP, Ganz P: Systemic nature of endothelial dysfunction in atherosclerosis. Am J Cardiol 1995, 75:71B-74B.
24. Neunteufl T, Heher S, Katzenschlager R, Wolfl G, Kostner K, Maurer G, Weidinger F: Late prognostic value of flow-mediated dilation in the brachial artery of patients with chest pain. Am J Cardiol 2000, 86(2):207-210.
25. Suwaidi JA, Hamasaki S, Higano ST, Nishimura RA, Holmes DR Jr, Lerman A: Long-term follow-up of patients with mild coronary artery disease and endothelial dysfunction. Circulation 2000, 101(9):948-954.
26. Hutcheson IR, Griffith TM: Central role of intracellular calcium stores in acute flow- and agonist-evoked endothelial nitric oxide release. Br J Pharmacol 1997, 122(1):117-125.
27. Celermajer D, Sorensen KE, Bull C, Robinson J, Deanfield JE: Endothelium-dependent dilation in the systemic arteries of asymptomatic subjects relates to coronary risk factors and their interaction. J Am Coll Cardiol 1994, 24(6):1468-1474.

28. Kritchevsky SB: A review of scientific research and recommendations regarding eggs. J Am Coll Nutr 2004, 23(6 Suppl):596S-600S.
29. McNamara DJ: Eggs and heart disease risk: perpetuating the misperception. Am J Clin Nutr 2002, 75(2):333-335.
30. McNamara DJ: The impact of egg limitations on coronary heart disease risk: do the numbers add up? J Am Coll Nutr 2000, 19(5 Suppl):540S-548S.
31. Katz DL, Evans M, Nawaz H, Njike V, Chan W, Comerford B, Hoxley M: Egg consumption & endothelial function in healthy adults: A randomized controlled crossover trial. International Journal of Cardiology 2005, 99(1):65-70.
32. Corretti MC, Plotnick GD, Vogel RA: Technical aspects of evaluating brachial artery vasodilatation using high-frequency ultrasound. Am J Physiol 1995, 268(4 Pt 2):H1397-1404.
33. Celermajer DS, Sorensen KE, Gooch VM, Spiegelhalter DJ, Miller OI, Sullivan ID, Lloyd JK, Deanfield JE: Non-invasive detection of endothelial dysfunction in children and adults at risk of atherosclerosis. Lancet 1992, 340(8828):1111-1115.
34. Plotnick GD, Corretti MC, Vogel RA: Effect of antioxidant vitamins on the transient impairment of endothelium-dependent brachial artery vasoactivity following a single high-fat meal. JAMA 1997, 278(20):1682-1686.
35. Mannion TC, Vita JA, Keaney JF Jr, Benjamin EJ, Hunter L, Polak JF: Non-invasive assessment of brachial artery endothelial vasomotor function: The effect of cuff position on level of discomfort and vasomotor responses. Vasc Med 1998, 3(4):263-267.
36. Sorensen KE, Celermajer DS, Spiegelhalter DJ, Georgakopoulos D, Robinson J, Thomas O, Deanfield JE: Non-invasive measurement of human endothelium dependent arterial responses: Accuracy and reproducibility. Br Heart J 1995, 74:247-253.
37. Faridi Z, Njike V, Dutta S, Ali A, Katz DL: Acute dark chocolate and cocoa ingestion and endothelial function: A randomized, placebo controlled, cross-over trial. Am J Clin Nutr 2008, 88(1):58-63.
38. Pyke KE, Tschakovsky ME: Peak vs. total reactive hyperemia: which determines the magnitude of flow-mediated dilation? J Appl Physiol 2007, 102(4):1510-1519.
39. Friedewald WT, Levy RI, Fredrickson DS: Estimation of the concentration of low-density lipoprotein cholesterol in plasma, without use of the preparative ultracentrifuge. Clin Chem 1972, 18(6):499-502.
40. Volek JS, Ballard KD, Silvestre R, Judelson DA, Quann EE, Forsythe CE, Fernandez ML, Kraemer WJ: Effects of dietary carbohydrate restriction versus low-fat diet on flow-mediated dilation. Metabolism 2009, 58(12):1769-1777.
41. Tyldum GA, Schjerve IE, Tjonna AE, Kirkeby-Garstad I, Stolen TO, Richardson RS, Wisloff U: Endothelial dysfunction induced by post-prandial lipemia: complete protection afforded by high-intensity aerobic interval exercise. J Am Coll Cardiol 2009, 53(2):200-206.
42. Preli RB, Klein KP, Herrington DM: Vascular effects of dietary L-arginine supplementation. Atherosclerosis 2002, 162(1):1-15.
43. Anderson TJ, Uehata A, Gerhard MD, Meredith IT, Knab S, Delagrange D, Lieberman EH, Ganz P, Creager MA, Yeung AC: Cardiovascular Division, Brigham and Women's Hospital, Harvard Medical School, Boston, Massachusetts, USA: Close

relation of endothelial function in the human coronary and peripheral circulation. J Am Coll Cardiol 1995, 26(5):1235-1241.

44. McNamara D: Cholesterol intake and plasma cholesterol: an update. J Am Coll Nutr 1997, 16(6):530-534.

45. Beynen AC, Katan MB: Effect of egg yolk feeding on the concentration and composition of serum lipoproteins in man. Atherosclerosis 1985, 54(2):157-166.

46. Caggiula AW, Mustad VA: Effects of dietary fat and fatty acids on coronary artery disease risk and total and lipoprotein cholesterol concentrations: epidemiologic studies. Am J Clin Nutr 1997, 65(5 Suppl):1597S-1610S.

47. Hu F, Stampfer M, Rimm E, Manson J, Ascherio A, Colditz G, et al.: A prospective study of egg consumption and risk of cardiovascular disease in men and women. JAMA 1999, 281:1387-1394.

48. Gramenzi A, Gentile A, Fasoli M, Negri E, Parazzini F, La Vecchia C: Association between certain foods and risk of acute myocardial infarction in women. Bmj 1990, 300(6727):771-773.

49. Djousse L, Gaziano JM: Egg consumption and risk of heart failure in the Physicians' Health Study. Circulation 2008, 117(4):512-516. Epub 2008 Jan 2014

50. Qureshi AI, Suri FK, Ahmed S, Nasar A, Divani AA, Kirmani JF: Regular egg consumption does not increase the risk of stroke and cardiovascular diseases. Med Sci Monit 2007, 13(1):CR1-8. Epub 2006 Dec 2018

51. Nakamura Y, Okamura T, Tamaki S, Kadowaki T, Hayakawa T, Kita Y, Okayama A, Ueshima H: NIPPON DATA80 Research Group: Egg consumption, serum cholesterol, and cause-specific and all-cause mortality: the National Integrated Project for Prospective Observation of Non-communicable Disease and Its Trends in the Aged, 1980 (NIPPON DATA80). Am J Clin Nutr 2004, 80:58-63.

52. Djousse L, Gaziano JM, Buring JE, Lee IM: Egg consumption and risk of type 2 diabetes in men and women. Diabetes Care 2009, 32(2):295-300.

53. Djousse L, Gaziano JM: Egg consumption in relation to cardiovascular disease and mortality: the Physicians' Health Study. Am J Clin Nutr 2008, 87(4):964-969.

54. Knopp RH, Retzlaff BM, Walden CE, Dowdy AA, Tsunehara CH, Austin MA, Nguyen T: A double-blind, randomized, controlled trial of the effects of two eggs per day in moderately hypercholesterolemic and combined hyperlipidemic subjects taught the NCEP step I diet. J Am Coll Nutr 1997, 16(6):551-561.

This chapter was originally published under the Creative Commons Attribution License. Nijke, V., Faridi, Z., Dutta, S., Gonzalez-Simon, A. L., and Katz, D. L. Daily Egg Consumption in Hyperlipidemic Adults: Effects on Endothelial Function and Cardiovascular Risk. Nutrition Journal 2010, 9:28. doi:10.1186/1475-2891-9-28. http://www.nutritionj.com/content/9/1/28

CHAPTER 8

PROTEIN-ENRICHED MEAL REPLACEMENTS DO NOT ADVERSELY AFFECT LIVER, KIDNEY OR BONE DENSITY: AN OUTPATIENT RANDOMIZED CONTROLLED TRIAL

ZHAOPING LI, LEO TREYZON, STEVE CHEN, ERIC YAN, GAIL THAMES, and CATHERINE L. CARPENTER

8.1 BACKGROUND

Obesity and overweight have reached epidemic proportions in the U.S. and increasingly around the world [1,2]. A number of studies have suggested that protein is the most satiating macronutrient and promotes the retention of lean body mass. Meals with increased protein to carbohydrate ratios have been demonstrated to increase satiety and decrease food intake [3,4] by comparison to standard protein intake. Increased protein intake results in both improved weight loss and improved maintenance of weight loss [5,6]. Therefore, protein-enriched or supplemented meal replacements have found their way into weight management practice.

There has been some concern that the long-term use of high protein diets may damage liver function, renal function, or reduce bone density [7,8]. While there are studies of the effects of increased intake of animal protein in the diet, protein-enriched meal replacements have not been evaluated in comparison to standard meal replacements in terms of effects on liver function, renal function, and bone mineral density in free-living populations.

Meal replacement (MR) is an important strategy in designing structured diets for weight management due to their simplicity, low cost, and convenience of protein-enriched meal replacement shakes by comparison to fast food meals [5,9,10]. Noakes et al [11] have shown that meal replacements are as effective as structured weight-loss diets. MR simplifies the weight loss plan by replacing one or two meals a day with a product of defined nutrient and calorie content. MR leads to increased weight losses over twelve weeks compared to simply restricting the intake of favorite food, and weight losses have been shown to be maintained for up to 4 years with the inclusion of one MR per day [12]. The present study was designed to recommend isocaloric weight management programs through the inclusion of either a protein or a carbohydrate supplement to a standard meal replacement powder to make either a standard or protein-enriched meal replacement.

8.2 METHODS

The study protocol was approved by the University of California Los Angeles Institutional Review Board. Healthy volunteers were recruited by public advertisement. Subjects over 30 years of age with a body mass index (BMI) between 27 to 40 kg/m^2, and in good health by history, physical examination, and basic laboratory screening (complete blood count, serum chemistries, liver panel, and lipid panel) were selected for the study. Subjects with type 2 diabetes or glucose intolerance were excluded as were individuals, who regularly drank more than one alcoholic beverage daily.

One hundred men and women who met the selection criteria were randomly assigned to either the HP (high protein) or SP (standard protein) treatment. This was a single-blinded study. Subjects were randomized in a 1:1 manner to either HP or SP diet using a computerized random proportion model.

Caloric intake to achieve weight loss was based on a 500 Kcal deficit of the participants' estimated resting metabolic rate as determined by body composition analysis by DEXA. Diet plans were individualized per subject by the research dietitian. Subjects were instructed to add to their meal replacements a set number of scoops of either protein or carbohydrate from powder canisters labeled as either A or B. The protein powder was measured with a calibrated scoop and subjects were instructed regarding how many scoops to use for their particular meal plan. Participants in the HP group received a diet plan that provided 2.2 grams of protein per kg of

LBM while the diet for the SP group provided 1.1 grams of protein per kg of LBM. The meal energy macronutrient composition in the HP group was approximately 30% protein, 30% fat, and 40% carbohydrate. The macronutrient composition in the SP diet was approximately 15% protein, 30% fat, and 55% carbohydrate. Both groups received the same isocaloric MR (Formula 1, Herbalife Intl., Los Angeles) with either a protein supplement for the HP group (Performance Protein Powder, Herbalife Intl., Los Angeles) or a carefully matched carbohydrate placebo containing maltodextrin and flavoring for SP group.

Instructions were provided for preparation of the MR. Subjects were advised to consume one MR in place of a meal and the other as a snack daily for 12 weeks, then one MR a day for an additional 40 weeks. All participants met individually with a registered dietitian at baseline for dietary instruction, and at week 2, month 1, 2, 3, 6, 9 and 12 to provide counseling and follow-up. Qualitative food logs including the servings of macronutrients and meal replacements were collected and reviewed with subjects at each visit. Participants were weighed and protein powder meal replacement products were dispensed at each visit. Subjects were given general advice for increasing their activity level with a goal of 30 minutes of aerobic exercise per day.

8.2.1 BODY WEIGHT

Subjects were weighed at each visit (Detecto-Medic; Deteco-Scales; Brooklyn, NY) while wearing no shoes after an overnight fast. Height was measured with a stadiometer (Detecto-Medic; Deteco-Scales; Brooklyn, NY) at week 0. BMI was calculated as weight (kg)/height squared (m).

8.2.2 BIOCHEMISTRY

Blood samples after >10 hours of overnight fasting were collected at months 0, 3, 6, and 12 for measurement of lipid profiles, electrolytes, liver and renal function tests. Twenty-four hour urine samples were collected at baseline and week 52 for urinary urea nitrogen, creatinine, calcium, phosphate excretions.

Plasma cholesterol was determined using standard enzymatic methods. Reagents, standards and calibrators were purchased from Pointe Scientific (Lincoln Park, MI). The HDL or alpha cholesterol is derived from the measurement on the supernatant following the precipitation of apo B containing lipoproteins with Heparin and $MnCl2$. The so-called LDL or beta lipoprotein cholesterol is estimated from these data using the Friedewald equation. All other tests were completed at Ronald Reagan Medical Center clinical laboratory using standard methods. Urinary urea nitrogen was measured with an enzymatic method of Talke and Schubert [13].

8.2.3 BONE DENSITY

Bone density was measured at baseline and 12 months by Dual Energy X-ray Absorptiometry by a Lunar Prodigy DEXA (GE Medical Systems, Waukesha, Wisconsin).

8.2.4 STATISTICAL ANALYSIS

All variable transformations and statistical analyses were performed using SAS version 9.2 [14]. We evaluated effectiveness of the subject random allocation by comparing patient characteristics and baseline measurements of the two study groups using t-tests (for continuous variables) and Chi-square tests (for categorical variables).

We computed t-tests within each treatment group using matched pair Analysis of Variance (ANOVA). Univariate and multivariate Repeated Measures ANOVA described within subject effects of changes over time for the total study sample; between treatment group effects; and changes over time by treatment group interactions. Because outcome data was not available for participants who dropped out of the study, we did not conduct intention to treat analysis. All data are presented as means ± standard deviation of the mean (SD).

8.3 RESULTS

100 obese men and women were randomly assigned to either a HP or SP MR diet plan. Fifteen subjects withdrew from the study within the first week after randomization due to inability to comply with the meal plan (6 in the HP group and 9 in the SP group) and those subjects were excluded from data analysis. Fifteen more subjects (9 in the HP group and 6 in the SP group) dropped out of the study during the 12 month trial due to loss of follow-up and personal reasons. No subject suffered any severe adverse event. Seventy subjects, (thirty-five subjects in each group) completed the 12-month study. Subject characteristics in the two treatment arms at baseline were not significantly different (Table 1). Mean age was 49.4 ± 11.0 years. Mean BMI at baseline was 34.43 ± 6.36 for HP group and 32.57 ± 4.10 kg/m^2 for SP group.

TABLE 1: Baseline characteristics of study participants

Characteristic		HP (N = 44)	SP (N = 41)
Demographic			
Women, No. (%)		36 (81.8)	26 (63.4)
Age, mean (SD)		48.9 11.8)	49.7 (9.1)
Race, No. (%)			
	Asian	4 (9.1)	1 (2.4)
	Black	9 (20.5)	8 (19.5)
	Caucasian	26 (59.1)	28 (68.3)
	Hispanic	4 (9.1)	2 (4.9)
	Others	1 (2.2)	2 (4.9)
Weight Factors, mean (SD)			
Body Weight, kg		93.5 (14.0)	92.7 (15.9)
BMI, kg/m^2		34.7 (6.8)	34.3(10.3)

All values are mean ± SD. Subjects who dropped out of the study after randomization were excluded. There were no significant differences between groups. HP: high protein group; SP: standard protein group.

8.3.1 WEIGHT LOSS

Subjects were weighed at baseline, 2 weeks, and monthly thereafter. Baseline body weight was not significantly different between these two groups. Both groups lost significant amounts of weight at 12 months (4.29 ± 5.90 kg; SP -4.66 ± 6.91 kg, $p < 0.01$). (Figure 1) After controlling for baseline weight, gender, and time period, there was no significant difference between the two treatment groups. For both dietary groups, BMI was significantly lower at 12 months (HP = -1.53 ± 2.17; SP = -1.77 ± 2.89 kg/m2). There were no significant differences in BMI changes between the two dietary groups.

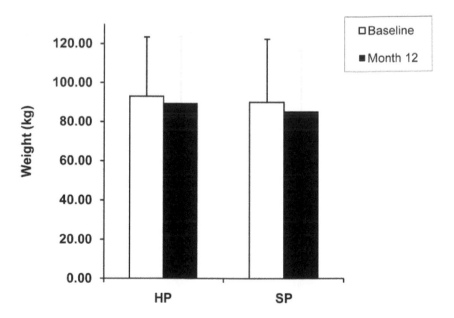

FIGURE 1: Body weight change in 12 months. Mean ± SD. Subjects in HP and SP group both lost significant amount weight in 12 months compared with baseline. Open square: baseline, black square: month 12, *P < 0.01 compared with baseline

8.3.2 CHOLESTEROL, HDL, LDL, TRIACYLGLYCEROL

There were significant reductions in total cholesterol for the HP group at 3 months (-15.20 ± 35.84 mg/dL, $p < 0.05$) and 6 months (-10.47 ± 30.46 mg/dL, $p < 0.05$) but not for the SP group (-4.98 ± 25.14; -9.31 ± 30.26 mg/dL, $p > 0.05$). The LDL concentration was significantly lowered at 3 months and 6 months (-7.74 ± 21.92; -7.83 ± 23.06 mg/dL, $p < 0.05$) for the HP group but not for SP group. There was significant elevation of HDL at month 6 for HP group only (2.53 ± 7.45 mg/dL, $p < 0.05$). The triacylglycerol concentration was reduced significantly only for the HP group at 3 months (-29.73 ± 58.22 mg/dL, $p < 0.05$). The difference between the two groups was not significant for any of the parameters. (Table 2)

TABLE 2: Blood lipid concentrations

	Cholesterol (mg/dL)		Triacylglycerol (mg/dL)		LDL (mg/dL)		HDL (mg/dL)	
	HP	SP	HP	SP	HP	SP	HP	SP
Baseline	198.85 ± 311.00	203.04 ± 39.08	136.07 ± 105.61	115.16 ± 55.57	116.88 ± 35.92	128.07 ± 36.28	54.34 ± 13.82	52.00 ± 10.56
Month 3	184.93 ± 272.00*	196.68 ± 36.36	106.24 ± 52.66*	113.59 ± 63.17	108.88 ± 36.54*	121.29 ± 36.36	54.83 ± 15.11	52.73 ± 11.19
Month 6	190.71 ± 279.00*	195.05 ± 38.38*	116.74 ± 74.19	106.64 ± 52.44	111.73 ± 26.79 *	119.74 ± 34.86 *	56.47 ± 15.47 *	53.95 ± 11.77
Month 12	188.33 ± 292.00	201.24 ± 37.60	119.33 ± 53.41	109.42 ± 63.78	132.07 ± 34.09	125.33 ± 32.78	54.63 ± 13.48	54.00 ± 11.48

All values are mean ± SD. There were no significant differences between groups. HP, high protein group; SP, standard protein group. * $p < 0.05$ compare with baseline

8.3.3 LIVER FUNCTION

All subjects had normal ranges of AST, ALT, bilirubin, and alkaline phosphatase at baseline. All those markers remained in the normal range and did not change significantly through the study (Table 3). No subject had any liver markers out of the normal range during any time of the study.

TABLE 3: Liver function tests

	ALT (U/L)		AST (U/L)		Alkaline Phosphatase (U/L)		Total Billirubin (mg/dL)	
	HP	SP	HP	SP	HP	SP	HP	SP
Baseline	27.07 ± 13.97	28.84 ± 15.00	25.2 ± 11.44	23.33 ± 13.00	66.2 ± 37.00	71.88 ± 41.00	0.75 ± 0.26	0.77 ± 0.40
Month 3	25.4 ± 7.51	28.53 ± 15.00	24.23 ± 7.30	24.39 ± 16.00	69.64 ± 39.00	72.85 ± 18.58	0.75 ± 0.20	0.82 ± 0.20
Month 6	24.87 ± 11.22	27.23 ± 16.00	24.26 ± 14.00	23.44 ± 5.17	69.89 ± 19.13	72.21 ± 17.78	0.80 ± 0.25	0.83 ± 0.30
Month 12	24.1 ± 12.00	26.91 ± 14.23	23.00 ± 13.00	23.76 ± 14.00	71.48 ± 40.00	69.18 ± 18.65	0.76 ± 0.36	0.84 ± 0.40

All values are mean ± SD. There were no significant differences between groups. HP, high protein group; SP, standard protein group

8.3.4 RENAL FUNCTION

No significant differences were found when comparing 12 month mean concentrations of serum creatinine, urea nitrogen and urine nitrogen and creatinine clearance within the groups and between the groups (Table 4). Urinary protein excretion significantly increased in the SP group but not in the HP group at month 12 (HP: 27.18 ± 105.33, mg/24 hours, p = 0.410; SP: 54.82 ± 83.35 mg/24 hours, p = 0.02). There was not any difference between the groups.

TABLE 4: Renal function, calcium, phosphate excretion and bone mineral density.

	HP		SP	
	Baseline	Month 12	Baseline	Month 12
Serum Creatinine (mg/dL)	0.82 ± 0.20	1.13 ± 1.85	0.87 ± 0.20	0.82 ± 0.18
Serum urea nitrogen (mg/dL)	12.37 ± 3.06	14.13 ± 5.77	12.14 ± 3.77	11.97 ± 3.73
Creatinne Clearance (mL/min)	129.78 ± 60.06	138.69 ± 40.39	116.89 ± 44.43	116.89 ± 42.84
Urine urea nitrogen (g/24 hr)	10.91 ± 4.49	12.22 ± 4.64	10.89 ± 4.73	9.58 ± 3.95
Urine Calcium (mg/24 hr)	184.68 ± 119.10	153.46 ± 77.07	25.2 ± 103.60	23.33 ± 75.74
Urine Protein (mg/24 hr)	141.25 ± 71.23	158.55 ± 88.82	114.39 ± 38.25	180.00 ± 86.56*
Bone mineral density (g/cm^2)	1.00 ± 0.00	1.04 ± 1.19	1.03 ± 0.17	1.01.00 ± 0.03

*All values are mean ± SD. There were no significant differences between groups. HP, high protein group; SP, standard protein group. *p < 0.05 compare with baseline*

8.3.5 BONE MINERAL DENSITY

No significant differences (p > 0.05) were observed at 12 months in total bone mineral density within-group or between groups (Table 3).

8.4 DISCUSSION

In this study, the energy deficit meal plan including meal replacements resulted in significant weight loss typical of meal replacement plans in both groups [15]. Since there was no run-in period, early dropouts were significant but 70 out of 85 subjects were retained after that point of the study. Because both diets were isocaloric the amounts of weight loss were the same enabling a meaningful comparison of the effects of the dietary intervention on liver function, renal function, and bone density in an outpatient setting. No special efforts were made to assess compliance which could be considered a limitation of the study. Compliance with diets is known to decrease on an outpatient basis and is an unmeasured effect that may account for the lack of findings of adverse events in our study. Nonetheless, this was a practical applied test of the issue as it would be encountered in people undertaking a weight management regimen.

Concerns that diets high in protein may have deleterious effects on renal function were not supported by the results of this study. There was no difference in creatinine clearance with either dietary pattern during weight reduction over one year. A previous study also reported that creatinine clearance was not altered by dietary protein in the context of weight loss while nitrogen balance was more positive in subjects who consumed a high protein diet than in those who consumed a high carbohydrate diet [16]. Skov et al [8] assessed changes in renal function by measuring the glomerular filtration rate (GFR) during high-protein and high-carbohydrate diets over a 6-month period and found that the high protein diet had no adverse effects on kidney function. More recently, Knight et al. determined whether protein intake influences the rate of renal function change in women prospectively studied over an 11-year period [7]. The Nurses'

Health Study evaluated 1624 enrolled women between the ages of 42 to 68 years in 1989 who provided blood samples in 1989 and 2000. Ninety-eight percent of women were white, while 1% were African American. In multivariate linear regression analyses, high protein intake was not significantly associated with change in estimated GFR in women with normal renal function (defined as an estimated GFR $\geq$ 80 mL/min per 1.73 m^2).

It has been suggested that a high protein diet may generate acidosis because of the presence of ketone bodies in the blood promoting calcium mobilization from bone to buffer the blood and maintain pH. This could promote urinary calcium loss [17,18]. There were no deleterious effects of increased protein intake at 2.2 g/kg LBM on markers of bone turnover in our study. In a 12- week study [19], a high protein diet increased the bone turnover markers while calcium excretion was decreased by 0.8 mmol/d. Evidence also indicates that high protein intake particularly higher animal protein intake is associated with decreased bone loss in older persons [20].

The trend of reduction in urinary calcium in this study was also unusual because dietary protein metabolism is associated with increased urinary calcium [21]. The high vegetable consumption with both dietary patterns may prevent this because high vegetable intakes have been shown to decrease urinary calcium [22]. An increase in calcium excretion was observed with the consumption of a high protein diet in the study by Johnston et al [16] which stated that this was due to the high calcium content of the high protein diet in their study. However, we did not observe this high protein pattern in which dietary calcium was very high.

Non-alcoholic fatty liver disease (NAFLD) is now the most common liver disease and is strongly linked to obesity and metabolic syndrome [23]. In middle aged women in the UK, Liu and colleagues [24] found that the relative risk of liver cirrhosis increased by 28% for every 5 unit increase in BMI above 22.5 in each stratum of alcohol consumption and estimated 17% of incident or fatal liver cirrhosis is attributable to excess body weight. Hart and colleagues [25] also show that being overweight or obese and drinking alcohol has a synergistic effect, which amplifies the insult to the liver and greatly increases the risk of liver related morbidity and mortality. Therefore, it is important to demonstrate that an effective weight management program does not elevate liver function tests and add insult to the liver. In this study, there were no adverse effects on liver function tests at either level of protein intake.

The one-year duration of the study may have led to reduced compliance to the meal plans. The study subjects met for a total of 8 sessions with our dietitian. These sessions were designed to support and encourage participants to follow the meal plan including the MR. At each visit, qualitative food logs for macronutrients and meal replacement were collected and reviewed. While we did not measure biochemical compliance, the overall weight loss we observed suggest relatively good compliance to our meal plans.

Noakes et al [19] reported that subjects with high serum triacylglycerol (>1.5 mmol/L) lost more fat mass with the high protein diet than with the high carbohydrate diet and suggesting a variation in responsiveness to diet based on other metabolic factors such as the presence of insulin resistance which was not measured in the current study.

As in many outpatient diet interventions long-term compliance is undercut by some unmeasured factors likely unrelated to the demonstrated satiety effects of added protein. Therefore, the expected effects on increased weight loss resulting from a high protein diet were not seen in this study. In our previous study, protein-enriched meal led to increased fat mass loss based on bioelectrical impedance analysis in spite of similar overall weight loss as the standard protein meal plan over 12 weeks [26]. The use of MR may have been the major influence on the weight loss by simplifying their weight loss efforts so that the power of the MR intervention may have obscured the difference between the weight losses of subjects using protein-enriched meal plans by comparison to standard meal plans [5].

The Institute of Medicine (IOM) of the National Academy of Sciences [27] has set acceptable macronutrient distribution ranges for carbohydrate (45%-65% of energy), protein (10%-35% of energy), and fat (20%-35% of energy; limit saturated and trans fats). These proportions provide a range broad enough to cover the macronutrient needs of most active individuals, but specific carbohydrate and protein recommendations are also typically made based on a g/kg body weight formula. These ranges are 5 to 12 g of carbohydrate/kg body weight and 1.2 to 1.8 g/kg body weight for protein depending on the level of physical activity. Clearly, for both the HP and SP group dietary protein intakes were within this recommended range for protein intake. Therefore, our research can only be applied to structured meal plans using protein-enriched shakes for their ability to increase satiety and should not be interpreted as a blanket endorsement of very high protein diets popular with some athletes exceeding the IOM recommendations by including pure protein supplements, high fat animal meats or other sources of organic acids and hidden fat which could adversely affect liver function, renal function, or bone density.

8.5 CONCLUSIONS

In summary, both the HP and SP diets resulted in the expected weight loss typical of an MR diet plan in free-living individuals at 12 months. Both diets were well tolerated, sustainable, and did not result in any adverse effects. There were no changes of liver function, renal function or bone mineral density based on routine clinical assessments.

REFERENCES

1. Ford ES, Mokdad AH: Epidemiology of obesity in the Western Hemisphere. J Clin Endocrinol Metab 2008, 93:S1-S8.
2. Yach D, Stuckler D, Brownell KD: Epidemiologic and economic consequences of the global epidemics of obesity and diabetes. Nat Med 2006, 12:62-66.
3. Lejeune MP, Westerterp KR, Adam TC, Luscombe-Marsh ND, Westerterp-Plantenga MS: Ghrelin and glucagon-like peptide 1 concentrations, 24-h satiety, and energy and substrate metabolism during a high-protein diet and measured in a respiration chamber. Am J Clin Nutr 2006, 83:89-94.
4. Weigle DS, Breen PA, Matthys CC, Callahan HS, Meeuws KE, Burden VR, et al.: A high-protein diet induces sustained reductions in appetite, ad libitum caloric intake, and body weight despite compensatory changes in diurnal plasma leptin and ghrelin concentrations. Am J Clin Nutr 2005, 82:41-48.
5. Heymsfield SB, van Mierlo CA, van der Knaap HC, Heo M, Frier HI: Weight management using a meal replacement strategy: meta and pooling analysis from six studies. Int J Obes Relat Metab Disord 2003, 27:537-49.
6. Paddon-Jones D, Westman E, Mattes RD, Wolfe RR, Astrup A, Westerterp-Plantenga M: Protein, weight management, and satiety. Am J Clin Nutr 2008, 87:1558S-61S.

7. Knight EL, Stampfer MJ, Hankinson SE, Spiegelman D, Curhan GC: The impact of protein intake on renal function decline in women with normal renal function or mild renal insufficiency. Ann Intern Med 2003, 138:460-467.
8. Skov AR, Toubro S, Bulow J, Krabbe K, Parving HH, Astrup A: Changes in renal function during weight loss induced by high vs low-protein low-fat diets in overweight subjects. Int J Obes Relat Metab Disord 1999, 23:1170-1177.
9. Ashley JM, Herzog H, Clodfelter S, Bovee V, Schrage J, Pritsos C: Nutrient adequacy during weight loss interventions: a randomized study in women comparing the dietary intake in a meal replacement group with a traditional food group. Nutr J 2007, 6:12.
10. Heber D, Ashley JM, Wang HJ, Elashoff RM: Clinical evaluation of a minimal intervention meal replacement regimen for weight reduction. J Am Coll Nutr 1994, 13:608-14.
11. Noakes M, Foster PR, Keogh JB, Clifton PM: Meal replacements are as effective as structured weight-loss diets for treating obesity in adults with features of metabolic syndrome. J Nutr 2004, 134:1894-99.
12. Flechtner-Mors M, Ditschuneit HH, Johnson TD, Suchard MA, Adler G: Metabolic and weight loss effects of long-term dietary intervention in obese patients: four-year results. Obes Res 2000, 8:399-402.
13. Talke H, Schubert GE: Enzymatic urea determination in the blood and serum in the warburg optical test. Klin Wochenschr 1965, 43:174-75.
14. Statistical Analysis System: SAS Statistics Software. version 9.1.3.
15. Ditschuneit HH: Do meal replacement drinks have a role in diabetes management? Nestle Nutr Workshop Ser Clin Perform Programme 2006, 11:171-79.
16. Johnston CS, Day CS, Swan PD: Postprandial thermogenesis is increased 100% on a high-protein, low-fat diet versus a high-carbohydrate, low-fat diet in healthy, young women. J Am Coll Nutr 2002, 21:55-61.
17. Kerstetter JE, Allen LH: Dietary protein increases urinary calcium. J Nutr 1990, 120:134-36.
18. Heaney RP, Recker RR: Effects of nitrogen, phosphorus, and caffeine on calcium balance in women. J Lab Clin Med 1982, 99:46-55.
19. Noakes M, Keogh JB, Foster PR, Clifton PM: Effect of an energy-restricted, high-protein, low-fat diet relative to a conventional high-carbohydrate, low-fat diet on weight loss, body composition, nutritional status, and markers of cardiovascular health in obese women. Am J Clin Nutr 2005, 81:1298-306.
20. Hannan MT, Tucker KL, wson-Hughes B, Cupples LA, Felson DT, Kiel DP: Effect of dietary protein on bone loss in elderly men and women: the Framingham Osteoporosis Study. J Bone Miner Res 2000, 15:2504-12.
21. Hegsted M, Linkswiler HM: Long-term effects of level of protein intake on calcium metabolism in young adult women. J Nutr 1981, 111:244-51.
22. Appel LJ, Moore TJ, Obarzanek E, Vollmer WM, Svetkey LP, Sacks FM, et al.: A clinical trial of the effects of dietary patterns on blood pressure. DASH Collaborative Research Group. N Engl J Med 1997, 336:1117-24.

23. Vanni E, Bugianesi E, Kotronen A, De MS, Yki-Jarvinen H, Svegliati-Baroni G: From the metabolic syndrome to NAFLD or vice versa? Dig Liver Dis 2010, 42:320-330.
24. Liu B, Balkwill A, Reeves G, Beral V: Body mass index and risk of liver cirrhosis in middle aged UK women: prospective study. BMJ 2010, 340:c912.
25. Hart CL, Morrison DS, Batty GD, Mitchell RJ, Davey SG: Effect of body mass index and alcohol consumption on liver disease: analysis of data from two prospective cohort studies. BMJ 2010, 340:c1240.
26. Treyzon L, Chen S, Hong K, Yan E, Carpenter CL, Thames G, et al.: A controlled trial of protein enrichment of meal replacements for weight reduction with retention of lean body mass. Nutr J 2008, 7:23.
27. Trumbo P, Schlicker S, Yates AA, Poos M: Dietary reference intakes for energy, carbohydrate, fiber, fat, fatty acids, cholesterol, protein and amino acids. J Am Diet Assoc 2002, 102:1621-30.

This chapter was originally published under the Creative Commons Attribution License. Li, Z., Treyzon, L., Chen, S., Yan, E., Thames, G., and Carpenter, C. L. Protein-Enriched Meal Replacements Do Not Adversely Affect Liver, Kidney or Bone Density: An Outpatient Randomized Controlled Trial. Nutrition Journal 2010, 9:72. doi:10.1186/1475-2891-9-72. http://www.nutritionj.com/content/9/1/72

FUNCTIONAL FOOD AND ORGANIC FOOD ARE COMPETING RATHER THAN SUPPORTING CONCEPTS IN EUROPE

JOHANNES KAHL, ANETA ZAŁĘCKA, ANGELIKA PLOEGER, SUSANNE BÜGEL, and MACHTELD HUBER

9.1 INTRODUCTION

Functional and organic foods are both segments of the food market in Europe and have been growing constantly within the last decade [1]. Whereas organic food covers the entire manner of production [2,3], functional food describes nutrition and/or health related product attributes [4,5]. Both foods are labeled and the labels send quality and/or health related messages that point towards high quality food and/or extra health benefits [6,7]. Although consumers perceive these messages in both types of foods [8,9], they are associated differently. Furthermore, the conceptual background of functional and organic food is different with respect to quality concepts [10,11], which reflect the difference in the underlying paradigms of these two different food segments [12]. The aim of this paper is to analyze the similarities and differences of functional and organic food in order to determine if the concept of functional food supports or contradicts organic food production. The work was done among members of the international association Food Quality and Health (FQH).

9.2 ORGANIC FOOD IN EUROPE

The market of organic food is constantly growing [13]. In 2009, the European turnover of the organic food market was 18.4 billion Euros, wherein Germany and France have the highest turnover of organic foods (in total 5.8 billion Euros and 3 billion Euros, respectively [14]), while the highest market shares were reached in Denmark with 7.2 percent of the total food turnover, followed by Austria with 6 percent [14]. In Europe, consumers buy organic food because they expect that it is "good" for their health, has less impact on the environment and the production respects animal welfare principles [15–18]. Results from scientific investigations on the quality of the food focus on organic versus non-organic production method comparisons. Process measurements (e.g., impact on environment) indicate less impact on environmental issues measured by Life-Cycle-Assessment [19–22]. Comparisons based on food constituents showed lower levels of residues in organic foods [23–25] but had no consistent answer related to nutrients and health related compounds [26]. Recent reviews on vegetables, fruits and milk showed significantly higher amounts of bioactive compounds in the organic food products [27,28]. The compounds found at a higher level in organic food are, for example, defense related secondary metabolites in fruits and vegetables and poly-unsaturated fatty acids (PUFAs) in milk. The effect of those compounds on health is still a matter of controversy [29–31]. Therefore expected health benefits can only be claimed on the basis of models [28].

Organic agriculture is defined as a "holistic production management system" [3], which takes into account soil, plant and animal health by applying natural methods and working in natural cycles [2,32]. Agricultural products should be further processed in order to maintain the "organic integrity and vital qualities" of the product in all stages [2,3]. Biological and mechanical methods should be applied in food production "opposed to using synthetic materials" [3]. The production of organic food has been regulated in Europe since 1991 with the actual regulations Nr. 834/2007 and 889/2008. Organic food products are labeled with a standard European logo. According to the EC regulations, processing should be done with care and all kinds of synthetic substances used during processing should be kept to a small number compared with non-organic food production [33,34]. Primary production, as well as processing and retailing, is certified according to the EC-regulations. Furthermore, import from third countries is regulated according to the European organic requirements and should guarantee the same quality of organic food as under EC regulation 834/2007.

Kahl et al. [35] analyzed the different guidelines and regulations relating to organic food quality issues and concluded that the quality of the food is described according to process issues, rather than concrete product attributes. Even so, one goal of organic food production is high quality of the products [2] as well as maintenance of their "true nature" during the production process [2]. Kahl et al. [10] identified health and sustainability as underlying goals in organic food production. Furthermore, they defined a combination of process and product aspects necessary for the definition of organic food quality. This is supported by results from consumer surveys [17,36,37]. The conceptual background refers to a system approach that considers the food as a whole rather than just the sum of its constituents. Therefore, one of the major future challenges for organic food quality research is to develop systemic tools for food evaluation.

9.3 FUNCTIONAL FOOD IN EUROPE

The European market for functional food is constantly growing [4,7]. The UK has the highest turnover of functional foods, reaching 3.3 billion Euros followed by Germany (3.02 billion Euros) and France (1.8 billion Euros). The main factor for market success is high quality, wherein a "holistic health image" of the product increases the market success [7]. Consumers buy functional food, because they believe it is healthy, while functional ingredients known by the consumers stimulate more market success than ingredients not well known to the general public (e.g., PUFAS versus selenium). Moreover, perceived relevance influences the willingness to buy products with health claims [38].

Diplock et al. [11] presented a first scientific concept of functional food in Europe. Functional food should contribute significantly either to enhancing health and well being or reducing risk of diseases. Functional foods can be divided into three categories according to the nature of the functional ingredients:

- natural food containing high levels of the functional ingredient or with high functionality;
- food to which functional ingredients were added or removed;
- food in which the nature of functional ingredients has been changed.

Functional food is not specifically regulated in Europe; neither does a uniform label exist. However, there are EC-regulations that cover this part of the food market: Firstly, the general food regulation in Europe (178/2002). This regulation describes the basic quality and safety requirements of European consumers and protects consumers against fraud. When the food is changed in its molecular structure or consists of ingredients not previously on the European market, the food falls under the European Novel Food Regulation (258/97). The most important regulation regarding functional food in Europe is the Health Claim Regulation from 2006 (1924/2006). All marketing referring to possible health implications of food is prohibited, unless the Commission, after recommendation by the European Food Safety Agency (EFSA), has authorized a defined and substantiated health claim through scientific evidence. Until 2011 only a few hundred health claims were authorized [39]. EFSA does not allow health claims for vegetables and fruits in general and for their processed products (e.g., apple juice). Furthermore, health claims cannot be made for fruits or vegetables due to a specific functional ingredient in the fruit or vegetable [39]. The main reason is that the food consumption patterns will be responsible for the potential health effect rather than specific varieties or amounts of the fruits and vegetables. Health claims are only authorized for processed products with one or more specific compounds and their evidenced effects on human health. Therefore it is still a major challenge for scientific research to identify and test the functionality of food compounds [4,5] and to make this information available to consumers [40].

9.4 SIMILARITIES AND DIFFERENCES

Both organic and functional foods intend to offer high quality products to the European consumer with added value related to health. Whereas functional food offers human health specific targets, organic food intends to combine the health of the consumer and the health of environment, animals and society. Both are defined as food according to the European food regulation (EC 178/2002). Claims relating to positive health effects, which should be labeled to inform the consumer, are regulated for both food segments under EC-regulation 1924/2006. Therefore in general, studies showing scientific evidence for a proposed health benefit must be carried out. Whereas for functional food a specific compound or food attribute can be tested (different levels of the compound in question), it is difficult to prove this in organic food, wherein the whole food instead of constituent parts has to be compared (what should the placebo be?). Organic and functional foods are both labeled. Functional food is labeled according to the potential positive human health function or product attribute (when authorized by EFSA). Here, specific claims are formulated for different products. Organic food is labeled according to the process of production by different labeling schemes [6] but there is also a standard European organic logo [41]. The effects of organic food are not limited to human health, but are also claimed to work on society and the environment (soil, plants, animals, atmosphere, etc.). For the marketing of functional food as well as organic food, consumer knowledge of the label or the specific product attributes is necessary [42–45].

One of the reasons that the European market for functional food is behind the US and Japanese markets may be due to the distrust of the consumers [1]. Consumers often do not believe in the additional health benefit, which may justify the higher price for these products [1]. Bech-Larsen and Scholderer [7] identified a paradox, wherein consumers on the one hand do not believe in these added values but on the other hand they feel unsupported by detailed research results. This paradox may become even sharper after the last decisions of the Commission (following EFSA recommendations), which restricts food claims to food products where substantial scientific evidence exists for a beneficial effect, based on single constituents or attributes.

Consumers believe that organic food has positive effects on health, environment and animal welfare [1,15,17], although only the production process is regulated, not the effects (e.g., levels of health related constituents). Eden [8] investigated consumers' awareness of organic and functional food. The study underlines that organic food is associated with naturalness, whereas functional food is associated with scientifically and technologically complicated processes. When consumer behavior is analyzed, the decision to buy organic food tends to be emotional, whereas there seems to be a rational decision process in buying functional food [46,47]. Furthermore, some organic consumers do not accept specially highlighted functions of food [9,43]. The question whether or not functional foods would compete with organic foods still cannot be addressed because of a lack of study material. Stolz et al. [42] indicate that conventional food with a communicated additional benefit will compete with the conventional counterpart, not the organic one.

When the underlying paradigms of functional versus organic food are analyzed, they seem to reflect the different visions of the European Knowledge-Based Bio-Economy [12]. Whereas functional food, being part of the dominant industrial life science vision, connects food quality with decomposability, organic food connects food production with integral product integrity. Allaire and Wolf [48] identified these divergent socio-technical paradigms as a re-arrangement of food constituents versus a holistic quality view recognizable by consumers. Whereas functional food is focused on product related attributes represented by single constituents or functions, organic food also takes sustainability of the process of production into account.

9.5 CONCLUSIONS

Based on this review of recent literature we conclude that the concept of functional food does not support organic food production in Europe.

Based on consumer behavior and the organic regulation, it seems full of contradictions. Although organic and functional foods can be grouped together with respect to the naturally increased levels of health related compounds, such as PUFAs and polyphenols, there will hardly be any health claims possible, as we discussed, with respect to the experimental design and recent decisions of the Commission. When food ingredients are added or changed in organic foods, it should be done within the frame of the European organic regulation, which is very strict. Only a few food additives are permitted so far and health related compounds such as bio-active compounds, antioxidants, etc. are not included (with the exception of vitamin C for specific purposes). Organic food may lose its "natural" image for consumers if producers follow the technological approach of re-designing food, as is done for functional foods. From their viewpoint, organic food derived from nature by a naturally sound agricultural system already contains all substances needed for human health and respects the health of the environment and society along the food chain. A key challenge for scientific research in both areas is the identification and validation of relevant indicators and parameters for the process (only organic) and product aspect (organic as well as functional) [10]. Furthermore, food seems to be the fundamental unit in nutrition and not only single constituents [49]. Health studies testing the effect of functional as well as organic food need approaches that reflect the dynamic reaction of a very complex system [50].

REFERENCES

1. The Nielsen Company. Organics and functional foods: A global Nielsen consumer report. Technical Report. Available online: http://pt.nielsen.com/documents/tr_0710_OrganicsFood.pdf (accessed on 3 July 2012).
2. Council Regulation (EC) No 834/2007. Available online: http://eur-lex.europa.eu/LexUriServ/ LexUriServ.do?uri=OJ:L:2007:189:0001:0023:EN:PDF (accessed on 3 July 2012).
3. Guidelines for the Production, Processing, Labelling and Marketing of Organically Produced Foods (GL 32-1999); Codex Alimentarius: Rome, Italy, 2010; Adopted 1999, Revisions 2001, 2003, 2004 and 2007, Amendments 2008, 2009 and 2010.

4. Kaur, S.; Das, M. Functional foods: An overview. Food Sci. Biotechnol. 2011, 20, 861–875.

5. Roberfroid, M.B. What is beneficial for health? The concept of functional food. Food Chem. Toxicol. 1999, 37, 1039–1041.

6. Janssen, M.; Hamm, U. Product labelling in the market for organic food: Consumers preferences and willingness-to-pay for different organic certification logos. Food Qual. Pref. 2012, 25, 9–22.

7. Bech-Larsen, T.; Scholderer, J. Functional foods in Europe: Consumer research, market experiences and regulatory aspects. Food Sci. Technol. 2007, 18, 231–234.

8. Eden, S. Food labels as boundary objects: How consumers make sense of organic and functional foods. Public Underst. Sci. 2011, 20, 179–194.

9. Mesias, F.J.; Martinez-Carrasco, F.; Martinez, J.M.; Gaspar, P. Functional and organic eggs as an alternative to conventional production: A conjoint analysis of consumer' preferences. J. Sci. Food Agric. 2011, 91, 532–538.

10. Kahl, J.; Baars, T.; Bügel, S.; Busscher, N.; Huber, M.; Kusche, D.; Rembialkowska, E.; Schmid, O.; Seidel, K.; Taupier-Letage, B.; Velimirov, A.; Zalecka, A. Organic food quality: A framework for concept, definition and evaluation from the European perspective. J. Sci. Food Agric. 2012, doi:10.1002/jsfa.5640.

11. Diplock, A.T.; Aggett, P.J.; Ashwell, M.; Bornet, F.; Fern, E.B.; Roberfroid, M.B. Scientific concepts of functional foods in Europe: Consensus document. Brit. J. Nutr. 1999, 81, 1–27.

12. Levidow, L.; Birch, K.; Papaioannou, T. Divergent paradigms of European Agro-Food Innovation: The Knowledge-Based Bio-Economy (KBBE) as an R&D agenda. Sci. Technol. Human Values 2012, doi:10.1177/0162243912438143.

13. Sahota, A. Global Organic Food & Drink Market. Available online: http://www.fibl. org/ fileadmin/documents/en/news/2011/sahota-2011-biofach-market.pdf (accessed on 3 July 2012).

14. OSEC Website. Available online: http://www.sippo.ch/internet/osec/en/home/import/publications/food.-ContentSlot-98296-ItemList-61735-File.File.pdf/SIPPO_Manual_18.04.2011_final.pdf (accessed on 3 July 2012).

15. Kriwy, P.; Mecking, R.-A. Health and environmental consciousness, costs of behaviour and the purchase of organic food. Int. J. Consumer Stud. 2012, 36, 30–37.

16. Pino, G.; Peluso, A.M.; Guido, G. Determinants of regular and occasional consumers' intentions to buy organic food. J. Consumer Aff. 2012, 46, 157–169.

17. Zagata, L. Consumers' beliefs and behavioural intentions towards organic food. Evidence from the Czech Republic. Appetite 2012, 59, 81–89.

18. Torjusen, H.; Sangstad, L.; O'Doherty-Jensen, K.; Kjaernes, U. European Consumers' Conceptions of Organic Food: A Review of Available Research; Project Report 4-2004 for National Institute for Consumer Research: Oslo, Norway, 2004.

19. Mondelaers, K; Aertsens, J.; van Huylenbroeck, G. A meta-analysis of the differences in environmental impacts between organic and conventional farming. Brit. Food J. 2009, 111, 1098–1119.

20. Gomiero, T.; Paolettia, M.G.; Pimentel, D. Energy and environmental issues in organic and conventional agriculture. Crit. Rev. Plant Sci. 2008, 27, 239–254.

21. Wood, R.; Lenzen, M.; Dey, C.; Lundie, S. A comparative study of some environmental impacts of conventional and organic farming in Australia. Agric. Syst. 2006, 89, 324–348.

22. Cederberg, C.; Mattsson, B. Life cycle assessment of milk production—A comparison of conventional and organic farming. J. Clean. Prod. 2000, 8, 49–60.

23. Lairon, D. Nutritional quality and safety of organic food: A review. Agron. Sustain. Dev. 2010, 30, 33–41.

24. Baker, P.B.; Benbrook, C.M.; Groth, E., III; Benbrook, L. Pesticide residues in conventional, integrated pest management (IPM)-grown and organic foods: Insights from three US data sets. Food Addit. Contam. 2002, 19, 427–446.

25. Kouba, M. Quality of organic animal products. Livest. Prod. Sci. 2003, 80, 33–40.

26. Dangour, A.D.; Dodhia, S.K.; Hayter, A.; Allen, E.; Lock, K.; Uauy, R. Nutritional quality of organic foods: A systematic review. Am. J. Clin. Nutr. 2009, 90, 680–685.

27. Palupi, E.; Jayanegara, A.; Ploeger, A.; Kahl, J. Comparison of nutritional quality between conventional and organic dairy products: A meta-analysis. J. Sci. Food Agric. 2012, doi:10.1002/jsfa.5639.

28. Brandt, K.; Leifert, C.; Sanderson, R.; Seal, C.J. Agroecosystem management and nutritional quality of plant foods: The case of organic fruits and vegetables. Crit. Rev. Plant Sci. 2011, 30, 177–197.

29. Hollman, P.C.H.; Cassidy, A.; Comte, B.; Heinonen, M.; Richelle, M.; Richling, E.; Serafini, M.; Scalbert, A.; Sies, H.; Vidry, S. The biological relevance of direct antioxidant effects of polyphenols for cardiovascular health in human is not established. J. Nutr. 2011, 141, 989–1009.

30. Hoefkens, C.; Sioen, I.; Baert, K.; de Meulenaer, B.; de Henauw, S.; Vandekinderen, I.; Devlieghere, F.; Opsomer, A.; Verbeke, W.; van Camp, J. Consuming organic versus conventional vegetables: The effect on nutrient and contaminant intakes. Food Chem. Toxicol. 2010, 48, 3058–3066.

31. Erdman, J.W., Jr.; Ford, N.A.; Lindshield, B.L. Are the health attributes of lycopene related to IST antioxidants function? Archiv. Biochem. Biophys. 2009, 483, 229–235.

32. IFOAM Basic Standards; International Federation of Organic Agriculture Movements: Bonn, Germany, 2008.

33. Food Standards Agency. Current EU approved additives and their E numbers. Available online: http://www.food.gov.uk/safereating/additivesbranch/enumberlist (accessed on 3 July 2012).

34. EU Web site. Available online: http://eur-lex.europa.eu/LexUriServ/LexUriServ.do?uri=OJ:L:2008:250:0001:0084:EN:PDF (accessed on 3 July 2012).

35. Kahl, J.; van den Burgt, G.-J.; Kusche, D.; Bügel, S.; Busscher, N.; Hallmann, E.; Kretzschmar, U.; Ploeger, A.; Rembialkowska, E.; Huber, M. Organic food claims in Europe. Food Technol. 2010, 3, 38–46.

36. Hjelmar, U. Consumers' purchase of organic food products: A matter of convenience and reflexive practices. Appetite 2011, 56, 336–344.

37. Hughner, R.S.; McDonagh, P.; Prothero, A.; Shultz, C.J., II; Stanton, J. Who are organic consumers? A compilation and review of why people purchase organic food. J. Consumer Behav. 2007, 6, 94–110.

38. Dean, M.; Lampila, P.; Shepherd, R.; Arvola, A.; Saba, A.; Vassallo, M.; Claupein, E.; Winkelmann, M.; Lähteenmäki, L. Perceived relevance and foods with health-related claims. Food Qual. Pref. 2012, 24, 129–135.

39. EU Web site. Available online: http://ec.europa.eu/nuhclaims/ (accessed on 3 July 2012).

40. Sahidi, F. Neutraceuticals and functional foods: Whole versus processed foods. Trends Food Sci. Technol. 2009, 20, 376–387.

41. Janssen, M.; Hamm, U. The mandatory EU logo for organic food: Consumer perceptions. Brit. Food J. 2012, 114, 335–352.

42. Stolz, H.; Stolze, M.; Janssen, M.; Hamm, U. Preferences and determinants for organic, conventional and conventional-plus products—The case of occasional organic consumers. Food Qual. Pref. 2011, 22, 772–779.

43. Markosyan, A.; McCluskey, J.J.; Wahl, T.I. Consumer response to information about a functional food product: Apples enriched with antioxidants. Can. J. Agric. Econ. 2009, 52, 325–341.

44. Verbeke, W.; Scholderer, J.; Lähteenmäki, L. Consumer appeal of nutrition and health claims in three existing product concepts. Appetite 2009, 52, 584–692.

45. Van Kleef, E.; van Trijp, H.C.M.; Luning, P. Functional foods: Health claim-food product compatability and the impact of health claim framing on consumer evaluation. Appetite 2005, 44, 299–308.

46. Szakaly, Z.; Szente, V.; Köver, G.; Polereczik, Z.; Szigeti, O. The influence of lifestyle on health behaviour and preference for functional food. Appetite 2012, 58, 406–413.

47. Padel, S.; Foster, C. Exploring the gap between attitudes and behaviour: Understanding why consumers buy or do not buy organic food. Brit. Food J. 2005, 107, 606–625.

48. Allaire, G.; Wolf, S.A. Cognitive representations and institutional hybridity in Agro-food Innovation. Sci. Technol. Human Values 2004, 29, 431–458.

49. Jacobs, D.R.; Tapsell, L.C. Food, not nutrients, is the fundamental unit in nutrition. Nutr. Rev. 2007, 65, 439–450.

50. Huber, M.; Knottnerus, J.A.; Green, L.; van der Horst, H.; Jadad, A.R.; Kromhout, D. How should we define health? Brit. Med. J. 2011, doi: http://dx.doi.org/10.1136/bmj.d4163.

This chapter was originally published under the Creative Commons Attribution License. Kahl, J., Załęcka, A., Ploeger, A., Bügel, S., and Huber, M. Functional Food and Organic Food Are Competing Rather Than Supporting Concepts in Europe. Agriculture 2012, 2(4), 316-324. doi:10.3390/agriculture2040316.

PART II

FUNCTIONAL FOODS: THE CONNECTION BETWEEN HEALTH AND FOOD SCIENCE

CHAPTER 10

EXERCISE AND FUNCTIONAL FOODS

WATARU AOI, YUJI NAITO, and TOSHIKAZU YOSHIKAWA

10.1 INTRODUCTION

Appropriate nutrition is essential for the proper performance of exercise. In particular, correct nutrition is critically important for improvement of athletic performance, conditioning, recovery from fatigue after exercise, and avoidance of injury. Although athletes need to eat a well-balanced basic diet, there are several nutritional factors that are difficult to obtain at a sufficient level from a normal diet since athletes require more nutrients than the recommended daily allowances. Thus, nutritional supplements containing carbohydrates, proteins, vitamins, and minerals have been widely used in various sporting fields, partly because these supplements are easily taken before, during, and/or after exercise. Several natural food components have also been shown to exert physiological effects, and some of them are considered to be useful (when ingested at high doses or continuously) for improving athletic performance or for avoiding the disturbance of homeostasis by strenuous exercise. Recently, food components with physiological actions have been called "functional foods" and the effects of such foods have been scientifically investigated. This article introduces some functional foods, including basic nutrients, which have been demonstrated to have a beneficial influence on the physiological changes that occur during exercise.

10.2 REPLENISHMENT OF WATER

Water is the main constituent of the human body, and it plays an essential role in circulatory function, chemical reactions involved in energy metabolism, elimination of waste products, and maintenance of the body temperature and plasma volume. When the body temperature rises due to the intense exercise or a high ambient temperature, sweating occurs in order to radiate heat [1-3], leading to the loss of a large amount of water and electrolytes such as sodium. This loss of body fluid impairs thermoregulation and circulatory system, leading to a decline of athletic performance [4,5]. Therefore, to maintain homeostasis and athletic performance, replenishment of water and electrolytes is essential before and during or after exercise. Generally, it is believed to be useful to drink isotonic fluid that contains electrolytes such sodium, potassium, and chloride at concentrations close to those in body fluids. It has also been suggested that intake of hypotonic fluid may exert a similar or more rapid effect on replenishment of body water [6,7] because it is rapidly absorbed from the small intestine along an electrochemical gradient. Also, the sodium concentration (and the osmolality) of sweat is lower than that of extracellular fluid, so the loss of water with sweating is much greater than the loss of electrolytes, leading to an increase in the plasma osmotic pressure. On the other hand, replenishment of water alone is unlikely to maintain homeostasis of body fluid in prolonged exercise that produces high sweat rates. Taking in only water in prolonged exercise leads to hyponatremia and a decrease in the osmotic pressure of body fluids and inhibit the release of antidiuretic hormone resulting in that water intake is suppressed and the urine output is increased (spontaneous dehydration) [8]. Latzka et al. [9] suggested that during prolonged exercise lasting longer than 90 minutes, fluid drink containing electrolytes and carbohydrate, not water alone, should be considered to provide to sustain carbohydrate oxidation and endurance performance. Furthermore, several studies have suggested that glycerol loading has been advocated as one of methods which prevent high temperature and dehydration in exercise [10,11]. Oral administration of 1.0–1.2 g/kg B.W. glycerol with water temporarily results in an increase of 300–700 ml body fluid [10,11] and improves endurance performance compared with placebo [12-14]. Glycerol acts as an osmolyte in body fluid, which would lead to an elevation of plasma osmolality [10]. Consequently, water reabsorption in the kidney is increased and urine excretion is decreased, which is considered as one of mechanisms of the effect.

10.3 IMPROVEMENT OF ENDURANCE

Energy consumed during exercise is mainly supplied by carbohydrates and lipids, so it is important for improvement of endurance to regulate the metabolism of these two substrates. During endurance exercise, glycogen (an energy substrate for muscle contraction) is gradually depleted, making it difficult to continue exercising. An effective way to improve endurance is to increase the glycogen stores in skeletal muscle and the liver before the commencement of exercise. When tissue glycogen stores are depleted, glycogen synthetase activity is transiently increased, leading to an increase of glycogen storage by conversion from carbohydrates [15,16]. For instance, it has been reported that glycogen stores can be increased by eating a low-carbohydrate diet for 3 days from 6 days prior to competition, followed by a high-carbohydrate diet for the next 3 days, resulting in the storage of 1.5 times more glycogen than normal [17]. If citrate, which inhibits glycolysis, is taken concurrently with a high-carbohydrate diet, glycogen stores will be further increased due to the inhibition of glycolysis [18,19]. It is also important for athletes to replenish the glycogen stores during post-exercise training to provide sufficient energy for the next training session or competition. For rapid replenishment of glycogen stores, a high-carbohydrate diet can be effective [19,20]. Intake of protein along with carbohydrate can be more effective for the rapid replenishment of muscle glycogen after exercise compared with carbohydrate supplements alone [21,22].

When prolonged exercise will be performed, such as a marathon, taking carbohydrates immediately before or during exercise is also an effective method of improving endurance. Under such conditions, it is desirable for the athlete to ingest monosaccharides or oligosaccharides, because these are rapidly absorbed and transported to the peripheral tissues. On the other hand, intake of carbohydrates inhibits the degradation of fat, which is another energy substrate, by stimulating insulin secretion [23,24]. This leads to impairment of energy production via lipid metabolism and accelerates glycolysis as alternate energy production pathway. As a result, the consumption of muscle glycogen will increase, and the intramuscular pH will decrease due to increased lactic acid production, which may lead to impairment of muscle contraction. Therefore, it is necessary to ingest carbohydrates that will not inhibit lipid metabolism. It has been suggested that supplements containing fructose, which cause less stimulation of insulin secretion and are unlikely to inhibit lipolysis, rather than common carbohydrates such as glucose and sucrose, may be better for improving endurance [25]. Furthermore, simultaneous intake of citrate can be expected to promote energy consumption from lipids through inhibition of glycolysis [26]. This will spare glycogen and inhibit lactic acid production, so that the weakening of muscle contraction will be delayed. An amino acid, arginine, has been reported to modulate hormones that control the blood glucose level without inhibiting lipid metabolism, and to delay glycogen depletion during exercise [26,27]. Therefore, intake of both citrate and arginine along with carbohydrates that cause little stimulation of insulin secretion before or during exercise may be an effective way to improve energy metabolism and to supply the optimum energy sources for prolonged exercise.

If there is a shift from predominantly glucose-based energy consumption to lipid-based energy consumption, this may lead to improvement of endurance by maintaining glycogen stores and inhibiting the decrease of intramuscular pH that results from generation of lactate during exercise. Several authors have reported about various factors that can stimulate lipid metabolism, although there is insufficient evidence about their efficacy. Carnitine is an intracellular enzyme that is required for fatty acid transport across the mitochondrial membrane into the mitochondria, and it promotes the β-oxidation of fatty acids [28,29]. Carnitine supplementation is expected to activate lipid metabolism in the skeletal muscles, and to also achieve the sparing of glycogen stores. In persons performing aerobic training, intake of 2–4 g of carnitine before exercise or on a daily basis was reported to increase the maximum oxygen consumption (anaerobic threshold) and also inhibited the accumulation of lactate after exercise [30,31]. The effect of caffeine on endurance has also been studied.

Caffeine inhibits phosphotidiesterase by promoting catecholamine release and increases hormone-sensitive lipase (HSL) activity, which leads to an increase of circulating free fatty acids and further improvement of endurance [32,33]. Capsaicin, obtained from hot red peppers, is likely to enhance fat metabolism by altering the balance of lipolytic hormones and promoting fat oxidation in skeletal muscle [34,35].

10.4 ENHANCEMENT OF MUSCLE STRENGTH

It is well known that the strength of a muscle is generally proportional to its cross-sectional area, and it is necessary to increase muscle bulk in order to enhance strength. Muscle tissue is mainly composed of proteins (such actin and myosin) and water, and it is important to increase the protein content by modulating protein metabolism when increasing muscle bulk. In other words, muscle bulk and strength can be increased by promoting protein synthesis or by inhibiting protein degradation. Resistance exercise is aimed at increasing muscle bulk, and it enhances the secretion and production of growth hormone and various growth factors [36]. Thus, resistance exercise promotes protein synthesis and an increment of muscle mass more strongly compared with aerobic exercise. In order to maximize the effect of resistance exercise, it is important to maintain the muscular pool and blood levels of various amino acids that are substrates for the synthesis of muscle proteins. For this purpose, it is necessary to maintain a positive nitrogen balance by increasing the dietary protein intake. Several studies have shown that protein requirements of strength training athletes are higher than those of sedentary individuals [37-40]. The daily recommended protein intake is estimated to be 1.4 – 1.8 g/kg for performing resistance exercise when the intake of calories and carbohydrate is adequate [41,42] although 1.0 g/kg of protein is generally sufficient for endurance athletes excluding an elite minority [43]. It may be difficult to maintain such a high dietary protein intake, but ingestion of protein supplements can be effective. A wide variety of raw materials are utilized for the production of powdered protein supplements, and products derived from soy beans, eggs, or whey (milk protein) are commercially available. All of these products contain a good balance of essential amino acids, and often achieve an amino acid score of 100. In particular, whey protein is believed to be an ideal source for building muscles, since such protein is easily digested and absorbed, resulting in a rapid increase in the blood level of amino acids [44,45]. In addition, branched-chain amino acids and glutamine, which promote the synthesis of muscle protein, have a high content in whey protein [44,46]. Not only the amount of protein intake, but also the timing of intake are important for building muscles efficiently. Eating a meal immediately after resistance exercise may contribute to a greater increase of muscle mass compared with ingesting a meal several hours later [47-49]. Also, intake of carbohydrates with protein can accelerate the synthesis of muscle protein via the actions of insulin, which increases protein synthesis and inhibits its catabolism [50,51].

In addition, it has been reported that the intake of amino acids and peptides is beneficial. Free amino acids and peptides do not need to be digested, so rapid absorption can be expected. Amino acids are not only utilized for the synthesis of muscle protein, and some of these molecules also exert a variety of physiological effects. Attention has been focused on the effects of branched-chain amino acids (BCAAs), including valine, leucine, and isoleucine, which are known to have a relatively high content in both muscle proteins and food proteins [52]. Most amino acids are metabolized in the liver, but BCAAs are metabolized in the muscles via special processes [52,53]. BCAAs are utilized as energy substrates and their oxidation is enhanced during exercise by activation of branched-chain-α-keto acid dehydrogenase (BCKDH) complex [54]. Furthermore, BCAAs modulate muscle protein metabolism to promote the synthesis and inhibit the degradation of proteins [54-56], resulting in an anabolic effect on the muscles. Glutamine has also been reported to promote muscle growth by inhibiting protein degradation [57-59]. It is the most abundant free amino acid in muscle tissue [60] and its intake leads to an increase of myocyte volume, resulting in stimulation of muscle growth [57-59]. Glutamine is also found at relatively high concentrations in many other human tissues and has an important homeostatic role [60]. Therefore, during catabolic states such as exercise, glutamine is released from skeletal muscle into the plasma to be utilized for maintenance of the glutamine level in other tissues [61]. Arginine is a precursor of nitric oxide and creatine, and its injection promotes the secretion of growth hormone [62,63], which may lead to an increase of muscle mass and strength. Although the effect of oral arginine on protein synthesis is equivocal, recent studies have indicated that combined intake of arginine with other compounds improves exercise performance [64-66].

Various other food components have also been studied to determine their effects on muscle strength and mass. A meta-analysis of studies done between 1967 and 2001 supported the use of two supplements, creatine and β-hydroxy-β-methylbutyrate (βHMB), to augment lean body mass and strength when performing resistance exercise [67]. The human body contains more than 100 g of creatine, almost all of which is stored in the skeletal muscles as creatine phosphate. This is used to produce ATP by degradation to creatine under anaerobic conditions, so improvement of anaerobic metabolism can be expected by increasing the stores of creatine. Intake of creatine also stimulates water retention and protein synthesis [68,69]. It has been reported that the intake of ≥ 3 g/day of creatine increases the intramuscular content of creatine phosphate and improves endurance, especially during activities with a high power output (such as short distance running or resistance exercise), as well as improving muscle strength [70-72]. In addition, it has been reported that intake of creatine accelerates the increase of lean body mass and muscle strength during resistance training [72-74]. βHMB is a metabolite of the branched-chain amino acid leucine, and it increases muscle bulk by inhibiting the degradation of protein via an influence on the metabolism of branched-chain amino acids [75,76]. It has been reported that intake of 1.5 to 3.0 g/day of βHMB for 3 to 8 weeks achieved a greater increase of muscle mass and power compared with the intake of placebo [77-79].

10.5 PREVENTION OF INJURY AND FATIGUE

Strenuous physical activity or unaccustomed exercise causes injury to the muscles, release of muscular protein, and muscle pain. The mechanism underlying delayed the muscle damage after intense physical activity is not fully understood, but it has been suggested that such delayed injury is due to an inflammatory reaction induced by phagocyte infiltration that is triggered by excessive mechanical stress [80,81], an increased intracellular

Ca^{2+} concentration [82,83], and oxidative stress [84]. There are several reports which examined whether antioxidants attenuate the muscle damage since a significant increase of oxidative products is noted in the exercised muscles and in the blood in post-exercise parallel to other parameters of delayed-onset muscle damage. Oxidative injury after acute exercise can be prevented by the intake of antioxidants, such as vitamins C and E, carotenoids, or polyphenols, not only during exercise, but also on a daily basis [84-91]. In contrast, several studies have indicated that antioxidants do not affect muscle damage and the inflammatory response caused by strenuous exercise [92-94]. One possibility on reason of the different results is that the effect of antioxidants is likely to be differences of exercise conditions, such intensity of mechanical stress and oxygen uptake. Reactive oxygen species (ROS) could be related to initiation of the muscle damage. ROS are generated from mitochondria and endothelium during exercise via elevation of the oxygen uptake of myocytes and ischemia-reperfusion process, which leads to invasion of phagocytes into the muscles after exercise via redox-sensitive inflammatory cascade. Therefore, the inflammatory response may be inhibited if ROS production during exercise is decreased just in large contribution of ROS on the initiation of muscle damage such endurance prolonged exercise not resistance exercise. Additionally, it would be better to take several antioxidants at the same time because different organelles are affected by each kind of antioxidant, such water-soluble or lipid-soluble compounds, and they can provide electrons to each other to prevent a change to pro-oxidant status.

Glucosamine and chondroitin are substances that protect the joints. Glucosamine is an amino acid synthesized in the body that is a component of synovial fluid, tendons, and ligaments in the joints. Chondroitin is mainly contained in cartilage, tendons, and the connective tissues of the skin, and plays an important role as a shock absorber due to its hygroscopic action. Supplementary oral intake of these substances is suggested to be effective for preventing or promoting recovery from osteoarthritis associated with exercise and aging [95,96] while the effect of supplementation in exercise is not clear.

There are various kinds of factors expressing fatigue condition induced by exercise such glycogen depletion and accumulation of lactic acid during exercise, and hyperactivation of sympathetic nerve in post-exercise.

As mentioned above, recovery of glycogen storage in muscle is promoted by high-carbohydrate diet. At the same time, it is more effective to take a factor with an inhibitory effect on glycolysis such as citrate and consider the timing of carbohydrate intake. Also, lactate accumulation in muscle inhibits capacity of muscular contraction associated with pH decrease in muscle, which could be one of fatigue conditions. Thus, dietary supplementation regulating production or clearance lactate may be effective. Dipeptides that are abundant in skeletal muscle, carnosine and anserine, are known to have a pH-buffering effect [97]. Supplementation of these dipeptides is also possible to inhibit the decline of intramuscular pH by exercise via the buffering action of these dipeptides [98-100].

10.6 MAINTENANCE OF IMMUNITY

It is generally believed that moderate exercise enhances immunocompetence and is effective for the prevention of inflammatory diseases, infection, and cancer, while excessive physical activity leads to immunosuppression and an increase of inflammatory and allergic disorders [101-103]. Susceptibility to infections following excessive physical activity is ascribed to an increase in the production of immunosuppressive factors such as adrenocortical hormones and anti-inflammatory cytokines, leading to a decrease in the number and activity of circulating natural killer cells and T cells as well as a lower IgA concentration in the saliva [104]. Therefore, athletes performing high-intensity training are exposed to the risk of impaired immunocompetence. Intake of carbohydrates during prolonged exercise at submaximal intensity attenuates the increase of plasma cortisol and cytokine levels after exercise, which could lead to the inhibition of immunosuppression [105-107]. Vitamin C and vitamin E have actions that promote immunity, and are essential for T cell differentiation and for maintenance of T cell function [104,108,109]. However, there is limited evidence about the effects of vitamins supplementation on immune function in relation to exercise. Glutamine is an important energy source for lymphocytes, macrophages, and neutrophils, and is also an essential amino acid for the differentiation and growth of these cells [57,110]. Intense exercise decreases the plasma glutamine concentration and this may be related to immunosuppression [111]. Castell et al. [112] reported that athletes who ingested glutamine had a lower infection rate after a marathon compared with the placebo group. They also demonstrated that intake of glutamine resulted in an increase of the T-helper/T-suppressor cell ratio [113]. Furthermore, glutamine enhances the activity of intestinal enterobacteria and inhibits the production of cytokines involved in inflammation or immunosuppression [110].

10.7 CONCLUSION

Due to a social background that includes changes of dietary habits, an aging population, and increased medical costs, people have shown a growing interest in health and have come to expect complex and diverse actions of foods. In recent years, various food factors that fulfill such requirements have been evaluated scientifically to determine whether they are any physiological effects like prevention of diseases. In the sports market, a variety of functional foods are available, but among these functional foods, some have not clearly demonstrated any efficacy and others are advertised with inappropriate and exaggerated claims, so consumers are often confused. Some of the food components described in this article should be studied further because of differing views with regard to their efficacy in different reports. Furthermore, the effectiveness of the components may differ according to gender, between individuals, and with the mode of ingestion, so that the optimum method of intake the quantity and quality of foods to be ingested, and the timing of their intake need to be established in accordance with the purpose of using each food or food component, after understanding the physiological changes by exercise. In the future, guidelines for the use and evaluation system of sports functional foods should be established with backing by clear scientific evidence related to the individual foods.

TABLE 1: Exercise and functional foods.

Physiological functions	A	B	C
		Hypotonic drinks	
Replenishment of water	Isotonic drinks	Glycerol	
		Arginine	
	High-carbohydrate	Caffeine	
Improvement of endurance	Citric acid	Carnitine	Capsaicine
	Protein		
	BCAA		
	Creatine		
Enhancement of muscle strength	β-HMB	Glutamine	Arginine
		Vitamins C and E	
Prevention of muscle/joint injuries or fatigue	High-carbohydrate	Carotenoids, Flavonoids	Glucosamine
	Citric acid	Carnosine, Anserine	Chondroitin
Prevention of a decrease in immunocompetence	Carbohydrate	Vitamins C and E	
		Glutamine	

A: The factors in this group has been shown adequate scientific evidence.
B: The factors in this group has been shown suggestive evidence.
C: The factors in this group has been shown no evidence while possible to effective

REFERENCES

1. Buono MJ, Wall AJ: Effect of hypohydration on core temperature during exercise in temperate and hot environments. Pflugers Arch 2000, 440:476-480.
2. Kenny GP, Periard J, Journeay WS, Signal RJ, Reardon FD: Effect of exercise intensity on the postexercise sweating threshold. J Appl Physiol 2003, 95:2355-2360.
3. Kondo N, Tominaga H, Shibasaki M, Aoki K, Okuda S, Nishiyasu T: Effect of exercise intensity on the sweating response to a sustained static exercise. J Appl Physiol 2000, 88:1590-1596.
4. Nose H, Takamata A, Mack GW, Oda Y, Kawabata T, Hashimoto S, Hirose M, Chihara E, Morimoto T: Right atrial pressure and forearm blood flow during prolonged exercise in a hot environment. Pfluger Arch 1994, 426:177-182.

5. Fortney SM, Wenger CB, Bove JR, Nadel ER: Effect of hyperosmolality on control of blood flow and sweating. J Appl Physiol 1984, 57:1688-1695.
6. Decombaz J, Gmunder B, Daget N, Munoz-Box R, Howald H: Acceptance of iso-tonic and hypotonic rehydrating beverages by athletes during training. Int J Sports Med 1992, 13:40-46.
7. Castellani JW, Maresh CM, Armstrong LE, Kenefick RW, Riebe D, Echegaray M, Kavouras S, Castracane VD: Endocrine responses during exercise-heat stress: ef-fects of prior isotonic and hypotonic intravenous rehydration. Eur J Appl Physiol 1998, 77:242-248.
8. Takamata A, Mack GW, Gillen CM, Nadel ER: Sodium appetite, thirst, and body fluid regulation in humans during rehydration without sodium replacement. Am J Physiol 1994, 266:R1493-R1502.
9. Latzka WA, Montain SJ: Water and electrolyte requirements for exercise. Clin Sports Med 1999, 18:513-524.
10. Robergs RA: Glycerol. Biochemistry, pharmacokinetics and clinical and practical applications. Sports Med 1998, 26:145-167.
11. Riedesel ML: Hyperhydration with glycerol solutions. J Appl Physiol 1998, 63:2262-2268.
12. Lyons TP: Effects of glycerol-induced hyperhydration prior to exercise in the heat on sweating and core temperature. Med Sci Sports Exerc 1990, 22:477-483.
13. Montner P: Pre-exercise glycerol hydration improves cycling endurance time. Int J Sports Med 1996, 17:27-33.
14. Anderson MJ: Effect of glycerol-induced hyperhydration on thermoregulation and metabolism during exercise in heat. Int J Sport Nutr Exerc Metab 2001, 11:315-333.
15. Hulyman E, Nilsson LH: Liver glycogen in men. Effect of different diets and muscu-lar exercise. Advances in Experimental Medicine and Biology. Muscle Metabolism During Exercise. Volume 2. Edited by Pernow B, Saltin B. Plenum Press, New York; 1971:143-151.
16. Forgac MT: Carbohydrate loading-a review. J Appl Physiol 1980, 48:624-629.
17. Evans WJ, Hughes VA: Dietary carbohydrates and endurance exercise. Am J Clin Nutr 1985, 41:1146-1154.
18. Saito S, Yoshitake Y, Suzuki M: Enhanced glycogen repletion in liver and skeletal muscle with citrate orally fed after exhaustive treadmill running and swimming. J Nutr Sci Vitaminol 1983, 29:45-52.
19. Saito A, Tasaki Y, Tagami K, Suzuki M: Muscle glycogen repletion and pre-exercise glycogen content: effect of carbohydrate loading in rats previously fed a high fat diet. Eur J Appl Physiol 1994, 68:483-488.
20. Saitoh S, Shimomura Y, Suzuki M: Effect of a high-carbohydrate diet intake on muscle glycogen repletion after exercise in rats previously fed a high-fat diet. Eur J Appl Physiol 1993, 66:127-133.
21. Tarnopolsky MA, Bosman M, Macdonald JR, Vandeputte D, Martin J, Roy BD: Postexercise protein-carbohydrate and carbohydrate supplements increase muscle glycogen in men and women. J Appl Physiol 1997, 83:1877-1883.

22. Ivy JL, Goforth HW Jr, Damon BM, McCauley TR, Parsons EC, Price TB: Early postexercise muscle glycogen recovery is enhanced with a carbohydrate-protein supplement. J Appl Physiol 2002, 93:1337-1344.
23. Jungas RL: Effects of insulin and proinsulin. E. Metabolic effects on adipose tissue in vitro. Hand b, Exp Pharmacol 1975, 7:371-412.
24. Arner P, Engfeldt P: Fasting-mediated alteration studies in insulin action on lipolysis and lipogenesis in obese women. Am J Physiol 1987, 253:E193-E201.
25. Mitsuzono R, Okamura K, Igaki K, Iwanaga K, Sakurai M: Effects of fructose ingestion on carbohydrate and lipid metabolism during prolonged exercise in distance runners. Appl Human Sci 1995, 14:125-131.
26. Saitoh S, Suzuki M: Nutritional design for repletion of liver and muscle glycogen during endurance exercise without inhibitinglipolysis. J Nutr Sci Vitaminol 1986, 32:343-353.
27. Palmer JP, Walter RM, Ensinck JW: Arginine-stimulated acute phase of insulin and glucagon secretion. I. in normal man. Diabetes 1975, 24:735-740.
28. Brass EP: Supplemental carnitine and exercise. Am J Clin Nutr 2000, 72:618S-623S.
29. Cerretelli P, Marconi C: L-carnitine supplementation in humans. The effects on physical performance. Int J Sports Med 1990, 11:1-14.
30. Marconi C, Sessi G, Carpinelli A, Cerretelli P: Effects of L-carnitine loading on the aerobic and anaerobic performance of endurance athletes. Eur J Appl Physiol 1985, 10:169-174.
31. Vecchiet L, Di Lisa F, Pieralisi G, Ripari P, Menabo R, Giamberardino MA, Siliprandi N: Influence of L-carnitine administration on maximal physical exercise. Eur J Appl Physiol 1990, 61:486-490.
32. Tarnopolsky MA: Caffeine and endurance performance. Sports Med 1994, 18:109-125.
33. Ryu S, Choi SK, Joung SS, Suh H, Cha YS, Lee S, Lim K: Caffeine as a lipolytic food component increases endurance performance in rats and athletes. J Nutr Sci Vitaminol 2001, 47:139-146.
34. Lim K, Yoshioka M, Kikuzato S, Kiyonaga A, Tanaka H, Shindo M, Suzuki M: Dietary red pepper ingestion increases carbohydrate oxidation at rest and during exercise in runners. Med Sci sports Exerc 1997, 29:355-361.
35. Yoshioka M, Lim K, Kikuzato S, Kiyonaga A, Tanaka H, Shindo M, Suzuki M: Effects of red pepper diet on the energy metabolism in men. J Nutr Sci Vitaminol 1995, 41:647-656.
36. Kraemer WJ, Marchitelli L, Gordon SE, Harman E, Dziados JE, Mello R, Frykman P, McCurry D, Fleck SJ: Hormonal and growth factor responses to heavy resistance exercise protocols. J Appl Physiol 1990, 69:1442-1450.
37. Mccall GE, Byrnes WC, Fleck SJ, Dickinson A, Kraemer WJ: Acute and chronic hormonal responses to resistance training designed to promote muscle hypertrophy. Can J Appl Physiol 1999, 24:96-107.
38. Tarnopolsky MA, MacDougall JD, Atkinson SA: Influence of protein intake and training status on nitrogen balance and lean body mass. J Appl Physiol 1988, 64:187-193.

39. Tarnopolsky MA, Atkinson SA, MacDougall JD, Chesley A, Phillips S, Schwarcz HP: Evaluation of protein requirements for trained strength athletes. J Appl Physiol 1992, 73:1986-1995.

40. Lemon PW, Tarnopolsky MA, MacDougall JD, Atkinson SA: Protein requirements and muscle mass/strength changes during intensive training in novice bodybuilders. J Appl Physiol 1992, 73:767-775.

41. Lemon PW: Is increased dietary protein necessary or beneficial for individuals with a physically active lifestyle? Nutr Rev 1996, 54:S169-S175.

42. Phillips SM: Protein requirements and supplementation in strength sports. Nutrition 2004, 20:689-695.

43. Tarnopolsky MA: Protein requirements for endurance athletes. Nutrition 2004, 20:662-668.

44. Ha E, Zemal MB: Functional properties of whey, whey components, and essential amino acids: mechanisms underlying health benefits for active people (review). J Nutr Biochem 2003, 14:251-258.

45. Boirie Y, Dangin M, Gachon P, Vasson MP, Maubois JL, Beaufrere B: Slow and fast dietary proteins differently modulate postprandial protein accretion. Proc Natl Acad Sci USA 1997, 94:14930-14935.

46. de Wit JN: Marschell Rhone-Poulenc Award Lecture. Nutritional and functional-characteristics of whey proteins in food products. J Dairy Sci 1998, 81:597-608.

47. Suzuki M, Doi T, Lee SJ, Okamura K, Shimizu S, Okano G, Sato Y, Shimomura Y, Fushiki T: Effect of meal timing after resistance exercise on hindlimb muscle mass and fat accumulation in trained rats. J Nutr Sci Vitaminol 1999, 45:401-409.

48. Esmarck B, Andersen JL, Olsen S, Richter EA, Mizuno M, Kjaer M: Timing of postexercise protein intake is important for muscle hypertrophy with resistance training in elderly humans. J Physiol 2001, 535:301-311.

49. Flakoll PJ, Judy T, Flinn K, Carr C, Flinn S: Postexercise protein supplementation improves health and muscle soreness during basic military training in Marine recruits. J Appl Physiol 2004, 96:951-956.

50. Borsheim E, Aarsland A, Wolfe RR: Effect of an amino acid, protein, and carbohydrate mixture on net muscle protein balance after resistance exercise. Int J Sport Nutr Exerc Metab 2004, 14:255-271.

51. Roy BD, Tarnopolsky MA, MacDougall JD, Fowles J, Yarasheski KE: Effect of glucose supplement timing on protein metabolism after resistance training. J Appl Physiol 1997, 82:1882-1888.

52. Harper AE, Miller RH, Block KP: Branched-chain amino acid metabolism. Annu Rev Nutr 1984, 4:409-454.

53. Rennie MJ: Influence of exercise on protein and amino acid metabolism. In Handbook of Physiology. Sect. 12: Exercise: Regulation and Integration Multiple Systemas. Volume chapter 22. Edited by Rowell LB, Shepherd JT. American Physiolosical Society, Bethesda, MD; 995-1035.

54. Shimomura Y, Murakami T, Nakai N, Nagasaki M, Harris RA: Exercise promotes BCAA catabolism: effects of BCAA supplementation on skeletal muscle during exercise. J Nutr 2004, 134:1583S-1587S.

55. Buse MG, Reid SS: Leusine: a possible regulator of protein turnover in muscle. J Clin Invest 1975, 56:1250-1261.

56. Tischler ME, Desautels M, Goldberg AL: Does leucine, leucyl-tRNA, or some metabolite of leucine regulate protein synthesis and degradation in skeletal and cardiac muscle. J Biol Chem 1982, 257:1613-1621.
57. Antonio J, Street C: Glutamine: A potentially useful supplement for athletes. Can J Appl Physiol 1999, 24:1-14.
58. Hankard RG, Haymond MW, Darmaun D: Effect of glutamine on leucine metabolism in humans. Am J Physiol 1996, 271:E748-E754.
59. Vom Dahl S, Haussinger D: Nutritional state and the swelling-induced inhibition of proteolysis in perfused rat liver. J Nutr 1996, 126:395-402.
60. Felig P: Amino acid metabolism in man. Annu Rev Biochem 1975, 44:933-955.
61. Castell LM: Glutamine supplementation in vitro and in vivo, in exercise and in immunodepression. Sports Med 2003, 33:323-345.
62. Merimee TJ, Rabinowitz D, Fineberg SE: Arginine-initiated release of human growth hormone. N Engl J Med 1969, 280:1434-1438.
63. Alba-Roth J, Muller OA, Schopohl J, Werder K: Arginine stimulates growth hormone secretion by suppressing endogenous somatostatin secretion. J Clin Endocrinol Metab 1988, 67:1186-1189.
64. Paddon-Jones D, Borsheim E, Wolfe RR: Potential ergogenic effects of arginine and creatine supplementation. J Nutr 2004, 134:2888S-2894S.
65. Ohtani M, Sugita M, Maruyama K: Amino Acid mixture improves training efficiency in athletes. J Nut 2006, 136:538S-543S.
66. Flakoll P, Sharp R, Baier S, Levenhagen D, Carr C, Nissen S: Effect of beta-hydroxy-beta-methylbutyrate, arginine, and lysine supplementation on strength, functionality, body composition, and protein metabolism in elderly women. Nutrition 2004, 20:445-451.
67. Nissen S, Sharp R: Effect of dietary supplements on lean mass and strength gains with resistance exercise: a meta-analysis. J Appl Physiol 2003, 94:651-659.
68. Balsom P, Soderlund K, Shodin B, Ekblom B: Skeletal muscle metabolism during short duration high-intensity exercise: influence of creatine supplementation. Acta Physiol Scand 1995, 1154:303-310.
69. Balsom P, Soderlund K, Ekblom B: Creatine in humans with special references to creatine supplementation. Sports Med 1994, 18:268-280.
70. Terjung RL, Clarkson P, Eichner ER, Greenhaff PL, Hespel PJ, Israel RG, Kraemer WJ, Meyer RA, Spriet LL, Tarnopolsky MA, Wagenmakers AJ, Williams MH: American College of Sports Medicine roundtable. The physiological and health effects of oral creatine supplementation. Med Sci Sports Exerc 2000, 32:706-717.
71. Casey A, Constantin-Teodosiu D, Howell D, Hultman E, Greenhalf PL: Creatine ingestion favorably affects performance and muscle metabolism during maximal exercise in humans. Am J Physiol 1996, 271:E31-E37.
72. Kreider R, Ferreira M, Wilson M, Grindstaff P, Plisk S, Reinardy J, Centler E, Almada AL: Effects of creatine supplementation on body composition, strength and sprint performance. Med Sci Sport Exerc 1998, 30:73-82.
73. Volec JS, Duncan ND, Mazzetti SA, Staron RS, Putukian M, Gomez AL, Pearson DR, Fink WJ, Kraemer WJ: Performance and muscle fiber adaptations to creatine supplementation and heavy resistance exercise. Med Sci Sport Exerc 1999, 31:1147-1156.

74. Vandenburghe K, Goris M, Van Hecke P, Van Leemputte M, Vangerven L, Hespel P: Long-term creatine intake is beneficial to muscle performance during resistance-training. J Appl Physiol 1997, 83:2055-2063.

75. Alon T, Bagchi D, Preuss HG: Supplementing with beta-hydroxy-beta-methylbutyr-ate (HMB) to build and maintain muscle mass: a review. Res Commun Mol Pathol Pharmacol 2002, 111:139-151.

76. Slater GJ, Jenkins D: Beta-hydroxy-beta-methylbutyrate (HMB) supplementation and the promotion of muscle growth and strength. Sport Med 2000, 30:105-116.

77. Nissen S, Sharp R, Ray M, Rathmacher JA, Rice D, Fuller JC Jr, Connelly AS, Ab-umrad N: Effect of leucine metabolite beta-hydroxy-beta-methylbutyrate on muscle metabolism during resistance-exercise training. J Appl Physiol 1996, 81:2095-2104.

78. Gallagher PM, Carrithers JA, Godard MP, Schulze KE, Trappe SW: Beta-hydroxy-beta-methylbutyrate ingestion, Part 1: effects on strength and fat free mass. Med Sci Sport Exerc 2000, 32:2109-2115.

79. Panton LB, Rathmacher JA, Baier S, Nissen S: Nutritional supplementation of the leucine metabolite beta-hydroxy-beta-methylbutyrate (hmb) during resistance train-ing. Nutrition 2000, 16:734-739.

80. Komulainen J, Takala TE, Kuipers H, Hesselink MK: The disruption of myofibre structures in rat skeletal muscle after forced lengthening contractions. Pflugers Arch 1998, 436:735-741.

81. Proske U, Morgan DL: Muscle damage from eccentric exercise: mechanism, me-chanical signs, adaptation and clinical applications. J Physiol 2001, 537:333-345.

82. Gissel H, Clausen T: Excitation-induced Ca2+ influx and skeletal muscle cell dam-age. Acta Physiol Scand 2001, 171:327-334.

83. Tidball JG: Inflammatory cell response to acute muscle injury. Med Sci Sports Exerc 1995, 27:1022-1032.

84. Aoi W, Naito Y, Takanami Y, Kawai Y, Sakuma K, Ichikawa H, Yoshida N, Yo-shikawa T: Oxidative stress and delayed-onset muscle damage after exercise. Free Radic Biol Med 2004, 37:480-487.

85. Phillips T, Childs AC, Dreon DM, Phinney S, Leeuwenburgh C: A dietary supple-ment attenuates IL-6 and CRP after eccentric exercise in untrained males. Med Sci Sports Exerc 2003, 35:2032-2037.

86. Takanami Y, Iwane H, Kawai Y, Shimomitsu T: Vitamin E supplementation and en-durance exercise. Are there benefits? Sports Med 2000, 29:73-83.

87. Kanter MM, Nolte LA, Holloszy JO: Effects of an antioxidant vitamin mixture on lipid peroxidation at rest and postexercise. J Appl Physiol 1993, 74:965-969.

88. Aoi W, Naito Y, Sakuma K, Kuchide M, Tokuda H, Maoka T, Toyokuni S, Oka S, Yasuhara M, Yoshikawa T: Astaxanthin limits exercise-induced skeletal and cardiac muscle damage in mice. Antioxid Redox Signal 2003, 5:139-144.

89. Sumida S, Doi T, Sakurai M, Yoshioka Y, Okamura K: Effect of a single bout of ex-ercise and βcarotene supplementation on the urinary excretion of 8-hydroxydeoxy-guanosine in humans. Free Rad Res 1997, 27:607-618.

90. Marquez R, Santangelo G, Sastre J, Goldschmidt P, Luyckx J, Pallardo FV, Vina J: Cyanoside chloride and chromocarbe diethylamine are more effective than vitamin C against exercise-induced oxidative stress. Pharmacol Toxicol 2001, 89:255-258.

91. Kato Y, Miyake Y, Yamamoto K, Shimomura Y, Ochi H, Mori Y, Osawa T: Prepara-
     tion of a monoclonal antibody to N(epsilon)-(hexanonyl)lysine: application to the
     evaluation of protective effects of flavonoid supplementation against exercise-in-
     duced oxidative stress in rat skeletal muscle. Biochem Biophys Res Commun 2000,
     274:389-393.

92. Petersen EW, Osrowski K, Ibfelt T, Richelle M, Offord E, Halkjaer-Kristensen J,
     Pedersen BK: Effect of vitamin E supplementation on cytokine response and on mus-
     cle damage after strenuous exercise. Am J Physiol Cell Physiol 2001, 280:C1570-
     C1575.

93. Warren JA, Jekins RR, Packer L, Witt EH, Armstrong RB: Elevated muscle vitamin
     E does not attenuate eccentric exercise-induced muscle injury. J Appl Physiol 1992,
     72:2168-2175.

94. Beaton LJ, Allan DA, Tarnopolsky MA, Tiidus PM, Phillips SM: Contraction-in-
     duced muscle damage is unaffected by vitamin E supplementation. Med Sci Sports
     Exerc 2002, 34:798-805.

95. Leffler CT, Philippi AF, Leffler SG, Mosure JC, Kim PD: Glucosamine, chondroitin,
     and manganese ascorbate for degenerative joint disease of the knee or low back: a
     randomized, double-blind, placebo-controlled pilot study. Mil Med 1999, 164:85-
     91.

96. Beren J, Hill SL, Diener-West M, Rose NR: Effect of pre-loading oral glucosamine
     HCl/chondroitin sulfate/manganese ascorbate combination on experimental arthritis
     in rats. Exp Biol Med (Maywood) 2001, 226:144-151.

97. Quershi J, Wood T: The effect of carnosine on glycolysis. Biochem Biophys Acta
     1962, 60:190-192.

98. Suzuki Y, Nakao T, Maemura H, Sato M, Kamahara K, Morimatsu F, Takamatsu K:
     Carnosine and anserine ingestion enhances contribution of nonbicarbonate buffer-
     ing. Med Sci Sports Exerc 2006, 38:334-338.

99. Harada R, Taguchi H, Urashima K, Sato M, Omori T, Morimatsu F: Effects of a
     chicken extract on endurance swimming in mice. J Jpn Soc Nutr Food Sci 2002,
     55:73-78.

100. Harada R, Urashima K, Sato M, Omori T, Morimatsu F: Effects of carnosine and a
     chicken extract on recovery from fatigue due to swimming in mice. J Jpn Soc Nutr
     Food Sci 2002, 55:209-214.

101. Oconnor FG, Wilder RP: Textbook of running medicine. MaGraw-Hill Co; 2004.

102. Fitzgerald L: Overtraining increases the susceptibility to infection. Int J Sports Med
     1991, 12:S5-S8.

103. Shephard RJ, Rhind S, Shek PN: Exercise and the immune system. Natural killer
     cells, interleukins and related responses. Sports Med 1994, 18:340-369.

104. Gleeson M, Nieman DC, Pederson BK: Exercise, nutrition and immune function. J
     Sports Sci 2004, 22:115-125.

105. Nieman DC, Nehlsen-Cannarella SL, Fagoaga OR, Henson DA, Utter A, Davis
     JM, Williams F, Butterworth DE: Effects of mode and carbohydrate on the granulo-
     cyte and monocyte response to intensive, prolonged exercise. J Appl Physiol 1998,
     84:1252-1259.

106. Nieman DC, Henson DA, Smith LL, Utter AC, Vinci DM, Davis JM, Kaminsky DE,
     Shute M: Cytokine changes after a marathonrace. J Appl Physiol 2001, 91:109-114.

107. Nehlsen-Cannarella SL, Fagoaga OR, Nieman DC, Henson DA, Butterworth DE, Schmitt RL, Bailey EM, Warren BJ, Utter A, Davis JM: Carbohydrate and the cytokine response to 2.5 h of running. J Appl Physiol 1997, 82:1662-1667.
108. Peake JM: Vitamin C: Effects of exercise and requirements with training. Int J Sport Nutr Exerc Metab 2003, 13:125-151.
109. Moriguchi S, Muraga M: Vitamin E and immunity. Vitam Horm 2000, 59:305-336.
110. Castell LM: Glutamine supplementation in vitro and in vivo, in exercise and immunodepression. Sports Med 2003, 33:323-345.
111. Keast D, Arstein D, Harper W, Fry RW, Morton AR: Depression of plasma glutamine concentration after exercise stress and its possible influence on the immune system. Med J Aust 1995, 162:15-18.
112. Castell LM, Newsholme EA: The effects of oral glutamine supplementation on athletes after prolonged, exhaustive exercise. Nutrition 1997, 13:738-742.
113. Castell LM, Poortmans JR, Newsholme EA: Does glutamine have a role in reducing infections in athletes? Eur J Appl Physiol 1996, 73:488-490.

This chapter was originally published under the Creative Commons Attribution License. Aoi, W., Naito, Y., and Yoshikawa, T. Exercise and Functional Foods. Nutrition Journal 2006, 5:15 doi:10.1186/1475-2891-5-15. doi:10.1186/1475-2891-5-15. http://www.nutritionj.com/content/5/1/15

CHAPTER 11

EFFECTS OF MICRONUTRIENT FORTIFIED MILK AND CEREAL FOOD FOR INFANTS AND CHILDREN: A SYSTEMATIC REVIEW

KLAUS EICHLER, SIMON WIESER, ISABELLE RÜTHEMANN, and URS BRÜGGER

11.1 BACKGROUND

Micronutrient (MN) deficiency is a common public health problem, specifically for infants and children, in many low and middle income countries. For example, anemia (caused by iron deficiency) or increased infection rates and mortality (exacerbated by vitamin A and zinc deficiency) are serious threats for child development [1]. The first two years of life represent a narrow time window, which is of outstanding importance for child development [2]. During this time period future growth and vulnerable physiological capacities, such as cognitive function and motor development, are determined. Even with optimum breastfeeding, these steps depend on a an adequate quantity and quality of complementary feeding, leading to an adequate MN supply [2]. Negative health consequences resulting from suboptimal feeding, such as stunting (i.e. low height-for-age), are associated with higher morbidity and decreased function in later life [3].

Several strategies have been shown to be effective in resolving MN deficiencies for different target groups and are proposed in recommendations and guidelines [4-6]: Food based approaches (e.g. spreads to increase energy-density and MN content of food; MN powders for home fortification with sprinkles) and MN supplementation (e.g. vitamin A capsules administered at defined intervals). In addition, fortification of staple food (e.g. fortified salt, flour or oil) is widely used to resolve MN deficiencies of general populations.

Fortified complementary feeding after 6 months of age, in combination with continued breastfeeding [7], typically comprises milk or cereals products (e.g. porridge or gruel) for infants. This type of food, however, is often not covered by programs that provide fortified staple food for the general population. Primary studies have assessed the effects of fortified milk or cereals for infants and children [8,9] and some countries, such as Mexico, have introduced country wide food programs, where fortified milk is one component [10]. However, the overall evidence of the effect of fortified milk and cereals on children has not been systematically assessed.

Thus, we performed a Systematic Review to specifically assess the impact of micronutrient fortified milk and cereal food on the health of infants and children compared to non-fortified food in randomized controlled trials.

11.2 METHODS

We performed our review in accordance with current guidelines for performing [11,12] and reporting of systematic reviews [13] and established a scientific advisory board (see Acknowledgments for participating experts). A review study protocol was developed in advance, though not published.

According to our research question we defined the following inclusion criteria: Population: Infants and children from 6 months to 5 years of age. While our primary focus was on age groups up to 2 years, we decided to set an upper age limit at 5 years, in order not to miss suitable studies with mixed age groups. Intervention: Micronutrient fortified milk or cereal food. Control intervention: Non-fortified food; additional other nutritional approaches, if such approaches were applied in the intervention and control group. Outcome: At least one of the following health related outcomes: surrogate measures (such as MN serum levels, hematological parameters), functional outcome (e.g. motor development), measures of morbidity (such as disease rates) or mortality. Study designs: Randomized controlled trials of any follow-up time.

We excluded studies with infants and toddlers younger than 6 months [14] or applying infant formula [15], studies addressing adolescents or adult women, interventions based on supplementation, home fortification, bare food based approaches, fortification with components other than micronutrients, and studies testing absorption of MN. A priori, we also excluded studies with fortification of staple food as provided for larger population groups to isolate the effect of fortified milk and cereals.

TABLE 1: Medline electronic search strategy

Step	Search Medline 1	Search Medline 2	Search Medline 3
	"Infant Formula"[MeSH]ª OR		
1	"Milk"[MeSH]	"economics" [MeSH] "micronutrients"	nutrition disorders[MeSH]
2	fortif*[TIAB]ᵇ	[MeSH] "Nutrition	child* OR infant* OR toddl*[TIAB]
3	1 AND 2	Disorders"[MeSH]	"cost*"[TIAB] OR "economics"[MeSH]
4	"Cereals"[MeSH]	1 AND 2 AND3	1 AND 2 AND 3
			"india*"[TIAB] OR "pakistan*"[TIAB] OR "philippine*"[TIAB] OR
5	fortif*[TIAB]	"cost*"[TIAB] "micronutrients"	"asia*"[TIAB] OR "africa*"[TIAB]
6	4 AND 5	[MeSH] "nutrition	4 AND 5
7	3 OR 6 child*[TIAB] OR infant*[TIAB] OR	disorders"[MeSH]	
8	toddler*[TIAB]	5 AND 6 AND 7	
9	7 AND 8	4 OR 8	

Three Medline searches were performed and retrieved references were cumulated. As this review was part of a larger project that evaluates the economic effects of micronutrient fortification as well, we included also search terms such as "cost" and "economics".
a MeSH: Medical Subject Heading.
bTIAB: Title/Abstract.

We systematically searched for studies using electronic databases (Medline [search strategy Table 1, Cochrane library; from 1966 to February 2011; no language restriction). As this review was part of a larger project, that evaluates the economic effects of MN fortification as well, we also included search terms such as "cost" and "economics". We screened reference lists of included papers and contacted experts in the field for additional references. In addition, we screened homepages of relevant organizations (e.g. WHO, United Nations [World Food Programme, Unicef, Millennium Development Goals], The World Bank, Pakistan National Nutrition Survey; International Clinical Epidemiology Network [16]; Global Alliance for Improved Nutrition, GAIN [17]; The Micronutrient Initiative [18]; Bill & Melinda Gates Foundation [19]). We also contacted a manufacturer (Nestlé) for further material and performed hand searches in relevant journals with developing countries issues (such as The Lancet). All references were stored in an EndNote X4 database (Thomson/ISI ResearchSoft Berkeley, CA, USA).

11.2.1 STUDY SELECTION AND DATA EXTRACTION

Three reviewers screened titles and abstracts for relevance and assessed potentially relevant studies for inclusion by full text. Teaching sessions were held in advance to improve conceptual consistency between reviewers. Disagreements were resolved by consensus meetings. If data of a specific population were published in several papers or if follow-up data were presented, we included each population only once. Using a predefined form, data were extracted by one reviewer in an Excel database and checked independently by a second reviewer.

We extracted data on general study information (e.g. study region; length and completeness of follow up), study setting (e.g. level of population recruitment), population details, intervention (e.g. daily amount of fortified MN, determined as daily difference between intervention and control group; composition of MN; comparator food) and outcome (e.g. morbidity rates; hemoglobin levels [g/dl; conversion to g/L with factor 10]).

One reviewer assessed risk of bias in individual studies with a component approach exploring methodological quality on the study level (adequate generation of random sequence, concealment of allocation, blinding) as well as on the outcome level (incomplete outcome data due to attrition; selective outcome reporting) [12].

11.2.2 STATISTICAL ANALYSIS

First, we calculated pooled estimates. For continuous variables we computed weighted mean differences (WMD) and 95%-confidence intervals (CI). For example, for analysis of hemoglobin change we used the mean change in the intervention and in the control group and their pooled standard deviation (SD). If the sample size decreased during the study, we used the lower sample size at the end of the study. If mean hemoglobin change per group and SD were not reported, we calculated change as the difference of baseline and final values for intervention and control group and applied the SD of final values [20]. If 95%-CI of mean values were reported we converted them to SD assuming normal distribution [21]. To check results for robustness, we also calculated WMD for final hemoglobin values of both study groups, as this data was reported more often. Due to considerable heterogeneity between trials, we applied a random effects model [22]. When authors reported only medians for continuous data (e.g. for ferritin levels), we did not include those data in the meta-analysis. For binary data, we calculated risk ratios and 95%-CI. Heterogeneity between trials was calculated with I2, that is the percentage of the total variation in estimated effects that is due to heterogeneity rather than chance (where values of 25% are assigned low, 50% moderate and 75% high) [23].

Second, we divided our dataset into pre-specified subgroups to explore the influence of possible modifying factors on the outcome (fortified milk vs. cereal food; high vs. low/middle-income countries; single- vs. dual/multi-micronutrient fortification strategy).

Third, we performed a meta-regression analysis weighted for the inverse of the variance of the outcome [12]. With this approach we evaluated the unique contribution of other a priori chosen independent factors on the most often reported outcome (dependent variable: hemoglobin level; independent variables: hemoglobin levels before intervention; daily amount of fortified MN; length of follow-up; completeness of follow-up).

For parametric and non-parametric tests P-values <0.05 were considered significant. Analyses were performed using the STATA SE 9 software package (StataCorp. 2007. Stata Statistical Software, College Station, Texas, USA).

11.3 RESULTS

11.3.1 DESCRIPTION OF INCLUDED STUDIES

Our searches retrieved 1153 potentially relevant studies (Figure 1). Eighteen RCT [8-10,24-38] (n = 10 fortified milk; n = 8 fortified cereals) fulfilled inclusion criteria and were included for our main analysis (Table 2).

These 18 trials comprised 5468 infants and children from different regions (2 studies from Asia [8,37], 5 studies from Africa [9,33-36]; 5 Studies from South- and Middle-America [10,27,29,30,38]; 6 Studies from Europe [24-26,28,31,32]).

Study population sizes varied from n = 33 to n = 1120 participants (median 166; IQR 92 to 361). Most participants belonged to vulnerable groups and had been recruited from different settings (8 studies: medical or community care centers:, 7 studies: low income risk groups; 2 studies: general population of peri-urban and rural areas; 1 study: no information given). The most frequent exclusion criteria were chronic diseases, severe anemia, severe mal-/under-nutrition, and low birth weight. Mean age of participants ranged from 6 to 23 months at inclusion (upper age limit was 3 years in one study [8]) and the sex ratio was well balanced. Mean hemoglobin values of children at baseline varied between studies from 9.0 g/dl to 12.6 g/dl (median of study values: 11.1 g/dl). Follow up periods were generally short and did not exceed one year (mean follow up: 8.2 months; range: 2.3 to 12).

Fortified milk was prepared with centrally processed fortified milk powder in most of the studies. Fortified cereals comprised centrally processed weaning or complementary food, such as fortified porridge, gruel or weaning rusk to prepare a pap. Iron was the most frequently used MN for fortification (15 of 18 trials), followed by zinc (9 trials) and vitamin A (6 trials). Seven studies used a single-MN fortification strategy (6 studies with iron only; 1 study with zinc only), two studies a dual- and 9 studies a multi-micronutrient (MMN) strategy (i.e. 3 or more MN, for example additional fortification with vitamin C and E, selenium, copper).

TABLE 2: Details of included studies for fortification of milk and cereal food

Study	Population	Intervention	Control food	Outcome	Comment
Author, year: Brown, 2007 [29] Design: RCT	Country: Peru Target population: periurban area; Age (mean; range): 0.6; 0.5 to 0.7 years Males (%): 46 Exclusion criteria: risk of acute malnutrition; chronic diseases	Cereals, fortified (porridge); single MN strategy MN applied a: Zn Iron dosage b: n.a.c mg/day; Iron compound d: n.a.	porridge, unfortified	After 0.5 year: plasma zinc; anthropometry; infections	Both groups recieved iron fortification and vitamin supplements, thus net intervention was zinc fortification.
Author, year: Daly, 1996 [31] Design: RCT	Country: UK (74% white; 24% Afro-Caribbean; 2% Asian) Target population: poor innerurban Age (mean; range): 0.65; 0.5 to 0.7 years Males (%): 47 Exclusion criteria: preterm at birth	Milk, fortified; multi MN strategy MN applied: Fe, VitA, other Vitamins, other MN Iron dosage: 5.47 mg/day; Iron compound: no info	milk, un-fortified	After 1 year: hema-tological parameters; anthropometry	Functional outcome was extracted from related paper Williams_1999 [39].
Author, year: Faber, 2005 [33] Design: RCT	Country: South Africa Target population: rural area, low socio-economic status, Age (mean; range): 0.7; 0.6 to 0.9 years Males (%): 51 Exclusion criteria: birth weight <2500 g, severe anemia	Cereals, fortified (porridge); multi MN strategy MN applied: Fe, Zn, other Vitamins Iron dosage: 27.5 mg/day; Iron compound: FeFu	porridge, unfortified	After 0.5 year: hemato-logical parameters, serum retinol, zinc; growth; motor development	Population baseline characteristics only for infants who completed the study.
Author, year: Gibson, 2011 [35] Design: RCT	Country: Zambia Target population: middle income class Age (mean; range): 0.5; 0.5 to 0.5 years Males (%): 48 Exclusion criteria: "not in good health"	Cereals, fortified (porridge); multi MN strategy MN applied: Fe, Zn, other Vitamins, other MN Iron dosage: 5.36 mg/day; Iron compound: no info	porridge, unfortified	After 1 year: hemato-logical parameters; serum zinc, anthropometry; hospital referral; death; diarrhea; pneumonia; mental and motor development	All children received VitA and Iodine by a public supplementation program. Some outcomes extracted from related paper (Chilenje_2010) [40] and (Manno_2011) [41].
Author, year: Gill, 1997 [24] Design: RCT	Country: Ireland Target population: no info Age (mean; range): 0.5; 0.5 to 0.5 years Males (%): 51 Exclusion criteria: severe or chronic disaese, malnutrition, congenital anomalies	Milk, fortified; single MN strategy MN applied: Fe Iron dosage: 6.54 mg/day; Iron compound: FeSu	formula milk, un-fortified for iron	After 0.75 year: hema-tological parameters, anthropometry	

TABLE 2: *Cont.*

Study	Population	Intervention	Control food	Outcome	Comment
Author, year: Lartey, 1999 [34] Design: RCT	Country: Ghana Target population: urban area Age (mean; range) 0.5; 0.5 to 0.5 years Males (%): 48 Exclusion criteria: congenital abnormalities	Cereals, fortified (porridge); multi MN strategy MN applied: Fe, Zn, VitA, other Vitamins, other MN Iron dosage: 14.25 mg/day; Iron compound: electrFe	porridge, unfortified	After 0.5 year: hematological parameters; anthropometry; diarrhea; fever; respiratory illness	Intervention cereal with 2 formulations of fortification depending on daily cereal intake of infant to avoid potential toxicity problems.
Author, year: Liu, 1993 [37] Design: RCT	Country: China Target population: all population classes (90% of all children) Age (mean; range): 0.8; 0.5 to 1.1 years Males (%): 55 Exclusion criteria: no info	Cereals, fortified (rusk); multi MN strategy MN applied: Fe, Zn, VitA, other Vitamins, other MN Iron dosage: 5 mg/day; Iron compound: FeAmCi	rusk, unfortified	After 0.25 year: hematological parameters; MN-serum levels, anthropometry	
Author, year: Maldonado Lonzano, 2007 [25] Design: RCT	Country: Spain Target population: no info Age (mean; range): 1.9; (range: no info) years Males (%): 58 Exclusion criteria: iron supplementation	Milk, fortified; multi MN strategy MN applied: Fe, other Vitamins, other MN Iron dosage: 5.9 mg/day; Iron compound: no info	milk, unfortified (cows whole milk formula)	After 0.33 year: hematological parameters	No child with anemia at baseline.
Author, year: Morley, 1999 [26] Design: RCT	Country: UK (Indian ethnicity) Target population: mother with higher eduction, non-manual social class Age (mean; range): 0.78; (range: no info) years Males (%): 50 Exclusion criteria: relevant disease; iron supplementation	Milk, fortified; single MN strategy MN applied: Fe Iron dosage: 1.8 mg/day; Iron compound: FeSu	formula, unfortified	After 0.75 year: hematological parameters, antropometry, motor and mental development	Only data from Norwich cohort blood samples could be taken at baseline and were extracted for Hb outcome.
Author, year: Nesamvuni, 2005 [36] Design: RCT	Country: South Africa Target population: poor socio-economic status, undernourished children Age (mean; range): no info; 1 to 3 years Males (%): 0 Exclusion criteria: physical or mental disability, severe undernutrition	Cereals, fortified (maize porridge; dual MN strategy MN applied: VitA, other Vitamins Iron dosage: n.a.mg/day; Iron compound: n.a.	maize meal, unfortified	After 1 year: hematological parameters, retinol level, anthropometry	Children and family members received the food.

TABLE 2: *Cont.*

Study	Population	Intervention	Control food	Outcome	Comment
Author, year: Oelofse, 2003 [9] Design: RCT	Country: South Africa Target population: urban disadvantaged black community (low socioeconomic status) Age (mean; range): 0.5; (range: no info) years Males (%): 0 Exclusion criteria: birth weight < 2.5 kg; congenital abnormalities	Cereals, fortified (porridge); dual MN strategy MN applied: Zn, other Vitamins Iron dosage: ~0.8 mg/day; Iron compound: FePP	normal diet	After 0.5 year: hematological parameters, zinc level, retinol level, anthropometry, psychomotor development	90% of control group already recieved commercially prepared complementary food. The food concentration of iron did not relevantly differ between groups, but of Zinc and of VitA.
Author, year: Rivera, 2010 [10] Design: RCT (accounted for cluster randomisation)	Country: Mexico Target population: households living in poverty Age (mean; range): no info; 1 to 2.5 years Males (%): 50 Exclusion criteria: no info	Milk, fortified; multi MN strategy MN applied: Fe, Zn, other Vitamins, other MN Iron dosage: 7.82 mg/day; Iron compound: FeGlu	milk, non-cal parameters	After 1 year: hematological parameters	Study results are adjusted for cluster effect. Evaluation of a large scale program (Leche Lincosa) in Mexico.
Author, year: Sazawal, 2010 [8] Design: RCT	Country: India Target population: periurban area, illiteracy of parents Age (mean; range): 1.9; 1 to 3 years Males (%): 50 Exclusion criteria: severe malnutrition; severe illness	Milk, fortified; multi MN strategy MN applied: Fe, Zn, VitA, other Vitamins, other MN Iron dosage: 8.3 mg/day; Iron compound: FeSu	milk, un-fortified	After 1 year: hematological parameters, anthropometry, severe illnesses, diarrhoea, lower respiratory tract infections, pneumonia	Some data extracted from relating paper: Sazawal_2006 [42] Completeness relates to hematologic parameters.
Author, year: Schümann, 2005 [38] Design: RCT	Country: Guatemala Target population: low income; periurban settlement Age (mean; range): 1.7; 1 to 2 years Males (%): 52 Exclusion criteria: gastric or intestinal diseases; infections	Cereals, fortified (bean paste); single MN strategy MN applied: Fe Iron dosage: 17.1 mg/day; Iron compound: FeSu	beans, un-fortified	After 0.19 year: hematological parameters	All children recieved anthelmintic treatment; all families were compensated. Three arm trial: Only data for FeSu group (n = 31) vs. control group (n = 30) extracted.

TABLE 2: *Cont.*

Study	Population	Intervention	Control food	Outcome	Comment
Author, year: Stevens, 1998 [32] Design: RCT	Country: UK (mostly caucasian) Target population: lower social classes were overrepresented Age (mean; range): 0.5; 0 to 0 years Males (%): 0 Exclusion criteria: illness, major congenital malformation	Milk, fortified; single MN strategy MN applied: Fe Iron dosage: 6.87 mg/day; Iron compound: FeSu	milk, unfortified	After 1 year: hematological parameters	
Author, year: Villalpando, 2006 [27] Design: RCT	Country: Mexico Target population: poor periurban community Age (mean; range): 1.8; 0.8 to 2.5 years Males (%): 50 Exclusion criteria: no info	Milk, fortified; multi MN strategy MN applied: Fe, Zn, other Vitamins Iron dosage: 6.74 mg/day; Iron compound: FeGlu	milk, unfortified	After 0.5 year: hematological parameters	The results of the study lead to broadening of a fortified milk distribution program in Mexico.
Author, year: Virtanen, 2001 [28] Design: RCT	Country: Sweden Target population: urban area Age (mean; range): 1; 1 to 1 years Males (%): 39 Exclusion criteria: milk intolerance; poor health	Milk, fortified; single MN strategy MN applied: Fe Iron dosage: 4.53 mg/day; Iron compound: FeGlu, FeLac	milk, unfortified	After 0.5 year: hematological parameters	
Author, year: Walter, 1998 [30] Design: RCT	Country: Chile Target population: From four contiguos urban communities Age (mean; range): 0.5; 0 to 0 years Males (%): 52 Exclusion criteria: major birth or neonatal complications, chronic illness	Milk, fortified; single MN strategy MN applied: Fe Iron dosage: 6.5 mg/day; Iron compound: FeSu	formula, low iron fortified	After 1 year: hematological parameters, anthropometry	

a MN (micronutrient) applied: Fe: iron; Zn: zinc; VitA: Vitamin A; other Vitamins: e.g. Vitamin C; other MN (micronutirents): e.g. Vitamin C; other MN (micronutirents): e.g. selen, copper.

b Iron dosage: Determined as daily difference between intervention and control group.

c n.a.: not applicable

d Iron compound: FeSu: iron-sulfate; FePP: iron-pyrophosphate; NaFeEDTA: natrium-iron-EDTA; FeFu: iron-fumarate; FeGlu: iron-gluconate; FeAmCi: ferric-ammonium-citrate; FeLa: Ferrous lactate; electrFe: electrolytic iron.

11.3.2 EFFECT ON HEMOGLOBIN LEVELS

Hemoglobin blood level was the most frequently reported outcome parameter. Across 13 studies that tested iron fortification irrespective of other added MN, the mean increase of hemoglobin compared to the control group was 0.62 g/dl (95%-CI: 0.34 to 0.89) for children fed with fortified milk or cereals (Figure 2). Heterogeneity was high (I2 = 86%). Comparison of different subgroups showed a stronger effect of the iron MMN fortification approach (n = 8 studies; hemoglobin increase 0.87 g/dl (95%-CI: 0.57 to 1.16; I2 = 82%) compared to the iron single-fortification strategy (n = 5 studies; hemoglobin increase 0.20 g/dl (95%-CI: -0.05 to 0.45; I2 = 43%). The daily applied iron dosage was similar for the single-iron approach (median: 6.5 mg) and the MMN-approach (median 6.7 mg).

11.3.3 EFFECT ON ANEMIA PREVALENCE

Eleven trials provided data for anemia rates, all of them using iron as a single- or a MMN-fortification strategy. Applied thresholds for anemia varied between 10.5 g/dl and 11 g/dl and the median of anemia rates at baseline was 36% (IQR: 15% to 40%; 9 studies with data). Fortified milk or cereals reduced the risk of suffering from anemia by 50% (risk ratio 0.50, 95%-CI: 0.33 to 0.75; I2 = 71%; Figure 3). Again, a stronger effect of the MMN fortification approach emerged (n = 7 studies; risk ratio 0.43 (95%-CI: 0.26 to 0.71; I2 = 81%) compared to the iron single-fortification strategy (n = 4 studies; risk ratio 0.76 (95%-CI: 0.45 to 1.28; I2 = 0%). Overall, the absolute risk reduction (ARR) of suffering from anemia was 14% (un-weighted data of 11 trials), translating into a number needed to treat (NNT) with fortified milk or cereals of 7 (95%-CI: 6 to 9) participants over a period of 8 months (i.e. the mean follow-up time) to avoid one case of anemia. For the MMN approach these results are even more favorable (un-weighted data of 7 trials: ARR 22%; NNT 5 [95%-CI: 4 to 6]).

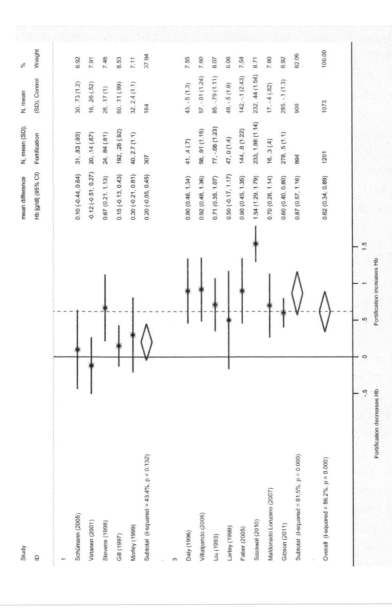

FIGURE 2: Effect of iron fortification of milk and cereals on hemoglobin (Hb) levels compared to non-fortified food. Only studies with iron fortification included (n = 13 RCT). Results are provided as weighted mean difference in hemoglobin (WMD: g/dl with 95%-CI; conversion to g/L with factor 10) between intervention and control group (iron single-fortification (1); iron multi micronutrient fortification (3); overall effect).

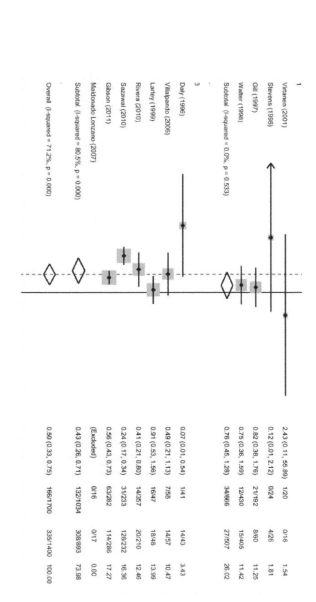

FIGURE 3: Effect of iron fortification of milk and cereals on anemia compared to non-fortified food. Only studies with iron fortification included (n = 11 RCT). Results are provided as risk ratio (RR, 95%-CI) of suffering from anemia in the intervention group compared to the control group (iron single-fortification (1); iron multi micronutrient fortification (3); overall effect).

TABLE 3: Risk of bias summary table

EN	Author	Year	Adequate sequence generation?	Allocation concealment?	Blinding?	Incomplete outcome data addressed	Are typical outcomers reported? (Selective outcome reporting)
675	Brown	2007	?	?	YES	YES	YES
818	Daly	1996	?	?	?	YES	YES
951	Faber	2005	NO	NO	YES	NO	YES
1058	Gibson	2011	?	?	YES	NO	YES
153	Gill	1997	?	YES	NO	NO	YES
1051	Lartey	1999	?	YES	NO	YES	YES
1154	Liu Maldonado	1993	?	?	?	NO	YES
257	Lonzano	2007	YES	YES	YES	YES	NO
282	Morley	1999	?	YES	YES	YES	YES
1149	Nesamvuni	2005	NO	NO	YES	YES	YES
297	Oelofse	2003	NO	NO	NO	NO	YES
333	Rivera	2010	?	?	YES	NO	NO
1	Sazawal	2010	YES	YES	YES	NO	YES
1172.2	Schumann	2005	NO	NO	YES	YES	NO
838	Stevens	1998	?	?	YES	NO	NO
403	Villalpando	2006	?	?	YES	YES	NO
404	Virtanen	2001	?	?	YES	NO	NO
797	Walter	1998	NO	YES	YES	NO	YES

11.3.4 EFFECT ON FERRITIN LEVELS

Ferritin is the most direct measure to conclude if iron stores increase by iron-fortified food consumption. Eleven trials provided data for ferritin serum levels. Ferritin levels were not adjusted for subclinical infections. Given the skewed distribution of ferritin values, authors often reported median estimates. Medians were significantly higher in the intervention groups (ranges of ferritin medians at end of study [micro-g/l]: intervention: 15.8 to 44.6; control: 6.5 to 28; $P < 0.01$). Only three studies provided mean values to be included in a meta-analysis, which showed an effect in the same direction. The mean ferritin increase with iron fortification was 11.3 micro-g/l (95%-CI: 3.3 to 19.2; $I2 = 79\%$) compared to control groups.

11.3.5 EFFECTS ON SERUM ZINC AND VITAMIN A LEVELS

Five studies provided data for change in serum zinc levels. MN fortification with zinc led to no relevant change in zinc serum levels (0.4 micro-g/dl (95%-CI: -1.7 to 2.6; I2 = 0%) compared to control groups. However, fortification increased vitamin A serum levels compared to control groups (four studies with data: Retinol increase by 3.7 micro-g/dl [95%-CI: 1.3 to 6.1; I2 = 37%]).

11.3.6 EFFECTS ON GROWTH, FUNCTIONAL MEASURES AND MORBIDITY

For three European studies, no relevant effect of fortification on height and weight was seen and morbidities were not an issue in this population.

All other results relate to non-European low-/middle income countries. Due to the short follow-up period in most of the studies, no meaningful conclusion can be drawn for possible effects of fortification on height or weight gain or z-scores (weight-for-age; height-for-age; weight-for-height). Of 9 studies with data, 7 trials reported no relevant differences between intervention and control group at the end of the study. In one study [36], more weight gain was seen in the intervention group after one year (4.6 kg vs. 2.0 kg; P < 0.05), in another study [8] children consuming fortified milk showed improvement in weight gain compared to control group (difference 0.21 kg/year [95%-CI 0.12 to 0.31) and height gain (difference 0.51 cm/year [95%-CI 0.27 to 0.75).

Of three studies with data for psychomotor development of children, two trials reported no relevant difference between groups [9,35] and one study [33] found slight improvements compared to the control group.

Of four studies with morbidity data of children, three trials reported no relevant differences between groups for infections [29], for diarrhea, fever and respiratory illness [34] and for referral to hospital or death in partly HIV exposed children [35]. In one study [8] fortified milk significantly reduced the probability of days with severe illness (by 15%), and the relative risk of diarrhea (by 18%) and lower respiratory illness (by 26%).

11.3.7 EXPLORING HETEROGENEITY

In our pre-specified subgroup analyses no relevant influence on the outcome was detected for the mode of fortified food (fortified milk vs. cereals). Hemoglobin change was somewhat higher in studies from low/middle income countries (0.78 g/dl (95%-CI: 0.41 to 1.15) compared to high income countries (0.42 g/dl (95%-CI: 0.10 to 0.73), but the difference was not statistically significant. The dual-/multi-micronutrient approach led to a significantly stronger effect of iron fortification on hemoglobin increase than the iron single-fortification strategy (Figure 3).

In our multivariable meta-regression analysis, none of the tested independent variables (mean hemoglobin level before intervention; daily amount of consumed iron, length of follow-up, completeness of follow up) was significantly associated with the change in hemoglobin.

11.3.8 SUMMARY ASSESSMENT OF RISK OF BIAS

Only two [8,25] of 18 trials provided enough information to conclude that both random sequence generation and allocation concealment was adequately performed (Table 3). For 11 trials this was unclear and inadequate procedures had been applied in 5 trials. Other criteria were fulfilled more often: Blinding was reported in 13 of 18 studies, incomplete outcome data were addressed in 8 of 18 trials and 12 of 18 studies showed no selective outcome reporting (i.e. besides serum markers also height/weight, functional measures or morbidities were reported).

In summary, the risk of bias for the most often reported outcomes hemoglobin change and anemia rates is unclear. However, a sensitivity analysis including only studies with low risk of bias led to similar results (three studies that fulfilled 4 of 5 quality criteria [8,25,26]: hemoglobin increase 0.87 g/dl (95%-CI: 0.09 to 1.65; I2 = 92%). Another sensitivity analysis showed that the result pattern remained basically unchanged after performing analyses using mean values of groups at the end of the study, instead of mean changes of groups.

TABLE 3: Risk of bias summary table

EN	Author	Year	Adequate sequence generation?	Allocation concealment?	Blinding?	Incomplete outcome data addressed	Are typical outcomers reported? (Selective outcome reporting)
675	Brown	2007	?	?	YES	YES	YES
818	Daly	1996	?	?	?	YES	YES
951	Faber	2005	NO	NO	YES	NO	YES
1058	Gibson	2011	?	?	YES	NO	YES
153	Gill	1997	?	YES	NO	NO	YES
1051	Lartey	1999	?	YES	NO	YES	YES
1154	Liu Maldonado	1993	?	?	?	NO	YES
257	Lonzano	2007	YES	YES	YES	YES	NO
282	Morley	1999	?	YES	YES	YES	YES
1149	Nesamvuni	2005	NO	NO	YES	YES	YES
297	Oelofse	2003	NO	NO	NO	NO	YES
333	Rivera	2010	?	?	YES	NO	NO
1	Sazawal	2010	YES	YES	YES	NO	YES
1172.2	Schumann	2005	NO	NO	YES	YES	NO
838	Stevens	1998	?	?	YES	NO	NO
403	Villalpando	2006	?	?	YES	YES	NO
404	Virtanen	2001	?	?	YES	NO	NO
797	Walter	1998	NO	YES	YES	NO	YES

The table presents each study by assessed methodological criterion in a cross-tabulation a. Studies are sorted for author name.
a Assessment categories: YES: criterion fulfilled; NO: criterion not fulfilled; "?": unclear, as no information given.

11.4 DISCUSSION

To our knowledge, this is the first systematic review, that has applied a meta-analysis to specifically weigh the overall evidence for the effects of fortified milk and cereal food suitable for complementary feeding of children. The evidence relates to study populations between 6 month and three years of age. Iron fortification leads to a clinically relevant increase in hemoglobin levels and reduction of anemia rates. For zinc and vitamin A fortification only surrogate parameters are reported, but the combination with iron (MMN approach) leads to a more pronounced effect on hemoglobin levels compared to an iron single-fortification strategy. The evidence for functional health outcomes is inconclusive.

11.4.1 STRENGTHS AND LIMITATIONS

We applied a thorough search strategy with a stepwise retrieval of studies using electronic databases and additional sources. We cannot exclude having missed references but we believe that we found a near complete sample of relevant papers for our specific research question.

Some limitations have to be mentioned. First, included studies showed short follow-up periods, thus the impact of fortified milk or cereal food on functional health outcomes (such as sustainable height and weight gain or mental and motor development) could not be assessed thoroughly. Second, for zinc and vitamin A fortification only surrogate outcomes as serum levels are available. However, the presence of additional fortified micronutrients seems to be important for the effects of iron on hemoglobin levels. The MMN approach is more effective than iron single fortification in our review, reflecting that complex micronutrient deficiencies are responsible for health problems [20]. Third, risk of bias is unclear mainly due to underreporting of the randomization procedure and incomplete outcome data. Finally, pooled estimates have to be interpreted cautiously as statistical heterogeneity between studies was considerable and meta-regression did not reveal significant associations of pre-specified study characteristics with study results. Possible sources for unexplained heterogeneity might be underreporting for co-interventions (e.g. public supplementation or food programs) or the diversity of applied MN preparations that have influence on MN absorption. For example, five different iron compounds were used (12 studies with data: six times FeSulfate; twice FeGluconate; one time, each, FePyrophosphate; FeFumarate; Ferric ammonium citrate; electrolytic iron). In addition, the difference in daily consumed iron between intervention and control group varied between 1.8 mg and 14.3 mg. Furthermore, molar ratios, a determinant for MN absorption, also showed variation (ranges of molar ratios: ascorbic acid/iron: 0.68 to 30; phytic acid/iron: 1.7 to 2.2; calcium/iron: 40 to 134).

11.4.2 EXISTING SYSTEMATIC REVIEWS AND RESEARCH NEEDS

Important contributions have been made in the recent years with other systematic reviews to evaluate the health effects of MN interventions. These reviews differ from ours: Dewey and Adu-Afarwuah gave a broad systematic overview of studies and programs aimed at improving biochemical and functional outcomes with complementary foods [43]. However, they did not perform a meta-analysis and presented results in a tabulated form or as averaged effect sizes. Some reviews have concentrated on MN supplementation only [44-48] or home fortification [49], other reviews have combined supplementation and fortification strategies for analysis [20,50], included children as well as adolescents or adult women [6,51,52], or included fortified staple food interventions [6,52].

The health effects found in our review are in line with effect sizes shown in some similar reviews above [20,47,52]. This underpins the validity of our findings and supports a strategy to intervene with fortified milk and cereal food for infants and children. Supplementation trials, for example with vitamin A, have been shown to reduce mortality and morbidity via improved nutritional status [47,48], even though serum level increases were small [53], similar to our review. Thus, some authors conclude that fortification would also have an impact on morbidity and mortality, although a conclusive answer cannot yet be given [52]. On the other hand, negative aspects of iron supplementation have been reported, such as increased morbidity and mortality in regions where malaria transmission is intense [54]. Thus, recommendations concerning iron supplementation have been formulated [55]. These adverse effects, however, may not be that relevant for fortified foods. Daily micronutrient dosages of fortified foods are much lower as compared to supplementation. Furthermore, children stop eating once they get saturated, which may also not be the case for high dosage sprinkles, that can be seen as a specific application of supplementation. Nevertheless, long term data concerning negative effects of iron fortified foods in regions with high prevalence of malaria and infectious diseases are lacking.

Further compiled evidence is needed to agree on the optimal MN preparation for fortified milk and cereals (such as composition of MN; suitable compounds; molar ratios for additives) to fully exhaust the potential of this approach. Future studies should also focus on health outcomes of MN fortification beyond the effect of iron on hematological results, for example via long-term follow-up of study populations.

11.4.3 IMPLICATIONS FOR DECISION MAKERS

There are multiple delivery mechanisms for fortified milk and cereal food. Production and distribution via government programs and local public agencies would be an obvious option to strengthen local structures. Implementation of effective strategies, however, does not always work well in the field due to logistical problems or inappropriate priority setting [56]. Thus, some have discussed the role of the business community in improving nutrition in developing countries [56-58]. Commercially distributed fortified foods (e.g. with iron) are already available in many markets, even in low-income countries. In a public private partnership, business partners can provide their professional knowledge and experience concerning technical problems with processing and fortification, supply and transport, or refrigeration and conservation issues (specifically important for milk) to get interventions more efficient.

A limitation of the market approach is that it may not reach the poorest of the poor. Thus, a combination of different delivery channels, as well as affordable prices, may be needed. Children with severe anemia, who may be overrepresented in very poor groups, are often excluded from trials due to ethical reasons. One may assume, that the positive effects on the hemoglobin levels may be even stronger for such children. Additional economic analyses are necessary to contribute to a deeper understanding of the health economic effects of such a strategy and to inform the priority setting of decision makers.

11.5 CONCLUSIONS

Multi micronutrient fortified milk and cereal products can be an effective option to reduce anemia of children up to three years of age in developing countries. On the basis of our data the evidence for functional health outcomes is still inconclusive.

REFERENCES

1. Horton S. Alderman H. Hunger and Malnutrition. Copenhagen Consensus Challenge Paper. Copenhagen Consensus Center, Rivera JA; 2008.
2. Black RE, Allen LH, Bhutta ZA, Caulfield LE, de Onis M, Ezzati M, Mathers C, Rivera J. Maternal and child undernutrition: global and regional exposures and health consequences. Lancet. 2008;371(9608):243–260. doi: 10.1016/S0140-6736(07)61690-0.
3. Dewey KG, Begum K. Long-term consequences of stunting in early life. Matern Child Nutr. 2011;7(Suppl 3):5–18.
4. Conclusions and recommendations of the WHO consultation on prevention and control of iron deficiency in infants and young children in malaria-endemic areas. Food and Nutrition Bulletin. 2007;28(4):S621–S631.
5. Allen L, de Benoist B. Dary O. Guidelines on food fortification with micronutritients. Edited by WHO Food and Agricultural Organization of the United Nations. Geneva, Hurrel R; 2006.
6. Bhutta ZA, Ahmed T, Black RE, Cousens S, Dewey K, Giugliani E, Haider BA, Kirkwood B, Morris SS, Sachdev HP. et al. What works? Interventions for maternal and child undernutrition and survival. Lancet. 2008;371(9610):417–440. doi: 10.1016/S0140-6736(07)61693-6.
7. Infant and young child nutrition. In WHA Resolution. 2001;542
8. Sazawal S, Dhingra U, Dhingra P, Hiremath G, Sarkar A, Dutta A, Menon VP, Black RE. Micronutrient fortified milk improves iron status, anemia and growth among children 1–4 years: a double masked, randomized, controlled trial. PLoS One. 2010;5(8):e12167. doi: 10.1371/journal.pone.0012167.
9. Oelofse A, Van Raaij JM, Benade AJ, Dhansay MA, Tolboom JJ, Hautvast JG. The effect of a micronutrient-fortified complementary food on micronutrient status, growth and development of 6- to 12-month-old disadvantaged urban South African infants. Int J Food Sci Nutr. 2003;54(5):399–407. doi: 10.1080/0963748031000092161.

10. Rivera JA, Shamah T, Villalpando S, Monterrubio E. Effectiveness of a large-scale iron-fortified milk distribution program on anemia and iron deficiency in low-income young children in Mexico. Am J Clin Nutr. 2010;91(2):431–439. doi: 10.3945/ajcn.2009.28104.

11. CRD's guidance for undertaking reviews in health care. University of York, York; 2008.

12. Higgins JP, Green S. Cochrane Handbook for Systematic Reviews of Interventions Version 5.1.0 [updated March 2011] The Cochrane Collaboration. 2011.

13. Moher D, Liberati A, Tetzlaff J, Altman DG. Preferred reporting items for systematic reviews and meta-analyses: the PRISMA statement. Ann Intern Med. 2009;151(4):264–269.

14. Heird WC. Progress in promoting breast-feeding, combating malnutrition, and composition and use of infant formula, 1981–2006. J Nutr. 2007;137(2):499S–502S.

15. Standard for infant formula and formulas for special medical purposes intended for infants. In CODEX STAN. 2007. pp. 72–1981.

16. International Clinical Epidemiology Network. http://www.inclentrust.org/

17. Global Alliance for Improved Nutrition. http://www.gainhealth.org/

18. The Micronutrient Initiative. http://www.micronutrient.org/english/view.asp?x=1.

19. http://www.gatesfoundation.org/Pages/home.aspx.

20. Allen LH, Peerson JM, Olney DK. Provision of multiple rather than two or fewer micronutrients more effectively improves growth and other outcomes in micronutrient-deficient children and adults. J Nutr. 2009;139(5):1022–1030. doi: 10.3945/jn.107.086199.

21. Altman DG, Machin D, Bryant TN, Gardner JM. Statistics with confidence. BMJ Books, Bristol; 2001.

22. Khan KS, Kunz R, Kleijnen J, Antes G. Systematic reviews to support evidence-based medicine. Royal Society of Medicine Press, London; 2003.

23. Higgins JP, Thompson SG, Deeks JJ, Altman DG. Measuring inconsistency in meta-analyses. BMJ. 2003;327(7414):557–560. doi: 10.1136/bmj.327.7414.557.

24. Gill DG, Vincent S, Segal DS. Follow-on formula in the prevention of iron deficiency: a multicentre study. Acta Paediatr. 1997;86(7):683–689. doi: 10.1111/j.1651-2227.1997.tb08568.x.

25. Maldonado Lozano J, Baro L, Ramirez-Tortosa MC, Gil F, Linde J, Lopez-Huertas E, Boza JJ, Gil A. Intake of an iron-supplemented milk formula as a preventive measure to avoid low iron status in 1–3 year-olds. An Pediatr (Barc) 2007;66(6):591–596. doi: 10.1157/13107394.

26. Morley R, Abbott R, Fairweather-Tait S, MacFadyen U, Stephenson T, Lucas A. Iron fortified follow on formula from 9 to 18 months improves iron status but not development or growth: a randomised trial. Arch Dis Child. 1999;81(3):247–252. doi: 10.1136/adc.81.3.247.

27. Villalpando S, Shamah T, Rivera JA, Lara Y, Monterrubio E. Fortifying milk with ferrous gluconate and zinc oxide in a public nutrition program reduced the prevalence of anemia in toddlers. J Nutr. 2006;136(10):2633–2637.

28. Virtanen MA, Svahn CJ, Viinikka LU, Raiha NC, Siimes MA, Axelsson IE. Iron-fortified and unfortified cow's milk: effects on iron intakes and iron status in young children. Acta Paediatr. 2001;90(7):724–731.

29. Brown KH, López de Romaña D, Arsenault JE, Peerson JM, Penny ME. Comparison of the effects of zinc delivered in a fortified food or a liquid supplement on the growth, morbidity, and plasma zinc concentrations of young Peruvian children. Am J Clin Nutr. 2007;2:538–547.

30. Walter T, Pino P, Pizarro F, Lozoff B. Prevention of iron-deficiency anemia: comparison of high- and low-iron formulas in term healthy infants after six months of life. J Pediatr. 1998;4:635–640.

31. Daly A, MacDonald A, Aukett A, Williams J, Wolf A, Davidson J, Booth IW. Prevention of anaemia in inner city toddlers by an iron supplemented cows' milk formula. Arch Dis Child. 1996;1:9–16.

32. Stevens D, Nelson A. The effect of iron in formula milk after 6 months of age. Arch Dis Child. 1995;3:216–220.

33. Faber M, Kvalsvig JD, Lombard CJ, Benadé AJ. Effect of a fortified maize-meal porridge on anemia, micronutrient status, and motor development of infants. Am J Clin Nutr. 2005;5:1032–1039.

34. Lartey A, Manu A, Brown KH, Peerson JM, Dewey KG. A randomized, community-based trial of the effects of improved, centrally processed complementary foods on growth and micronutrient status of Ghanaian infants from 6 to 12 mo of age. Am J Clin Nutr. 1999;70(3):391–404.

35. Gibson RS, Kafwembe E, Mwanza S, Gosset L, Bailey KB, Mullen A, Baisley K, Filteau S. A Micronutrient-Fortified Food Enhances Iron and Selenium Status of Zambian Infants but Has Limited Efficacy on Zinc. J Nutr. 2011;141:935–943. doi: 10.3945/jn.110.135228.

36. Nesamvuni AE, Vorster HH, Margetts BM, Kruger A. Fortification of maize meal improved the nutritional status of 1-3-year-old African children. Public Health Nutr. 2005;8(5):461–467.

37. Liu DS, Bates CJ, Yin TA, Wang XB, Lu CQ. Nutritional efficacy of a fortified weaning rusk in a rural area near Beijing. Am J Clin Nutr. 1993;57(4):506–511.

38. Schumann K, Romero-Abal ME, Maurer A, Luck T, Beard J, Murray-Kolb L, Bulux J, Mena I, Solomons NW. Haematological response to haem iron or ferrous sulphate mixed with refried black beans in moderately anaemic Guatemalan pre-school children. Public Health Nutr. 2005;8(6):572–581.

39. Williams J, Wolff A, Daly A, MacDonald A, Aukett A, Booth IW. Iron supplemented formula milk related to reduction in psychomotor decline in infants from inner city areas: randomised study. BMJ. 1999;318(7185):693–697. doi: 10.1136/bmj.318.7185.693.

40. Micronutrient fortification to improve growth and health of maternally HIV-unexposed and exposed Zambian infants: a randomised controlled trial. PloS one. 2010;6:e11165.

41. Manno D, Kowa PK, Bwalya HK, Siame J, Grantham-McGregor S, Baisley K, De Stavola BL, Jaffar S, Filteau S. Rich micronutrient fortification of locally produced infant food does not improve mental and motor development of Zambian infants: a randomised controlled trial. Br J Nutr. 2011. pp. 1–11.

42. Sazawal S, Dhingra U, Dhingra P, Hiremath G, Kumar J, Sarkar A, Menon VP, Black RE. Effects of fortified milk on morbidity in young children in north India: community based, randomised, double masked placebo controlled trial. BMJ. 2007;334(7585):140. doi: 10.1136/bmj.39035.482396.55.

43. Dewey KG, Adu-Afarwuah S. Systematic review of the efficacy and effectiveness of complementary feeding interventions in developing countries. Matern Child Nutr. 2008;4(Suppl 1):24–85.
44. Gera T, Sachdev HP, Nestel P, Sachdev SS. Effect of iron supplementation on haemoglobin response in children: systematic review of randomised controlled trials. J Pediatr Gastroenterol Nutr. 2007;44(4):468–486. doi: 10.1097/01. mpg.0000243440.85452.38.
45. Ojukwu JU, Okebe JU, Yahav D, Paul M. Oral iron supplementation for preventing or treating anaemia among children in malaria-endemic areas. Cochrane Database Syst Rev. 2009;3 CD006589.
46. Bhutta ZA, Black RE, Brown KH, Gardner JM, Gore S, Hidayat A, Khatun F, Martorell R, Ninh NX, Penny ME. et al. Prevention of diarrhea and pneumonia by zinc supplementation in children in developing countries: pooled analysis of randomized controlled trials. Zinc Investigators' Collaborative Group. J Pediatr. 1999;135(6):689–697.
47. Mayo-Wilson E, Imdad A, Herzer K, Yakoob MY, Bhutta ZA. Vitamin A supplements for preventing mortality, illness, and blindness in children aged under 5: systematic review and meta-analysis. BMJ. 2011;343:d5094. doi: 10.1136/bmj.d5094.
48. Haider Batool A, Bhutta Zulfiqar A. Neonatal vitamin A supplementation for the prevention of mortality and morbidity in term neonates in developing countries (Review) Cochrane Library. 2011;10:CD006980.
49. De-Regil LM, Suchdev PS, Vist GE, Walleser S, Pena-Rosas JP. Home fortification of foods with multiple micronutrient powders for health and nutrition in children under two years of age. Cochrane Database Syst Rev. 2010;12:CD008524. CD006980.
50. Ramakrishnan U, Aburto N, McCabe G, Martorell R. Multimicronutrient interventions but not vitamin a or iron interventions alone improve child growth: results of 3 meta-analyses. J Nutr. 2004;134(10):2592–2602.
51. Hess SY, Brown KH. Impact of zinc fortification on zinc nutrition. Food Nutr Bull. 2009;30(1):79–107.
52. Bhutta Zulfiqar A, Das JK, Dean SV, Salam RA. Effectiveness of Food Fortification with Micronutrients. A Review. Nestle, Geneva; 2012.
53. Imdad A, Herzer K, Mayo-Wilson E, Yakoob M, Bhutta Z. Vitamin A supplementation for preventiing morbidity and mortality in children six months to five years of age. Cochrane Database of Systematic Reviews. 2010;5
54. Sazawal S, Black RE, Ramsan M, Chwaya HM, Stoltzfus RJ, Dutta A, Dhingra U, Kabole I, Deb S, Othman MK. et al. Effects of routine prophylactic supplementation with iron and folic acid on admission to hospital and mortality in preschool children in a high malaria transmission setting: community-based, randomised, placebo-controlled trial. Lancet. 2006;367(9505):133–143. doi: 10.1016/S0140-6736(06)67962-2.
55. Iron supplementation of young children in regions where malaria transmission is intense and infectious disease higly prevalent. World Health Organisation, Geneva; 2007.
56. Bryce J, Coitinho D, Darnton-Hill I, Pelletier D, Pinstrup-Andersen P. Maternal and child undernutrition: effective action at national level. Lancet. 2008;371(9611):510–526. doi: 10.1016/S0140-6736(07)61694-8.

57. Darnton-Hill I, Nalubola R. Fortification strategies to meet micronutrient needs: successes and failures. Proc Nutr Soc. 2002;61(2):231–241. doi: 10.1079/PNS2002150.
58. Scaling up Nutrition: Road Map Implementation. http://www.unscn.org/sacling_up_nutrition_sun/

CHAPTER 12

EFFECT OF MINERAL-ENRICHED DIET AND MEDICINAL HERBS ON FE, MN, ZN, AND CU UPTAKE IN CHICKEN

DUCU SANDU STEF and IOSIF GERGEN

12.1 BACKGROUND

Chicken meat is widely consumed worldwide. Chicken meat and meat products are important for human diet because they provide a great part of nutrients (protein, lipids), including necessary minerals and trace elements. The main source of metals in these meat foodstuffs arises from processing and manipulation of feeds. Fe, Zn, Cu and Mn are the essential metals which are required in small quantities and occur naturally in various vegetables and meat foodstuffs. However, those essential metals are given special attention due to their toxic effect in the body when their concentrations exceed limits of safe exposure. The contents of these metals in vegetables and meat foodstuffs may vary depending on the general (varieties, maturity, genetics, and age) and environmental (soils, geographical locations, season, water source and use of fertilizers) conditions of plants and animals and on methods of handling and processing. Heavy metals contaminate the environment and enter the food chain. Contamination with these metals is serious threat because of their toxicity, bioaccumulation and biomagnification in the food chain [1-4].

To avoid deficiencies that can lead to a wide variety of clinical and pathological disorders diets for livestock are supplemented with minerals

[5]. Fe, Mn, Zn and Cu can be considered trace minerals with a central role in many metabolic processes throughout the body and are essential for correct growth and development of all animals, including humans. They predominantly act as catalysts in many enzyme and hormone systems which influence on growth, bone development, feathering, enzyme structure and function, and appetite. Deficiency symptoms are manifested as disturbances in metabolic processes, resulting in lower production performance, loss of appetite, reproductive disorders, and impaired immune response. Deficiencies can be caused by inadequate mineral intake or by the presence of antagonists in the diet, which interfere with or unbalance minerals uptake [2,6,7].

Generally, inorganic mineral salts (sulfates, oxides and carbonates) have been used within feed formulations, because they offer a cost-effective solution to meet the requirement of the animal for trace elements. Feeding studies demonstrated that chelated trace minerals are at least 30% more bioavailable than inorganic trace mineral salts when fed to broilers. However, it has yet to be conclusively proven that mineral chelates are better absorbed in the monogastric enterocyte. Several papers reported on metal binding to proteins in the cells [8,9]. The higher availability of chelates may be linked to the shielding of the minerals positive charge during chelation. This allows the mineral to withstand the binding activity of the negatively charged mucin layer and results in lower competition between minerals of similar charge in their resorption from the gut and transfer to the enterocyte. These phenomena, combined with lower complex formation in the intestinal lumen with compounds such as phytate, may contribute to the higher absorption of minerals from the gut. Feeding trials in mammalian species have shown that organic complexed trace minerals have higher relative bioavailability than inorganic ones and provide alternative pathways for absorption, thus leading to a reduction in the excretion of minerals [10-12].

Natural feed additives of plant origin are generally believed to be safer, healthier and less subject to hazards for humans and animals. Spice and many medicinal herbs and plant extracts have appetite- and digestion-stimulating properties, antioxidant and antimicrobial activities, influence the poultry productivity and health mainly by stabilization of normal gut microflora and prevention of pathogens colonization. Also they have beneficial effects on digestive enzymes production, improve and exert certain immunological consequences in bird's body. Mixtures of complex compounds, vitamins and minerals found in plants tend to work together synergistically. These combinations were more effective than they were each used in isolated form. These beneficial effects make them useful as potential natural animal feed additives [13-16].

Antioxidant effect of aromatic plants is due to the presence and specific arrangement of hydroxyl groups in their phenolic compounds. These polyphenol also have important biological activities in vitro such as anti-tumour, chemo-preventive and anti-inflammatory activities. It has been proposed that polyphenol from some medicinal plants may greatly increase the functionality of food in terms of health and wellness [17-19].

Antioxidant properties of compounds from some medicinal herbs can result from their free radical scavenging activity but their ability to chelate transition metal ions, especially Fe (II) and Cu (II), also plays an important role [20]. The metal chelating ability of polyphenol is related to the presence of ortho-dihydroxy polyphenol, i.e., molecules bearing catechol or galloyl groups and condensed tannins. Also since metal chelation can occur at physiological pH it has a physiological significance [21,22].

Therefore, the aim of our study was to evaluate, for the first time in the field, the effect of different medicinal herb rich in polyphenol (Lemon balm, Sage, St. John's wort and Small-flowered Willowherb) used as dietary supplements on bioaccumulation of some essential metals (Fe, Mn, Zn and Cu) in chicken liver and chicken meat from legs and breast.

TABLE 1: Total phenol in feeding diets (a, b, c: different letters within the same column indicate significant differences among levels ($p \leq 0.05$)

Feeding diet	Trial code	Total phenols, mg gallic acid/g fodder		Relative to control
		Average	± 95% confidence	%
Basal diet, minerals and 2% Lemon balm	Le	2.89a	0.15	146
Basal diet, minerals and 2% Sage	Sa	2.73a	0.15	138
Basal diet, minerals and 2% St. John's wort	Jo	2.39b	0.10	121
Basal diet, minerals and 2% Small-flowered Willowherb	Wi	2.31b	0.11	117
Basal diet and minerals	Min	1.98c	0.10	100
Only basal diet (control)	B	1.98c	0.10	100

12.2 RESULTS AND DISCUSSIONS

The total phenols contents, determined by Folin-Ciocalteu method, of experimental feeding diets are shown in Table 1.

Although the addition of medicinal herbs in the basic diet was the same (2%), due to various polyphenol contents of medicinal herbs [23], a different content of polyphenol in the final diet was identified. Higher content in polyphenol was identified in feeding diet supplemented with Lemon balm and Sage (Le, Sa), followed by feeding diets supplemented with St. John's wort and Small-flowered Willowherb (Jo, Wi). Relative to control feeding diet, content of polyphenol ranged between 117-121% for Jo and Wi diets and between 138-146% for Le and Sa feeding diets.

Except Small-flowered Willowherb all other herbs used in the experiment contain volatile oils consist mostly of volatile compounds of mono, tri- and sesquiterpenes. Besides different content of total polyphenol, herb and flowers of these plants contain various types of polyphenol.

In Lemon balm herb a lot of polyphenol were identified: flavonoids such as luteolin, quercetin, apigenin and kaempferol; phenylpropanoids including hydroxycinnamic acid derivatives (caffeic and chlorogenic acids) and in particular rosmarinic acid and other various condensed tannins [24].

In Sage herb were identified a lot of flavonoids, formed from flavones and their glycosides, phenolic glycosides, benzoic acid derivate, hydroxycinnamic acid derivate composed from free caffeic acid, caffeic acid dimer (rosmarinic acid), caffeic acid trimmers, caffeic acid tetramer, phenolic diterpenes [25].

St. John's wort herb and flowers contain various polyphenol: flavonoids (epigallocatechin, rutin, hyperoside, isoquercetin, quercitrin, quercetin, amentoflavone, astilbin, miquelianin), phenolic acids (chlorogenic acid, 3-O-coumaroylquinic acid), and various naphtodianthrones: (hypericin, pseudohypericin, protohypericin, protopseudohypericin), phloroglucinols (hyperforin, adhyperforin). The naphthodianthrones hypericin and pseudohypericin along with the phloroglucinol derivative hyperforin are thought to be the active components [26, 27].

Willowherb (Small-flowered Willowherb, Epilobium parviflorum) species contain flavonoids, especially derivatives of kaempferol, quercetin, and myricetin. In E. parviflorum and E. angustifolium, β-sitosterol, various esters of sitosterol, and sitosterol glucoside have been detected. Gallic-acid derivatives may be present. Two macrocyclic ellagitannins, oenothein A and oenothein B, have been identified as the main constituents responsible for the inhibition of 5-alpha-reductase and aromatase enzymes [28]. According to Kohlert et al. [29], most active principles of plant extract are absorbed in the intestine by enterocytes, and readily metabolized by the body. The products of this metabolism are transformed into polar compounds by conjugation with glucuronate and excreted in the urine. As the active compounds are readily metabolized and have a short half-life, the risk of tissue accumulation is minimal [30].

The diverse composition of medicinal herb is reflected in a complex influence on bioaccumulation of essential metals Fe, Mn, Zn and Cu in chicken meats. The content and correlation of the investigated metals of different chicken meats is shown in Table 2, Table 3, Table 4, Table 5, Table 6 and Table 7.

TABLE 2: Metals contents (mg/kg dry matter) in chicken liver, fed with different diets

Metals/ Feeding diet	Feeding diet code	Fe		Mn		Zn		Cu	
		Average	± Std	Average	± Std	Average	± Std	Average	± Std
Basal diet, minerals and 2% Lemon balm	Le	45.28	4.43	6.13	0.58	22.75	2.13	4.39	0.52
Basal diet, minerals and 2% Sage	Sa	58.65	6.01	7.35	0.81	28.09	3.22	7.35	0.81
Basal diet, minerals and 2% St. John's wort	Jo	47.76	4.39	5.08	0.49	23.70	2.17	4.16	0.37
Basal diet, minerals and 2% Small-flowered Willowherb	Wi	34.00	4.50	4.93	0.54	16.62	2.01	4.50	0.52
Basal diet and minerals	Min	45.80	3.99	3.78	0.43	17.31	1.57	4.62	0.44
Only basal diet (control)	B	31.00	3.11	2.69	0.25	13.64	1.16	2.96	0.36

TABLE 3: Correlations (Pearson) between metal-metal in chicken liver and metals-total phenols concentrations in diets (* significance, $p < 0.05$)

Compounds	Fe	Mn	Zn	Cu	Phen
Fe	1	0.78182	0.9192*	0.5922	0.57469
Mn	0.78182	1	0.91376*	0.66654	0.91365*
Zn	0.9192*	0.91376*	1	0.4994	0.80512
Cu	0.5922	0.66654	0.4994	1	0.42859
Phen	0.57469	0.91365*	0.80512	0.42859	1

TABLE 4: Metals contents (mg/kg dry matter) in meat from chicken legs, fed with different diets

Metals/ Feeding diet	Feeding diet code	Fe		Mn		Zn		Cu	
		Average	± Std	Average	± Std	Average	± Std	Average	± Std
Basal diet, minerals and 2% Lemon balm	Le	10.54	1.11	0.14	0.03	16.37	1.55	0.54	0.06
Basal diet, minerals and 2% Sage	Sa	15.62	1.73	0.33	0.05	15.32	2.02	0.62	0.07
Basal diet, minerals and 2% St. John's wort	Jo	12.49	1.22	0.17	0.03	16.09	1.57	0.33	0.03
Basal diet, minerals and 2% Small-flowered Willowherb	Wi	12.71	1.11	0.40	0.03	19.09	2.05	0.48	0.05
Basal diet and minerals	Min	12.85	1.11	0.36	0.03	11.00	1.51	0.33	0.03
Only basal diet (control)	B	6.94	0.58	0.15	0.02	9.00	1.11	0.25	0.02

TABLE 5: Correlations (Pearson) between metal-metal in chicken legs and metals-total phenols concentrations in diets (* significance, $p < 0.05$)

Compounds	Fe	Mn	Zn	Cu	Phen
Fe	1	0.77136	0.53473	0.64072	0.39564
Mn	0.77136	1	0.40392	0.41003	-0.04474
Zn	0.53473	0.40392	1	0.65861	0.64321
Cu	0.64072	0.41003	0.65861	1	0.85253*
Phen	0.39564	-0.04474	0.64321	0.85253*	1

TABLE 6: Metals contents (mg/kg dry matter) in chicken breast meat, fed with different diets

Metals/ Feeding diet	Feeding diet code	Fe		Mn		Zn		Cu	
		Average	± Std	Average	± Std	Average	± Std	Average	± Std
Basal diet, minerals and 2% Lemon balm	Le	4.88	0.55	0.01	0.001	7.25	0.82	0.44	0.04
Basal diet, minerals and 2% Sage	Sa	4.44	0.61	0.03	0.003	7.49	0.73	0.74	0.09
Basal diet, minerals and 2% St. John's wort	Jo	4.06	0.55	0.01	0.001	6.66	0.72	0.57	0.07
Basal diet, minerals and 2% Small-flowered Willowherb	Wi	3.10	0.43	0.01	0.001	5.80	0.65	0.34	0.03
Basal diet and minerals	Min	3.67	0.34	0.02	0.002	7.02	0.75	0.50	0.04
Only basal diet (control)	B	2.81	0.24	0.01	0.001	5.64	0.63	0.30	0.02

TABLE 7: Correlations (Pearson) between metal-metal in chicken breast and metals-total phenols concentrations in diets (* significance, $p < 0.05$)

Compounds	Fe	Mn	Zn	Cu	Phen
Fe	1	0.32347	0.90172*	0.48622	0.85683*
Mn	0.32347	1	0.64476	0.72181	0.19063
Zn	0.90172*	0.64476	1	0.79057	0.63427
Cu	0.48622	0.72181	0.79057	1	0.0621
Phen	0.85683*	0.19063	0.63427	0.0621	1

For better emphasizing of the influence of different feeding diets over the bioaccumulation of metals in chicken meats, for each metal were calculated the accumulation factors (AF). This factor is the ratio between concentrations of metal in meat from experimental trial to concentration of metal in meat from control trial:

AF metal = metal content in meat from X trial (mg/kg dry matter)/ metal content in meat from control trial (mg/kg dry matter)

Like control trial the trial with feeding diet composed from basal diet only was selected.

12.2.1 BIOACCUMULATION OF METALS IN CHICKEN LIVER

The contents of bioactive metals Fe, Mn, Zn and Cu in chicken liver from chicks fed with different diets are presented in Table 2. In chicken liver, Fe is the predominant metal (31 to 58.65 mg/kg dry matter), followed by Zn (13.64 to 28.09 mg/kg dry matter). Mn and Cu are in the same range of concentration (3 to 7 mg/kg dry matter) significantly lower than previous. These data are in accordance with other authors who found the same contents in same meat food stuffs and are under maximum acceptable limit imposed by national and international legislation[31-34].

A linear regression correlation test was performed to investigate correlations between metal-metal and metals-diet total phenols concentrations (Phen). The values of correlation coefficients are given in Table 3.

In chicken liver there are significant correlations between dominant metals: Fe-Zn (r = 0.9192) and Mn-Zn (0.9137) and relative good correlation between Fe-Mn (r = 0.7818). Only Mn and Zn have relative good correlations with total phenols content (r = 0.9136 and 0.8051 respectively).

The addition of metal salts in the feed significantly influences all metals accumulations in the liver, with minor differences to the type of metal. Under the influence of medicinal herbs, accumulation of metals in the liver, however, presents significant differences from the group that received a diet supplemented only with metal salts (Min diet) (Figure 1).

Size of AF is differentiated, depending both on the metal and diet types. Diets enriched with Lemon balm and Sage, rich in polyphenol, have the most positively pronounced effect on uptake of metals. Between the two diets, Sa diet presents the largest accumulation factors for all metals and Le diet only for Mn and Zn, comparatively with Min diet.

The lowest influences on accumulation factors of metals in the liver have diets enriched with St. John's wort and Small-flowered Willowherb,

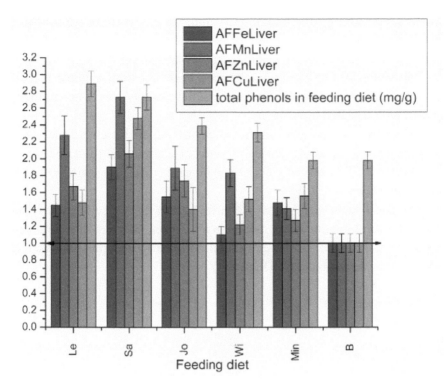

FIGURE 1: Metals accumulation factors (AF) in chicken liver under influence of different feeding diets (Error bar represent range for 95% confidence).

diets with lower polyphenol content (Table 1). These diets significantly positively influence only the Mn uptake. Jo diet significantly increases Zn accumulation while Wi diet significantly reduces the accumulation of Fe in chicken liver. Generally the diet enriched in Willowherb had a poor or even negative effect on metals accumulation in chicken liver, compared with other diets.

12.2.2 BIOACCUMULATION OF METALS IN MEAT FROM CHICKEN LEGS

In chicken legs meat (Table 4) Zn and Fe are the predominant metals and were found in the same range of concentration ($\approx$ 9 to 19 mg/kg dry matter for Zn and ≈ 7 to 16 mg/kg dry matter for Fe). These are followed by Cu and Mn in the same range of concentration (≈ 0.25 to 0.64 for Cu and ≈ 0.14 to 0.40 for Mn), significantly lower than previous.

The values of correlation coefficients obtained from a linear regression test between metal-metal and metals-diet total phenols concentrations (Phen) are presented in Table 5.

Generally poor correlations between investigated metals in chicken legs meat were found. A relatively good correlation was found between: Fe-Mn ($r = 0.7714$) and only Cu has significant correlation with total phenols content ($r = 0.8525$).

The addition of metal salts in the feed (Min diet) significantly influences the accumulation of all metals in meat from chicken legs, with significant differences of the type of metal. Mn and Fe accumulation factors were greater than Zn and Cu (Figure 2).

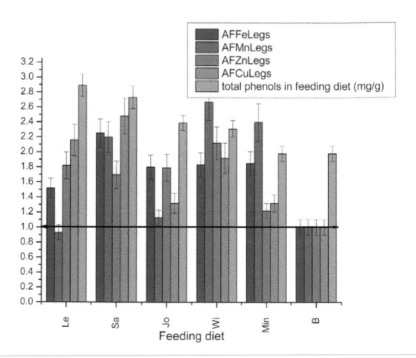

FIGURE 2: Metals accumulation factors (AF) in meat from chicken legs under influence of different feeding diets (Error bar represent range for 95% confidence).

Under the influence of medicinal herbs, accumulation of metals in meat from chicken legs differs significantly from the group that received a diet supplemented only with metals (Min diet). Especially accumulation of Cu was positively influenced by diets rich in polyphenol (Le, Sa). The addition of Sage and Small-flowered Willowherb in diet significantly influence accumulation of Zn and Cu and less Mn and Fe. Note that addition of Lemon Balm and St. John's wort herbs in the diet have significantly negative effects on accumulation of Mn in chicken legs meat.

12.2.3 BIOACCUMULATION OF METALS IN CHICKEN BREAST MEAT

In chicken breast meat (Table 6), Zn is the predominant metal ($\approx$ 5.6 to 7.5 mg/kg dry matter) followed by Fe ($\approx$ 2.8 to 4.9 mg/kg dry matter) in the same range of concentration and Cu ($\approx$ 0.3 to 0.74 mg/kg dry matter) at lower concentration. Mn concentration ($\approx$ 0.01 to 0.03 mg/kg dry matter) is significantly lower than previous.

A linear regression correlation test was performed to investigate correlations between metal-metal and metals- diet total phenols concentrations (Phen). The values of correlation coefficients are given in Table 7.

There are significant correlations between Fe-Zn concentrations (r = 0.90172) in chicken breast meat and relative good correlations were found between: Zn-Cu (r = 0.7906) and Mn-Cu (r = 0.7218). Only Fe has significant correlation with total phenols content (r = 0.8568).

The addition of metal salts in the feed significantly influences the accumulation of all metals in meat from chicken legs, with significant differences from the type of metal. In Min diet, Mn and Cu uptake factors were larger than Zn and Fe (Figure 3). Under the influence of medicinal plants, metals accumulation in chicken breast differ significantly from the group that received a diet supplemented only with metals salts (Figure 3).

Fe accumulation is significantly positively influenced by Lemon balm and Sage diets, rich in polyphenol, and less affected or even reduced in diets with St. John's herb and Small-flowered Willowherb, which has a lower concentration of polyphenol. Accumulation of Mn and Cu is more positively influenced by Sage diet; the other diets even cause a significant

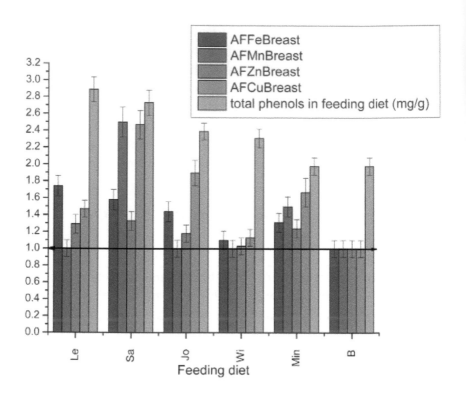

FIGURE 3: Metals accumulation factors (AF) in chicken breast meat under influence of different feeding diets (Error bar represent range for 95% confidence).

reduction of Mn accumulation. Accumulation of Zn in chicken breast meat is generally less influenced by all the diets used. Sage diet (Sa) clearly differ from other diets having the greatest positive influence on the accumulation of Mn, Cu and Fe and less on Zn.

12.3 CONCLUSIONS

The contents of essential metals (Fe, Mn, Zn and Cu) in chicken meats are different, depending mainly on the type of meat. So Fe is the predominant metal in liver and Zn is the predominant metal in legs and breast chicken meat. The addition of metal salts in the feed influences accumulations of all metals in the liver, legs and breast chicken meat with specific difference to the type of metal and meat. The greatest influences were observed in legs meat for Fe and Mn. However, under the influence of medicinal herbs (Lemon balm, Sage, St. John's wort and Small-flowered Willowherb) rich in different type of polyphenol (flavonoids, phenols acids, benzoic acid derivate, phenylpropanoids derivate, condensed tannins), accumulation of metals in the liver, legs and breast chicken meat presents significant differences from the group that received a diet supplemented only with metals salts. Each medicinal herb from diet had a specific influence on the metals accumulation and generally moderate or poor correlations were observed between total phenols and metal accumulation factors. Great influence on all metal accumulation factors was observed in diet enriched with sage, which had only significantly positive effect for all type of chicken meat. The effects of medicinal herb in the metals biochemistry of broiler are still unclear and complex. The differences regarding metals accumulations could appear due to the antagonism between metal ions or to the chemical composition of chicken diets, which comprise other chelating agents like amino acids and protein from soybean and corn. This fact leads to a competition between polyphenols and amino acids for metals complexation and bioaccumulation. Although some effects have already been demonstrated, the mechanisms involved in bioaccumulation of these are still widely unknown. More investigations on the action of the active compounds and their effects in vivo are required and significant improvement of animal performance must be shown before medicinal herbs or plants extracts are effectively adopted as feed additives.

12.3.1 EXPERIMENTAL

In our research the national and international behaviour codes were complied regarding the experiment with animals.

12.3.2 EXPERIMENTAL DESIGN

A total of 180 mixed one-day-old broiler chicks (Ross 308) were purchased from a local hatchery, weighed on arrival and randomly allocated to 18 pens (1 × 1 m) of 10 birds each, with equal numbers of male and females (tree replicates per each treatment) in a ventilated broiler house containing wood shavings as litter material. Water and feed were available ad libitum. All chickens were fed the similar starter (weeks 0-3 of age) and grower (weeks 4-6 of age) diets (Table 8), but receive different diet defined as: Le, Sa, Jo, Wi, Min and B treatments, respectively. Ingredients and nutrients composition of diets are shown in Table 8 and 9. Medicinal herbs, Lemon balm (Melissae folium), Sage herb (Salvia officinalis), St. John's wort (Hypericum perforatum) and Small-flowered Willowherb (Epilobium parviflorum), purchased from local specialized market, have been dried and ground to pass through a 2-mm screen. Minerals salts, p.a. purity (Merck Chemicals, Germany) used for metal sources were: Mohr salt as a source of iron, $MnSO_4$ $4H_2O$ as a source of manganese, $ZnSO_4$ $7H_2O$ as a source of zinc and $CuSO_4$ $5H_2O$ as a source of copper. Due to low metal uptake from medicinal plants [35], its contribution to experimental diets has not been taken into account. Experimental diets were prepared weekly by a special feed factory. Diets were formulated to meet or exceed the requirements of the National Research Council (NRC) [36] for broilers at this age. Composition of basal diet is presented in Table 9.

TABLE 8: Ingredients and nutrient composition of experimental diets

Nr.	Diets	Diet code
1	Basal diet + 2% Lemon Balm + minerals from salts, at NRC level (80 mg/kg Fe, 60 mg/kg Mn, 40 mg/kg Zn and 8 mg/kg Cu)	Le
2	Basal diet + 2% Sage herb + minerals from salts, at NRC level (80 mg/kg Fe, 60 mg/kg Mn, 40 mg/kg Zn and 8 mg/kg Cu)	Sa
3	Basal diet + 2% St. John's wort + minerals from salts, at NRC level (80 mg/kg Fe, 60 mg/kg Mn, 40 mg/kg Zn and 8 mg/kg Cu)	Jo
4	Basal diet + 2% Small-flowered Willowherb + minerals from salts, at NRC level (80 mg/kg Fe, 60 mg/kg Mn, 40 mg/kg Zn and 8 mg/kg Cu)	Wi
5	Basal diet + minerals from salts, at NRC level (80 mg/kg Fe, 60 mg/kg Mn, 40 mg/kg Zn and 8 mg/kg Cu)	Min
6	Basal diet only (kept as control)	B

TABLE 9: Ingredients and nutrient composition of basal diet

Composition	Stage 0-3 weeks (%)	Stage 4-6 weeks (%)
Corn	56.5	61,62
Soybean meal	34	31
Fish meal	5	2
Sunflower oil	1,5	2,5
CaCO3	0,8	1
Dicalcium phosphate	1,2	1
NaCl	0,2	0,2
DL Metionine	0,3	0,18
Vitaminic Premix	0,5	0,5
Nutritive characteristic		
E M (kcal/kg feed)	3204	3214
Crude protein (%)	22,91	20,03
Lizine (%)	1,27	1,06
Metionine + cistine (%)	0,95	0,72
Calcium (%)	1,03	0,89
Total Phosphor (%)	0,73	0,62
Crude Celuloze (%)	3,16	3,08

12.3.3 SAMPLES COLLECTION AND PREPARATION

12.3.3.1 MEAT SAMPLES

At days 43 of age, 6 birds per pen were selected, weighed and slaughtered to obtain the meat samples. Meat samples were first dried in oven at 105°C for 24 hours and then digested in the Muffle furnace by stepwise increase of the temperature up to 550°C till the white ash formed. The ash was dissolved in 0.5 N HNO3 and filtered through ash-free filter paper before analysis. Each sample solution was made up with dilute HNO_3 (0.5 N) to a final volume of 50 mL and analysed by flame atomic absorption spectrometry. Necessary dilutions were made [2, 3].

12.3.3.2 METALS ANALYSIS AND QUALITY CONTROL

The concentrations of Fe, Mn, Zn, Cu in the solutions were determined using a flame atomic absorption spectrophotometer with high resolution continuum source (Model ContrAA 300, Analytik Jena, Germany), fitted with a specific conditions of particular metal using appropriate drift blanks. Mix standard solutions of heavy metals (1000 mg/L), namely iron, manganese, zinc and copper - ICP Multielement Standard solution IV CertiPUR, were purchased from Merck Germany. Solutions of varying concentrations were prepared for all the metals by diluting the standards. HPLC water was used for the preparation of reagents and standards. All chemicals were trace metal grade (Suprapur). Concentrate nitric acid (HNO_3 65%), were obtained from Merck Germany. All glassware was treated with Pierce solution 20% (v/v), rinsed with cold tap water followed by 20% (v/v) nitric acid and then rinsed with double-distilled water. For quality control purposes, blanks and duplicates samples were analysed during the procedure. NCS Certified Reference Material-DC 85104a and 85105a (China National Analysis Center for Iron&Steel), was analysed for quality assurance. Per cent recovery means were: Fe (93%), Mn (94%), Zn (105%), and Cu (103%). The variation coefficients were below 10%. Detection limits (μg/mL) were determined by the calibration curve method: Fe (0.14), Mn (0.18), Zn (0.45), Cu (0.11). The blank reagent and standard reference soil materials were included in each sample batch to verify the accuracy and precision of the digestion procedure and also for subsequent analyses.

12.3.3.3 FEED ANALYSIS FOR TOTAL POLYPHENOL

The most used extraction solvent of polyphenol from vegetable matrix is methanol or ethanol and water-methanol/ethanol solutions [37]. The hydro-ethanol extracts for total polyphenol contents determination were prepared as following: 1 g feed samples, grounded and sieved at 0.3 mm was mixed with 20 mL 50% ethanol (v/v). The obtained hydro-ethanol extracts were rested for 30 minutes and finally filtered. The total polyphenol content was determinate using Folin & Ciocalteu method [38]. In this reagent Molybdenum is in superior oxidation state (+6) and has a yellow colour. The polyphenol are responsible for the reduction of Mo 6+ to the inferior state of oxidation (Mo 5+, Mo 4+) with blue colour. It was prepared 2.0 M Folin-Ciocalteu phenol reagent, 10 mM gallic acid standard solutions (50% ethanol) and 7.5% carbonate solutions. All chemicals and reagents were analytical grade or purest quality purchased from Merck and Fluka. For the preparation of calibration curve 0.5 mL aliquot of 0.2, 0.3, 0.4, 0.8 and 1.2 μM/mL gallic acid solution were mixed with 2.0 mL Folin-Ciocalteu reagent (diluted 1/10) and 2.0 mL sodium carbonate solution. The absorbance of solutions was read after 2 h at 750 nm using UV-VIS spectrophotometer SPECORD 205 by Analytik Jena. All determinations were realized in triplicate for each sample. Total polyphenol content was expressed like mg gallic acid/g feed. Squared correlation coefficient (R2) for calibration curve was 0.994 [39,40].

12.3.3.4 STATISTICAL TREATMENT OF DATA

Measurements of total polyphenol and metals contents in feed respectively meat samples were expressed in terms of means and standard deviation. The whole data were subjected to a statistical analysis and correlation matrices were produced to examine the inter-relationships between the investigated metals concentrations of the meat samples and between these metals and total polyphenol from different diets, Student's t-test was employed to estimate the significance of values. Statistical significance was computed using Pair-Samples T-Test, with a significance level of p < 0.05. The data were statistically analysed and plotted using a statistical package Origin Pro 8.5.1.

REFERENCES

1. Sabir SM, Khan SW, Hayat I: Effect of environmental pollution on quality of meat in district Bagh, Azad Kashmir. Pakistan Journal of Nutrition 2003, 2:98-101.
2. Birla Singh K, Taneja SK: Concentration of Zn, Cu and Mn in vegetables and meat foodstuffs commonly available in Manipur: a North Eastern state of India. EJEAF-Che 2010, 9:610-616.
3. Harmanescu M, Alda LM, Bordean DM, Gogoasa I, Gergen I: Heavy metals health risk assessment for population via consumption of vegetables grown in old mining area; a case study: Banat County Romania. Chem Centr J 2011, 5:64.

4. Murariu M, Gradinaru RV, Mihai M, Jurcoane S, Drochioiu G: Unexpected effect of nickel complexes of some histidine-containing peptides on Escherichia coli. Romanian Biotechnological Letters 2011, 16(3):6242-6246.
5. Drinceanu D, Julean C, Simiz E, Ştef L, Luca I, Sofian D: Effects of mineral supplements on bioproductive results in egg-laying hens farmed in organic systems. Scientific Papers: Animal Science and Biotechnologies 2011, 44:30-36.
6. Abdulkarimi R, Abdullahi A, Amini M: Mentha extract consumption (Mentha piperita L) reduced blood iron concentration and increased TIBC levels in broiler chickens. Journal of American Science 2011, 7:494-500.
7. Okoye COB, Ibeto CN, Ihedioha JN: Assessment of heavy metals in chicken feeds sold in south eastern, Nigeria. Advances in Applied Science Research 2011, 2:63-68.
8. Murariu M, Dragan ES, Drochioiu G: Electrospray ionization mass spectrometric approach of conformationally-induced metal binding to oligopeptides. Eur J Mass Spectrom 2010, 16:511.
9. Drochioiu G, Manea M, Dragusanu M, Murariu M, Dragan ES, Petre BA, Mezo G, Przybylski M: Interaction of beta-amyloid(1-40) peptide with pairs of metal ions: An electrospray ion trap mass spectrometric model study. Biophys Chem 2009, 144(1-2):9-20.
10. Bao YM, Choct M, Iji PA, Bruerton K: Effect of organically complexed Copper, Iron, Manganese and Zinc on broiler performance, mineral excretion and accumulation in tissues. J Appl Poult Res 2007, 16:448-455.
11. Nollet L, van der Klis JD, Lensing M, Spring P: The effect of replacing inorganic with organic trace minerals in broiler diets on productive performance and mineral excretion. J Appl Poult Res Winter 2007, 16:592-597.
12. Abdallah AG, El-Husseiny OM, Abdel-Latif KO: Influence of some dietary organic mineral supplementations on broiler performance. International Journal of Poultry Science 2009, 8:291-298.
13. Hernandez F, Madrid J, Garcıa V, Orengo J, Megıas MD: Influence of Two Plant Etracts on Broilers Performance, Digestibility, and Digestive Organ Size. Poultry Sci 2004, 83:169-174.
14. Stef L, Dumitrescu G, Drinceanu D, Stef D, Mot D, Julean C, Tetileanu R, Corcionivoschi N: The effect of medicinal plants and plant extracted oils on broiler duodenum morphology and immunological profile. Rom Biotechnol Lett 2009, 14:4606-4614.
15. Khaligh F, Sadeghi G, Karimi A, Vaziry A: Evaluation of different medicinal plants blends in diets for broiler chickens. Journal of Medicinal Plants Research 2011, 5:1971-1977.
16. Zaki AA, Elbarawy AM, Darwish AS: Biochemical studies on the effect of Nasturtium Officinalis plant extract in chickens fed raw soya bean meals. Aust J Basic & Appl Sci 2011, 5:755-761.
17. Jang A, Liu XD, Shin MH, Lee BD, Lee SK, Lee JH, Jo C: Antioxidative potential of raw breast meat from broiler chicks fed a dietary medicinal herb extract mix. Poultry Sci 2008, 87:2382-2389.
18. Brenes A, Viveros A, Goñi I, Centeno C, Saura-Calixto F, Arija I: Effect of grape seed extract on growth performance, protein and polyphenol digestibilities, and antioxidant activity in chickens. Span J Agric Res 2010, 8:326-333.

19. Polat U, Yesilbag D, Eren M: Serum biochemical profile of broiler chickens fed diets containing Rosemary and Rosemary volatile oil. J Biol Environ Sci 2011, 5:23-30.
20. Karamać M: Chelation of Cu(II), Zn(II), and Fe(II) by tannin constituents of selected edible nuts. Int J Mol Sci 2009, 10:5485-5497.
21. Khokhar S, Apenten RKO: Iron binding characteristics of phenolic compounds: some tentative structure-activity relations. Food Chem 2003, 81:133-140.
22. Andjelković M, Van Camp J, De Meulenaer B, Depaemelaere G, Socaciu C, Verloo M, Verhe R: Iron-chelation properties of phenolic acids bearing catechol and galloyl groups. Food Chem 2006, 98:23-31.
23. Vargias L, Harmanescu M, Gergen I: Mineral Content and Antioxidant Capacity of Aqueous and Methanolic Extract from Epilobium Hirsutum. Scientifical Researches. Agroalimentary Processes and Technologies 2005, XI(1):199-204.
24. Safra J, Pospíšilová M, Honegr J, Spilková J: Determination of selected antioxidants in Melissae herba by isotachophoresis and capillary zone electrophoresis in the column-coupling configuration. J Chromatogr A 2007, 1171:124-132.
25. European Medicines Agency: Assessment report on Salvia Officinalis L., Folium and Salvia Officinalis L., Aetheroleum. [http://www.emea.europa.eu/docs/en_GB/document_library/Herbal_-_HMPC_assessment_report/2010/02/WC500070850.pdf] 2010.
26. Franchi GG, Nencini C, Collavoli E, Massarelli P: Composition and antioxidant activity in vitro of different St. John's Wort (Hypericum perforatum L.) extracts. Journal of Medicinal Plants Research 2011, 5:4349-4353.
27. Orčić DZ, Mimica-Dukić NM, Franišković MM, Petrović SS, Jovin EĐ: Antioxidant activity relationship of phenolic compounds in Hypericum perforatum L. Chem Centr J 2011, 5:34.
28. Valimareanu S, Deliu C: Polymorphism detection of in vitro cultivated Epilobium (Onagraceae) species using RAPD method. Contribuţii Botanice 2008, XLII:113-120.
29. Kohlert C, Van Rensen I, Marz R, Schindler G, Graefe EU, Veit M: Bioavailability and pharmokinetics of natural volatile terpenes in animal and humans. Planta Medica 2000, 66:495-505.
30. Barreto MSR, Menten JFM, Racanicci AMC, Pereira PWZ, Rizzo PV: Plant extracts used as growth promoters in broilers. Brazilian Journal of Poultry Science 2008, 10:109-115.
31. Iwegbue CMA, Nwajei GE, Iyoha EH: Heavy metal residues of chicken meat and gizzard and turkey meat consumed in Southern Nigeria. Bulgarian Journal of Veterinary Medicine 2008, 11:275-280.
32. Uluozlu OD, Tuzen M, Mendil D, Soylak M: Assessment of trace element contents of chicken products from Turkey. Journal of Hazardous Materials 2009, 163:982-987.
33. Akan JC, Abdulrahman FI, Sodipo OA, Chiroma YA: Distribution of heavy metals in the liver, kidney and meat of beef, mutton, caprine and chicken from Kasuwan Shanu market in Maiduguri Metropolis, Borno State, Nigeria. Research Journal of Applied Sciences, Engineering and Technology 2010, 2:743-748.
34. Ordinance 975/16 XII 1998 of Romanian Ministry of Health [http://www.unilab.ro/downloads/O%20975%20norme%20igienico-sanitare%20-%20marieta.pdf]

35. Harmanescu M: Heavy Metals Determination in Selected Medicinal Plants. Annals of the Faculty of Engineering Hunedoara - Journal of Engineering 2007, V(1):63-68.

36. Subcommittee on Poultry Nutrition, National Research Council: Nutrient Requirements of Poultry. Ninth Revised edition. The National Academies Home, 500 Fifth St. N.W. Washington, D.C; 1994. 20001 http://www.nap.edu/catalog.php?record_id=2114

37. Ignat I, Volf I, Popa VI: A critical review of methods for characterisation of polyphenolic compounds in fruits and vegetables. Food Chem 2011, 126:1821-1835.

38. Folin O, Ciocalteu V: On tyrosine and tryptophane determination in proteins. Journal of Biological Chemistry 1927, 27:627-650.

39. Harmanescu M, Moisuc A, Radu F, Dragan SI, Gergen I: Total polyphenol content determination in complex matrix of medicinal plants from Romania by NIR spectroscopy. Bulletin UASVM Agriculture 2008, 65:123-128.

40. Moigradean D, Poiana MA, Gogoasa I, Harmanescu M, Gergen I, Lazureanu A: The correlations between total antioxidant capacity and total polyphenol content established for tomatoes. Lucrări stiinţifice medicină veterinară 2007, XL:486-489.

This chapter was originally published under the Creative Commons Attribution License. Stef, D. S., and Gergen, I. Effect of Mineral-Enriched Diet and Medicinal Herbs on Fe, Mn, Zn, and Cu Uptake in Chicken. Chemistry Central Journal 2012, 6:19. doi:10.1186/1752-153X-6-19.

CHAPTER 13

CONSUMING IODINE ENRICHED EGGS TO SOLVE THE IODINE DEFICIENCY ENDEMIC FOR REMOTE AREAS IN THAILAND

WIYADA CHAROENSIRIWATANA, PONGSANT SRIJANTR, PUNTHIP TEEYAPANT, and JINTANA WONGVILAIRATTANA

13.1 BACKGROUND

Iodine deficiency disorders (IDD) had been widely recognized as one of the important public health problems especially in developing countries throughout the world [1]. Thailand, one of the developing countries in South East Asia has started the public health activities on elimination of IDD endemic areas since 1989. In 1991, a national survey on total goitre prevalence (TGP) in 20,596 schools from 3,366,867 children was done in 53 provinces throughout the country with the mean TGP of 15.79% [2]. Later in 1994, World Health Organization (WHO) produced a document in collaboration with United Nations International Children's Emergency Fund (UNICEF) and International Council for Control of Iodine Deficiency Disorders (ICCIDD) for the guidance concerning the IDD surveillance indicators and salt iodization has been selected as a strategy to control and elimination of IDD [3]. Since then Thailand has emphasized various kind of strategic planning for elimination of IDD endemic areas such as iodization of drinking water for villagers, administration of iodine capsules every 6-10 months for population in remote areas where transportation was the main problem for the enrichment of iodine in communities, iodization of edible salt at 30 ppm for daily use in household etc. In 2006, a survey for the use of iodine salts from 819 households were done in iodine deficiency endemic areas in Udon Thani province and found that only 10.26% of the households consumed iodine edible salts which were passed the standard of edible iodized salts (not less than 30 ppm of iodine) declared by Thailand FDA [4]. In addition, the TSH index for monitoring IDD by WHO/UNICEF/ICCIDD guideline, showed that during the years 2003 - 2006 the number of neonates having TSH >5 mU/L were 13.54%, 15.28%, 21.55%, 19.56% respectively [5]. These TSH index showed that Thailand was exposed to iodine deficiency condition and also indicated that the current public health activities did not achieve the goal for the elimination of IDD.

A new pilot programme has been proposed to public health sector to introduce a more stable form of iodine in daily nutrition through the existence of biological active iodo-organic compounds from animal and plant origins in the natural food chain which will be sustainable in the long-term. The design of the programme was initiated with the aim to increase the content of iodine in eggs and vegetables. National Statistic Data on Thailand eggs consumption during 2006 and 2007 were 9,789 and 9,376 million eggs respectively or 142 and 150 eggs per person per year [6]. Thus, the hen eggs would be one of the reliable sources for iodine consumption in daily nutrition. This new programme would be implemented with the philosophy of Sufficient Economy Concept introduced by His Majesty the King of Thailand to conduct all Thai peoples living for better quality of life with the appropriate needs of socio-economic development in moderation, reasonableness and self-immunity for sufficient protection from both internal and external impacts arising with knowledge and spirit to supply their long-term necessities throughout the country starting from the level of the families, communities as well as the nation. It was believed that with this new concept of the biological active iodo-organic compound enrichment in the natural food chain through the Sufficient Economy Philosophy, the new programme for elimination of iodine deficiency endemic areas would be successfully implemented through the cooperation with all concerned sectors throughout the country.

14.2 METHODS

The design of the study was to determine the iodine status of communities by using urine iodine excretion concept before and after the implement of iodine enrichment in the natural food chain. The collecting of 858 urine of child bearing age women as first morning mid-stream urine from 5 districts in Udon Thani province were done in 2006 in order to establish base-line iodine status of the areas before implementing the programme. The urine specimens were transported to the laboratory immediately in ice boxes and kept frozen at -20°C until the analysis. The urine iodine was determined by Inductively Coupled Plasma Mass Spectrometry (ICP-MS) with the use of tellurium as internal standard. During the assay performance, the internal quality control samples prepared from urine iodine reference standard code 2670 from US-National Institute of Standards and Technology (NIST) were inserted in every ten samples of the running assay [7,8].

In 2008, the model programme was operated in cooperation with Napu Sub-district Municipality and the experimental design was approved from the Napu Sub-district Municipal Committee in compliance with the Helsinki Declaration. The propose programme on iodine enriched eggs in the form of biological active iodo-organic compounds from animal origins was started by selecting two neighbourhood villages in those base-line study areas: Ban Nong Nok Kean and Ban Kew in Napu sub-district of Udon Thani province. A hen egg farm located in these two village areas was selected as a model farm for the supply of iodine enriched eggs to the communities. The base-line of iodine content in eggs from regular feeding at this model farm was done by collecting 30 fresh eggs weight between 50-55 g, and determining the iodine content according to the TCM-040 based on Compendium of Methods for Food Analysis [9]. Then, the programme was initiated by replacing regular feed with the iodine enriched feeding formula prepared by the addition of iodine as potassium iodide (KI) to yield a final concentration of 4 mg of iodine per kilogram of poultry feed for the daily feeding process of the farm. The hens consumed the iodine poultry feed at the amount of 120-130 g per hen per day. After one month feeding, 30 eggs were collected and sent to the laboratory for the determination of iodine content.

TABLE 1: Median urine iodine contents from child bearing-age women in the study areas of Udon Thani province in August 2006

Province	District	Sub-district	Village	No. of samples	Median Urine Iodine (µg/dL)
Udon Thani	Nam Som	Nam Som		101	6.50
			Ban Phon	69	7.68
			Ban Na Muang Thai	32	3.86
Udon Thani	Ban Dung			174	4.97
		Ban Chantr	Ban Subsomboon	59	3.26
		Ban Chai	Ban Non Sa-ard	55	4.96
		Srisutho	Ban Sriburapa	60	6.45
Udon Thani	Pen	Napu		128	4.61
			Ban Kew	63	3.82
			Ban Nong Nok Khaen	65	5.24
Udon Thani	Muang			212	6.14
		Ban Jan	Ban Dong Keng	73	7.09
		Sam Praw	Ban Na Yaad	66	6.26
		Nong Na Kham	Ban Jampa	73	4.97
Udon Thani	Sri That	Tad Thong		238	4.64
			Ban Kud Na Khor	74	5.01
			Ban Pa Whai	67	4.67
			Ban Ratsomboon	97	4.07
Total in Udon Thani province				858	5.16

The biological active form of iodine from the iodine enriched eggs was evaluated by the determination of urine iodine excretion of volunteers before and after consuming a hard boiled iodine enriched egg continuously for five days, as one item in their breakfast and no other iodine enriched food was taken in the volunteers' meals during these five days. The first morning mid-stream urine from each volunteer was collected and the urine specimens were kept frozen at -20°C until the analysis. There were 124 women volunteers from these two villages, age between 20 - 63 years. All village volunteers were explained about the study details and consent forms were given to the volunteers for their agreement for study participation with an included statement that they could dropped out from this study at any time. The iodine content in urine was determined using the ICP-MS procedure. The median urine iodine was calculated and ANOVA was used for statistic analysis of the urine samples from the village volunteers for the treatment before and after eating iodine enriched eggs.

14.3 RESULTS

Table 1 showed that in August 2006 the median urine iodine of 858 child bearing-age women volunteers was 5.16 μg/dL (range from 3.26 - 7.68 μg/dL in all 5 districts) which meant that the villagers had an iodine deficiency condition even though the campaign from the provincial health office and district health centers emphasized their public health education programme on the use of iodized salt.

Table 2 showed the iodine content of hen eggs produced by the model farm. It was found that the average iodine content in eggs from the regular feeding formula was 25.31 μg per egg (or 75.96 μg per 100 gram of fresh weight) whereas the iodine enriched feeding formula yielded the mean iodine content in the range of 93.57 - 97.76 μg per egg (or 182.67 - 184.58 μg per 100 gram of fresh weight). The production batches that was provided to the village volunteers for their breakfast had an iodine content in the range of 90.97 - 104.14 μg per egg for egg size #2 (60 - 65 g per egg), and a range of 87.76 - 98.18 μg per egg for egg size #3 (55 - 60 g per egg).

TABLE 2: Iodine content in eggs from the experimental farm before and after the iodine enriched feeding

	Average iodine content in eggs	
	μg per egg	μg per 100 g.wt
Eggs from regular feeding (#4; 50-55 g)	25.31	75.96
Iodine enriched eggs (#3; 55-60 g)	93.57	182.67
Iodine enriched eggs (#2; 60-65 g)	97.76	184.58

Table 3 showed the level of urine iodine of the village volunteers. It was found that the base-line median urine iodine before consuming the iodine enriched eggs from 65 volunteers in Ban Nong Nok Kean was 7.04

μg/dL (SD = 8.54, n = 65), and from 59 volunteers in Ban Kew was 7.00 μg/dL (SD = 7.15, n = 59). The median urine iodine from 124 volunteers in these two villages was 7.03 μg/dL (SD = 7.89, n = 124) which indicated the mild iodine deficiency condition of these two village areas.

TABLE 3: Urine iodine contents of women volunteers in Ban Nong Nok Kean village and Ban Kew village before consuming iodine enriched eggs

Urine iodine content (μg/dL)	Ban Kew village	Ban Nong Nok Kean village	Total (2 villages)
maximum	37.30	45.51	45.51
minimum	1.00	1.51	1.00
Std.Dev	7.15	8.54	7.89
Median	7.00	7.04	7.03
Population of volunteers	59	65	124

Table 4 showed the results for urine iodine levels of the village volunteers after consuming a hard boiled iodine enriched egg continuously for five days as one item in their breakfast meals. The result of a median urine iodine from 55 volunteers of Ban Nong Nok Kean was 13.95 μg/dL (SD = 10.76, n = 55), and from 57 volunteers of Ban Kew was 20.76 μg/dL (SD = 13.63, n = 57). The median urine iodine from the volunteers in these two villages after having iodine enriched eggs was 16.57 μg/dL (SD = 12.56, n = 112).

TABLE 4: Urine iodine contents of volunteers in Ban Nong Nok Kean village and Ban Kew village after consuming iodine enriched eggs

Urine iodine content (μg/dL)	Ban Kew village	Ban Nong Nok Kean village	Total (2 villages)
maximum	61.2	50.93	61.2
minimum	3.94	3.53	3.53
Std.Dev	13.63	10.76	12.56
Median	20.76	13.95	16.57
Number of volunteers	57	55	112

Table 5 demonstrated the comparison using statistic ANOVA analysis of urine iodine content of the women volunteers from the two villages in the study: before and after consuming iodine enriched eggs. The result showed that there was a highly significant difference (P value < 0.001) in the urine iodine content of volunteers before and after consuming iodine enriched eggs.

TABLE 5: Comparison of urine iodine contents of women volunteers (as the same volunteers) with treatment before and after consuming iodine enriched eggs in Ban Nong Nok Kean village and Ban Kew village

	Median urine iodine contents (µg/dL)	
	Ban Kew village	Ban Nong Nok Kean village
Before consuming iodine enriched eggs	6.87	7.11
After consuming iodine enriched eggs	20.76	13.09
F-test	P < 0.001	P < 0.001
Statistical signification	**	**
Number of volunteers	57	53

** highly statistical significant as P value < 0.001
Note: There were 14 volunteers (12 in Ban Nong Nok Kean village and 2 in Ban Kew village) that dropped out during the study.

14.4 DISCUSSION

Before the implementation of the iodine enrichment programme in the natural food chain of study areas, the median urine iodine from 858 volunteers (Table 1) in 5 districts of Udon Thani province was 5.16 µg/dL (range 3.26 - 7.68 µg/dL). This demonstrated that the community urine iodine level was at a mild iodine deficiency status [10], even though there was an active campaign in the region to encourage consumption of edible iodized salts by the provincial health office.

The iodine enriched feeding formula was successfully prepared and yielded about 3.8 fold higher iodine content in the produced eggs than the regular formula (Table 2). Thus, the iodine enriched feeding formula could be used instead of the regular one and the cost of iodine added in the formula was only 0.33% of the cost per kg of poultry feed. This would not significantly affect the cost of hen egg production.

The result of urine iodine survey before the dietary enrichment of iodine diet in two villages of the study area: Ban Kew village and Ban Nong Nok Kean village in Napu sub-district, Udon Thani province was 7.03 µg/dL (Table 3) which confirmed that these two communities were still in the condition of mild iodine deficiency. These results also showed that during the period of 2006 - 2008 the diets consumed by the community were still lacking iodine and the concerned public health sectors had not created any awareness to eliminate this iodine deficiency crisis.

For the five day study of continuously consuming an iodine enriched egg as one item at breakfast, there were 124 volunteers that participated at the beginning of the study and 14 volunteers dropped out in later stage. There were 53 volunteers from Ban Nong Nok Kean village and 57 volunteers from Ban Kew village that participated through out the study programme. By using an ANOVA method for statistical assessment, the result showed that the median urine iodine level before and after consumption of iodine enriched eggs were highly significant different at $P < 0.001$ (Table 5). The effect of this innovative process produced a remarkable increase of the urine iodine level from the condition of iodine deficiency (median urine iodine content 6.87 - 7.11 µg/dL) to the optimal level of iodine (median urine iodine content 13.09 - 20.76 µg/dL) in the villagers of the study areas. The urinary iodine excretion could also be used as a valid marker for reporting the recent dietary iodine intake as well as one of the key indicators for monitoring the IDD situation at the community level and the sustained adherence to public health efforts to eliminated IDD.

14.5 CONCLUSIONS

In summary, the pilot model farm for the production of iodine enriched eggs supplied to the neighbouring communities was successfully developed which resulted in a self-support system of nutritional iodine enrichment for the communities. The WHO/UNICEF/ICCIDD has recommended that the daily iodine in take for various age groups be in the range between 90 - 200 μg. Since eggs could be consumed as daily food products in every Thai family, it would be possible to supply iodine eggs as the new iodine daily diet source to all Thai communities due to the fact that the cost for the addition of iodine as potassium iodide or potassium iodate to poultry feed was very low and did not significant increase the production cost of eggs from the farm. In addition, the preparation of iodine enriched formula poultry feed could be self processed at the farms and did not require any complicated equipments for the iodine enrichment steps. This innovative and inexpensive strategy could be easily applied to all remote areas throughout the country with the community programme of Sufficient Economy Concept to overcome the problem of iodine deficiency endemic in Thailand.

REFERENCES

1. Bruno De Benoist, Andersson Maria, Egli Ines, Takkouche Bahi, Allen Hnreitte: Iodine status worldwide: WHO Global Database on Iodine Deficiency. Geneva, World Health Organization, United Nations; 2004.
2. Pawabutr , Paichit : National Policy on the Control of Iodine Deficiency Disorders. Proceeding of National Seminar on the Control of Iodine Deficiency Disorders in Thailand: 3-5 March 1992; Chiang Mai, Thailand 1992, 19-29. (in Thai)
3. World Health Organization: WHO/UNICEF/ICCIDD Indicators for Assessing iodine deficiency disorders and their control through salt iodization. (WHO/NUT/94.6). Geneva, World Health Organization, United Nations; 1994.
4. Department of Health: Report on Monitoring of Iodine Deficiency Disorders in 2003 for 62 provinces in Thailand. Ministry of Public Health, Thailand; 2004. (in Thai)
5. Wiyada Charoensiriwatana, Srijantr Pongsant, Janejai Noppavan, Hasan Supaphan: Application of geographic information system in TSH neonatal screening for monitoring of iodine deficiency areas in Thailand. Southeast Asian J Trop Med Public Health 2008, 39:362-367.
6. Thailand industrial Statistic for hen eggs in 2008 [http://hebe.cpportal.net] (in Thai)
7. Date AR, Gray AL: Applications of inductively coupled plasma-mass spectrometry. New York, USA: Chapman and Hall; 1989.
8. Mulligan KJ, Davidson TM, Caruso JA: Feasibility of the direct analysis of urine by inductively coupled argon plasma-mass spectrometry for biological monitoring of exposure to metals. J Anal At Spectrom 1990, 5:301.
9. Department of Medical Sciences: Iodine in Food Spectro-photometric Method. Compendium of Methods for Food Analysis: Bangkok, Thailand 2003, 47-49.

10. World Health Organization: WHO/UNICEF/ICCIDD. Assessment of iodine deficiency disorders and monitoring their elimination: A guide for programme managers. 3rd edition. Geneva, World Health Organization, United Nations; 2007.

This chapter was originally published under the Creative Commons Attribution License. Charoensiriwatana, W., Srijantr, P., Teeyapant, P., and Wongvilairattana, J. Consuming Iodine Enriched Eggs to Solve the Iodine Deficiency Endemic for Remote Areas in Thailand. Nutrition Journal 2010, 9:68. doi:10.1186/1475-2891-9-68. http://www.nutritionj.com/content/9/1/68

CHAPTER 14

CINNAMON EXTRACT INHIBITS α-GLUCOSIDASE ACTIVITY AND DAMPENS POSTPRANDIAL GLUCOSE EXCURSION IN DIABETIC RATS

H. MOHAMED SHAM SHIHABUDEEN, D. HANSI PRISCILLA, and KAVITHA THIRUMURUGAN

14.1 BACKGROUND

In individuals with type 2 diabetes, nutrient intake related first-phase insulin response is severely diminished or absent resulting in persistently elevated postprandial glucose (PPG) throughout most of the day [1]. This is due to the delayed peak insulin levels which are insufficient to control PPG excursions adequately [2]. Postprandial hyperglycemia is a major risk factor for micro- and macro vascular complications associated with diabetes [3,4] and so controlling postprandial plasma glucose level is critical in the early treatment of diabetes mellitus and in reducing chronic vascular complications [5]. The acute glucose fluctuations during the postprandial period exhibits a more specific triggering effect on oxidative stress than chronic sustained hyperglycemia which suggests that the therapy in type 2 diabetes should target not only hemoglobin A1c and mean glucose concentrations but also acute glucose swings [6,7].

Mammalian α-glucosidase anchored in the mucosal brush border of the small intestine catalyzes the end step digestion of starch and sucrose that are abundant carbohydrates in human diet [8]. α-glucosidase inhibitors (AGI) delay the breakdown of carbohydrate in small intestine and diminish the postprandial blood glucose excursion in diabetic subjects [9,10] and thus have a lowering effect on postprandial blood glucose and insulin levels. Commercially available α-glucosidase inhibitors such as acarbose, miglitol and voglibose are widely used to treat patients with type 2 diabetes [11,12]. AGI is shown to reduce the insulin requirements for type 1 diabetes and it also improves reactive hypoglycemia [10]. As the α-glucosidase inhibitors exhibit therapeutic effect by restricting carbohydrate absorption, the undigested carbohydrate dislodged to the colon undergoes fermentation by colonic flora to result in adverse effects such as flatulence, abdominal discomfort and diarrhoea [13]. However the adverse effects are dose dependent and get reduced with the duration of therapy [14,15].

Several α-glucosidase inhibitors have been isolated from medicinal plants to develop as an alternative drug with increased potency and lesser adverse effects than the existing drugs [16]. Cinnamon is used in traditional medicine for treating diabetes and it was found to have insulin secretagogue property [17] and insulin sensitizing property [18]. Besides the antidiabetic effect, the cinnamon bark and cinnamon oil have been reported to possess antioxidant activity [19], antinociceptive property [20], acaricidal property [21], and activity against urinary tract infections [22]. In a human clinical trial, it was found that intake of cinnamon with rice pudding reduced postprandial blood glucose and delayed gastric emptying [23].

Ahmad Gholamhoseinian [24] screened 200 Iranian medicinal plants in vitro and reported that the cinnamon extract exhibited strong inhibition on yeast α-glucosidase. However, the nature of the enzyme inhibition was not studied in detail. As most of the plant derived inhibitors showing effective inhibition on yeast α-glucosidase do not effectively inhibit the mammalian α-glucosidase, we have prompted to evaluate the same. In addition, we have studied the effect of cinnamon extract on postprandial glucose excursion associated with disaccharides and monosaccharide challenge in normal and STZ induced diabetic rats.

14.2 METHODS

14.2.1 PLANT MATERIALS

Cinnamomum zeylanicum (CZ) bark was collected form Mailadumpara, Kerala and authenticated by Angelin Vijayakumari, Head, Department of Plant Biology and Biotechnology, Voorhees College, Vellore, India. A voucher specimen of the plant (ID: VRC001) was deposited in the Herbarium Center, Voorhees College, Vellore, India.

14.2.2 EXTRACTION METHODS

Shade dried bark (50 g) was milled and extracted using methanol (250 ml) in Soxhlet apparatus for 8 hours. Then, the extract was evaporated to dryness and the final dry chocolate colour crude extract was stored in dark at -20°C until used for the experiments.

14.2.3 PHYTOCHEMICAL ANALYSIS

The phytochemical analysis of cinnamon bark extract has been performed to find the presence of major secondary metabolites like flavonoids, tannins, saponins, steroid, glycosides, coumarins, anthraquinones and alkaloids. Standard protocols according to Trease and Evans [25] and Harborne [26] were followed to analyze tannins, flavonoids, glycosides, terpenoids, alkaloids, coumarins, and anthraquinones. Steroidal rings analysis was performed following method described by Sofowora [27]. Saponins were analysed by following the protocol described by Wall [28].

14.2.4 ENZYME ASSAY

p-Nitrophenyl-α-D-glucopyranoside (PNPG), Yeast α-glucosidase (EC 3.2.1.20), sodium phosphate salts and sodium carbonate were purchased from Sisco (SRL), India. Rat-intestinal acetone powder was obtained from Sigma (USA). Acarbose was bought from Bayer pharmaceuticals, India. α- glucosidase inhibitory activity was performed following the modified method of Pistia Brueggeman and Hollingsworth [29,30]. Mammalian α- glucosidase was prepared following the modified method of Jo [31]. Rat-intestinal acetone powder (200 mg) was dissolved in 4 ml of 50 mM ice cold phosphate buffer and sonicated for 15 minutes at 4°C. After vigorous vortexing for 20 minutes, the suspension was centrifuged (10,000 g,

4°C, 30 minutes) and the resulting supernatant was used for the assay. A reaction mixture containing 50 μl of phosphate buffer (50 mM; pH 6.8), 10 μl of yeast or Rat α-glucosidase (1 U/ml) and 20 μl of plant extract of varying concentrations was pre-incubated for 5 min at 37°C, and then 20 μl of 1 mM PNPG was added to the mixture as a substrate. After further incubation at 37°C for 30 min, the reaction was stopped by adding 50 μl of Na2CO3 (0.1 M). All the enzyme, inhibitor and substrate solutions were made using the same buffer. Acarbose was used as a positive control and water as negative control. Enzymatic activity was quantified by measuring the absorbance at 405 nm in a microtiter plate reader (Bio-TEK, USA). Experiments were done in triplicates. The percentage of enzyme inhibition by the sample was calculated by the following formula: % Inhibition = {[(AC - AS)/AC] ×100}, where AC is the absorbance of the control and AS is the absorbance of the tested sample. The concentration of an inhibitor required to inhibit 50% of enzyme activity under the mentioned assay conditions is defined as the IC50 value.

14.2.5 KINETICS OF α-GLUCOSIDASE INHIBITION BY CZ

The mode of inhibition of CZ extract against mammalian α-glucosidase activity was measured with increasing concentrations of PNPG (0.5,1,2 and 4 mM) as a substrate in the absence and presence of CZ at 0.5 mg/ml and 1 mg/ml. Optimal amounts of CZ used were determined based on the enzyme inhibitory activity assay. Mode of inhibition of CZ was determined by Lineweaver-Burk plot analysis of the data calculated following Michaelis-Menten kinetics [32,33].

14.2.6 DIALYSIS FOR REVERSIBILITY OF CZ ACTION

α-glucosidase (100 U/ml) was incubated with CZ (23.5 mg/ml) in 0.5 ml of sodium phosphate buffer (50 mM, pH 6.7) for 2 h at 37°C and dialyzed against sodium phosphate buffer (5 mM, pH 6.7) at 4°C for 24 h, changing the buffer every 12 h. Another premixed-enzyme solution (0.5 ml) was kept at 4°C for 24 h without dialysis for the control experiment. Reversibility of CZ has been determined by comparing the residual enzyme activity after dialysis with that of non-dialyzed one [34,35].

14.2.7 EXPERIMENTAL ANIMALS

Adult male Albino wistar rats were maintained during the experiments in the animal house, Center for Biomedical Research, VIT University, Vellore. 12-13 weeks old rats, weighing 160-210 g were kept in polycarbonate cage housed in a room with a 12-h light/12-h dark cycle at $25 \pm 2°C$, fed with standard rodent diet and water ad libitum. All animal procedures were approved by the ethical committee in accordance with our institutional Animal Ethics Committee, 1333/C/10/CPCSEA.

14.2.8 INDUCTION OF DIABETES

Rats previously fasted for 16 h were given single intraperitoneal injection of 45 mg/kg body wt. streptozotocin (Sigma, USA) dissolved in freshly prepared citrate buffer (0.1 M, pH4.5). Animals with fasting blood glucose over 250 mg/dl, three days after streptozotocin administration were considered diabetic and they received treatment similar to that of normal rats.

14.2.9 MALTOSE AND SUCROSE LOADING IN NORMAL RATS

Total of eighteen rats were segregated into three groups of six animals each. After 16 hours fasting, Group 1 had received maltose or sucrose (2 g/kg body wt; p.o.) as the diabetic control. Group 2 was coadministered with maltose or sucrose (2 g/kg body wt; p.o.) and CZ extract (300 mg/kg body wt; p.o.). Group 3 was coadministered with maltose or sucrose (2 g/kg body wt; p.o.) and acarbose (5 mg/kg body wt; p.o.). Selected dosages of cinnamon extract and acarbose were determined to be safe based on the previous studies [36-38]. Blood glucose level was measured before and 30, 60 and 120 minutes after the maltose or sucrose loading using a Glucometer (One touch Horizon™). The change in blood glucose from the basal level after the carbohydrate load was analysed and represented as delta blood glucose.

14.2.10 MALTOSE AND SUCROSE LOADING IN DIABETIC RATS

Total of 24 rats were sorted into four groups of six animals each. After 16 hours fasting, they were given single intraperitoneal injection of 45 mg/kg body wt. streptozotocin (Sigma, USA). Group 1 had received maltose or sucrose (2 g/kg body wt; p.o.) as the diabetic control. Group 2 was coadministered with maltose or sucrose (2 g/kg body wt; p.o.) and CZ extract (300 mg/kg body wt; p.o.). Group 3 was coadministered with maltose or sucrose (2 g/kg body wt; p.o.) and CZ extract (600 mg/kg body wt; p.o.); Group 4 was coadministered with maltose or sucrose (2 g/kg body wt; p.o.) and acarbose (5 mg/kg body wt; p.o.). Blood glucose level was measured at 0, 30, 60, and 120 minutes after the maltose or sucrose loading using a Glucometer (One touch Horizon™). Deviation in blood glucose concentration from the basal value was analysed and represented as delta blood glucose.

14.2.11 GLUCOSE LOADING IN NORMAL RATS

Total of twelve normal rats were segregated into two groups of six animals each. After 16 hours fasting, Group 1 had received glucose (2 g/kg body wt; p.o.) as the control. Group 2 was coadministered with glucose (2 g/kg body wt; p.o.) and CZ extract (300 mg/kg body wt; p.o.). Blood glucose level was measured before and 30, 60 and 120 minutes after the glucose loading using a Glucometer (One touch Horizon™). The change in blood glucose from the basal level after the oral load was analysed and represented as delta blood glucose.

14.2.12 GLUCOSE LOADING IN DIABETIC RATS

Total of twelve diabetic rats were segregated into two groups of six animals each. After 16 hours fasting, Group 1 had received glucose (2 g/kg body wt; p.o.) as the control. Group 2 was coadministered with glucose (2 g/kg body wt; p.o.) and CZ extract (300 mg/kg body wt; p.o.). Blood glucose level was measured before and 30, 60 and 120 minutes after the glucose loading using a Glucometer (One touch Horizon™). The change in blood glucose from the basal level after the oral load was analysed and represented as delta blood glucose.

14.2.13 STATISTICAL ANALYSES

Statistical analysis was performed using t-test or one-way analysis of variance (ANOVA) followed by Dunnett's Multiple Comparison Test using GraphPad Prism software. P-values of less than 0.05 were considered to be statistically significant. The delta blood glucose levels were expressed as mean ± SE for six animals in each group.

14.3 RESULTS

14.3.1 PHYTOCHEMICAL CONSTITUENTS OF CZ

Phytochemical analysis of the cinnamon extract indicated the presence of flavonoids, glycosides, coumarins, alkaloids, anthraquinone, steroids, tannins and terpenoids.

14.3.2 IN VITRO α-GLUCOSIDASE INHIBITION BY CZ

Yeast α-glucosidase inhibition potential of the CZ extract and acarbose was measured (Figure 1A). It displays effective inhibition of α-glucosidase by CZ extract with IC 50 value of 5.83 µg/ml. Acarbose used as the positive control showed IC 50 value of 36.89 µg/ml (Figure 1B), under similar assay conditions. CZ extract and acarbose inhibited rat-intestinal α-glucosidase with IC 50 value of 676 µg/ml and 34.11 µg/ml, respectively (Figure 2A and 2B).

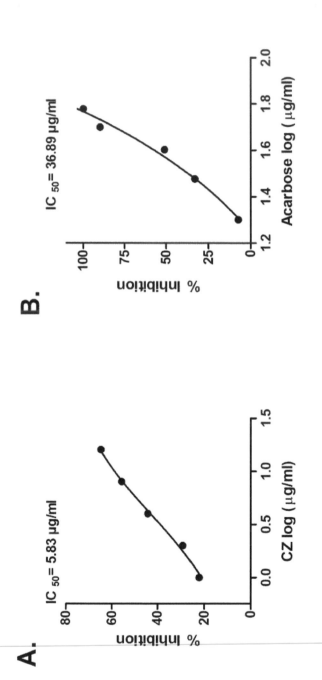

FIGURE 1: Inhibition of yeast α-glucosidase by CZ extract. A. Inhibition of α-glucosidase by CZ extract at various concentrations (1-16 μg/ml). B. Inhibition α-glucosidase by acarbose at various concentrations (1-60 μg/ml). The α-glucosidase inhibition was analyzed by measuring p-nitrophenol released from PNPG at 405 nm after 30 minutes of incubation at 37°C. Results are expressed as mean of percent inhibition ± S.E.M against log 10 concentration of inhibitor.

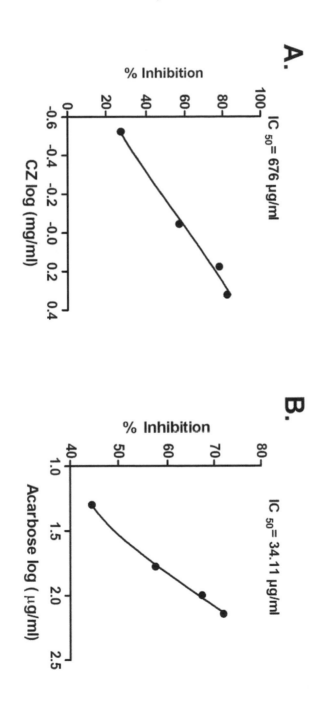

FIGURE 2: Inhibition of Mammalian α-glucosidase by CZ extract. A. Inhibition of mammalian α-glucosidase by CZ extract at various concentrations (0.3-2.1 mg/ml). B. Inhibition mammalian α-glucosidase by acarbose at various concentrations (20-140 µg/ml). The α-glucosidase inhibition was analyzed by measuring p-nitrophenol released from pNPG at 405 nm after 30 minutes of incubation at 37°C. Results are expressed as mean of percent inhibition ± S.E.M against log 10 concentration of inhibitor.

14.3.3 MODE OF α-GLUCOSIDASE INHIBITION BY CZ

The mode of inhibition of CZ extract on rat-intestinal α-glucosidase activity was analyzed using LB plot. The double-reciprocal plot displayed competitive inhibition of the enzyme activity (Figure 3). The Km value increased with increase in the CZ concentration and Vmax remained unaltered (Table 1).

TABLE 1: Kinetic analysis of α-glucosidase inhibition by CZ

CZ (mg/ml)	Vmax (mM/min)	Km (mM)
0	0.94	0.85
0.5	0.94	1.22
1	0.94	1.59

Table 1. α-glucosidase with different concentrations of PNPG (0.5-4 mM) was incubated in the absence and presence of CZ at two different concentrations (0.5 and 1 mg/ml) at 37°C for 30 min. Km and Vmax were calculated from Lineweaver-Burk plot

14.3.4 REVERSIBILITY OF CZ ACTION

The enzyme activity of α-glucosidase was almost completely recovered after the dialysis, shown by the enzyme mixed inhibitor curve (EID) that was similar to the curves of enzyme control without dialysis (EC) and with dialysis (ED) (Figure 4). Proximal running of ED as experimental control along with EC and EID ensures that dialysis alone does not greatly affect the enzyme activity. However, the non-dialyzed mixture of enzyme and extract (EIC) showed its inhibited activity.

14.3.5 MALTOSE LOADING IN NORMAL RATS

Postprandial blood glucose variation was measured after loading maltose to the normal rats with and without the coadministration of CZ extract. In the control group, blood glucose level increased by an average of 50 mg/dl at 30 minutes after the maltose load. In the group that received CZ extract

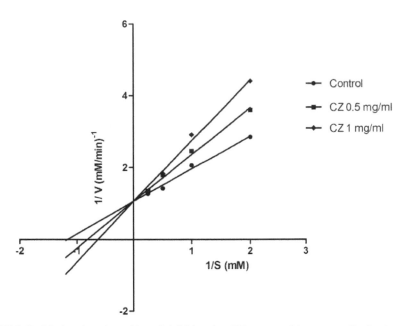

FIGURE 3: Mode of α-glucosidase inhibition by CZ extract. Lineweaver-Burk plot of α- glucosidase inhibition by CZ. α-glucosidase was treated with various concentrations of pNP-glycoside (0.5-4 mM) in the absence or presence of CZ at two different concentrations (0.5 and 1 mg/ml). The kinetics assay has been performed after incubating the mixture at 37°C for 30 min.

along with maltose, the 30 minutes post-load glucose level increased only marginally by 9 mg/dl on an average (Figure 5A). This indicates the potency of CZ extract to significantly suppress high maltose diet associated postprandial glucose elevation. Compared to control, the whole glycemic response is reduced by 65.1% on CZ treatment (Figure 5B).

14.3.6 MALTOSE LOADING IN DIABETIC RATS

As CZ extract exhibited appreciable postprandial blood glucose lowering effect in the normal rats, we examined its inhibitory effect on STZ induced diabetic rats. In the control group, blood glucose level increased to an average

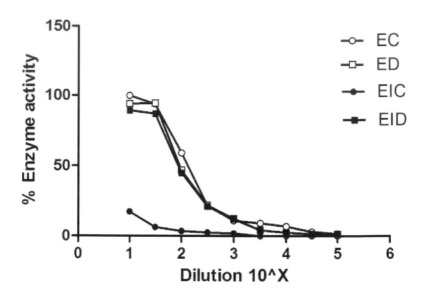

FIGURE 4: Reversibility of CZ action. α-glucosidase (100 U) was incubated with CZ (23.5 mg) in 0.5 ml of sodium phosphate buffer (50 mM; pH 6.7) for 2 h at 37°C and dialyzed against sodium phosphate buffer (5 mM; pH 6.7) at 4°C for 24 h. Reversibility of CZ was determined by comparing the residual enzyme activity after dialysis with that of non-dialyzed one. α-glucosidase alone (EC, ED) and the complex of α-glucosidase and CZ (EIC, EID) were dialyzed against 5 mM sodium phosphate buffer (pH 6.7) at 4°C (ED, EID) or were kept at 4°C (EC, EIC) for 24 h.

of 362 mg/dl above the basal level 30 min after CZ administration and decreased thereafter (Figure 6A). However, the rise of the post-load blood glucose has been significantly impeded in a dose dependent fashion on coadministering CZ with maltose at different doses (300, 600 mg/kg body wt.). Similar kind of suppression effect was observed in the group that received acarbose (5 mg/kg body wt.) as the positive control along with maltose. Compared to control, the whole glycemic response is reduced by 78.2%, 86.3% and 54.2% when treated with 300, 600 mg/kg body wt. of CZ and 5 mg/kg body wt. of acarbose, respectively (Figure 6B).

A.

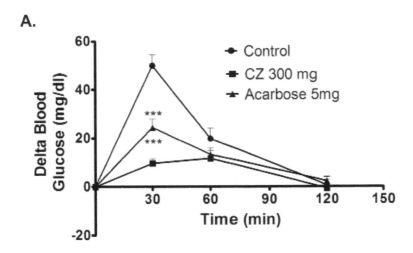

B.

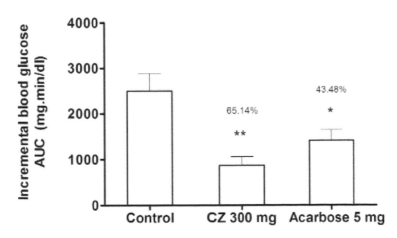

FIGURE 5: Inhibitory effects of CZ on blood glucose after maltose loading in normal rats. The normal rats fasted for 16 h received maltose (2 g/kg body wt; p.o.) and dose of CZ (300 mg/kg body wt; p.o.) by gastric intubation. Control group received maltose (2 g/kg body wt; p.o.) alone, and the drug control group received maltose (2 g/kg body wt; p.o.) plus acarbose (5 mg/kg). Blood glucose was measured at 0, 30, 60 and 120 min after food administration. A. The glycemic response curve in normal rats after maltose challenge. B. The incremental AUC_{0-120} min in normal rats after maltose administration. Data are expressed as the mean ± S.E, n = 6. *, $P < 0.05$ vs. control; **, $P < 0.01$ vs. control; ***, $P < 0.001$ vs. control.

A.

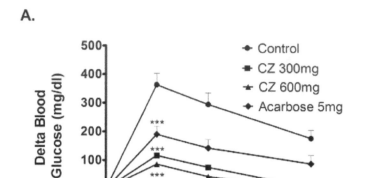

B.

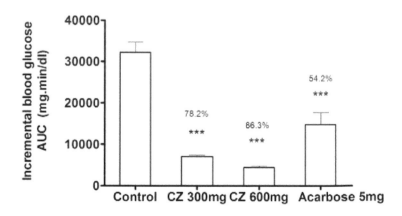

FIGURE 6: Inhibitory effect of CZ on blood glucose after maltose loading in diabetic rats. The diabetic rats fasted for 16 h received maltose (2 g/kg body wt; p.o.) and different doses of CZ (300 mg/kg body wt; p.o. and 600 mg/kg body wt; p.o.) by gastric intubations. Control animals were given only maltose (2 g/kg body wt; p.o.) and the drug control group received maltose (2 g/kg body wt; p.o.) plus acarbose (5 mg/kg). Blood glucose was monitored at 0, 30, 60 and 120 min after food administration. The result shows the significantly impeded 30 minutes post-load glucose level in the CZ 300 mg and CZ 600 mg treated group compared to control. Data are expressed as the mean ± S.E, n = 6. *, P < 0.05 vs. control; **, P < 0.01 vs. control; ***, P < 0.001 vs. control.

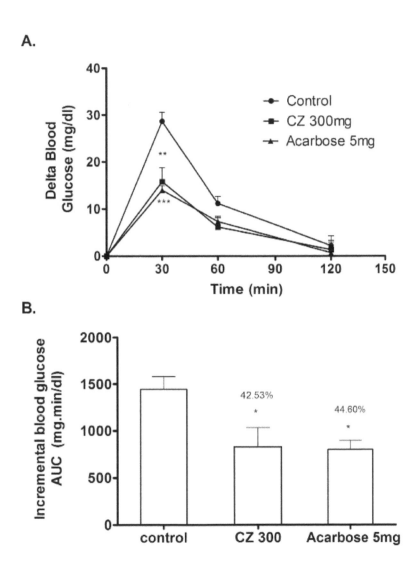

FIGURE 7: Inhibitory effects of CZ on blood glucose after sucrose loading in normal rats. The normal rats fasted for 16 h received sucrose (2 g/kg body wt; p.o.) and dose of CZ (300 mg/kg body wt; p.o.) by gastric intubation. Control group received sucrose (2 g/kg body wt; p.o.) alone and the drug control group received sucrose (2 g/kg body wt; p.o.) plus acarbose (5 mg/kg). Blood glucose was measured at 0, 30, 60 and 120 min after food administration. A. The glycemic response curve in normal rats after sucrose challenge. B. The incremental AUC0-120 min in normal rats after sucrose administration. Data are expressed as the mean ± S.E, n = 6. *, $P < 0.05$ vs. control; **, $P < 0.01$ vs. control; ***, $P < 0.001$ vs. control.

14.3.7 SUCROSE LOADING IN NORMAL RATS

Postprandial blood glucose variation was measured after loading sucrose to the normal rats with and without the coadministration of CZ extract. In the control group, blood glucose level increased by an average of 28.6 mg/dl at 30 minutes after the sucrose load. In the group that received CZ extract along with sucrose, the 30 minutes post-load glucose level increased only by 15.8 mg/dl on an average (Figure 7A). This indicates the potency of CZ extract to significantly suppress high sucrose diet associated postprandial glucose elevation. Compared to control, the whole glycemic response is reduced by 42.5% on CZ treatment and 44.6% on acarbose treatment (Figure 7B).

14.3.8 SUCROSE LOADING IN DIABETIC RATS

Postprandial blood glucose variation was measured after loading sucrose to the diabetic rats with and without the coadministration of CZ extract. In the control group, blood glucose level increased to an average of 151.6 mg/dl above the basal level 30 min after CZ administration and decreased thereafter (Figure 8A). However, the rise of the post-load blood glucose has been significantly impeded in a dose dependent fashion on coadministering CZ with sucrose at different doses (300, 600 mg/kg body wt.). Similar kind of suppression effect was observed in the group that received acarbose (5 mg/kg body wt.) as the positive control along with sucrose. Compared to control, the whole glycemic response is reduced by 52.0%, 67.5% and 70.7% when treated with 300, 600 mg/kg body wt of CZ and 5 mg/kg body wt. of acarbose, respectively (Figure 8B).

14.3.9 GLUCOSE LOADING IN NORMAL RATS

To affirm that the observed suppression of postprandial glucose is due to the inhibition of α-glucosidase, postprandial blood glucose variation was measured after loading glucose to the normal rats with and without the

A.

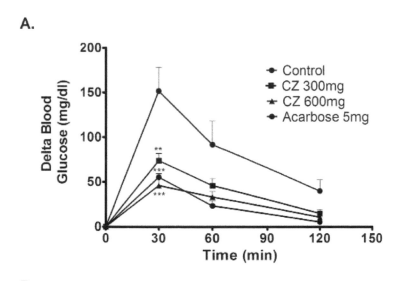

B.

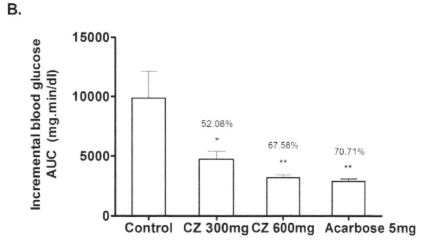

FIGURE 8: Inhibitory effect of CZ on blood glucose after sucrose loading in diabetic rats. The diabetic rats fasted for 16 h received sucrose (2 g/kg body wt; p.o.) and a dose of CZ (300 mg/kg body wt; p.o.) by gastric intubations. Control animals were given only sucrose (2 g/kg body wt; p.o.) and the drug control group received sucrose (2 g/kg body wt; p.o.) plus acarbose (5 mg/kg). Blood glucose was monitored at 0, 30, 60 and 120 min after food administration. A. The glycemic response curve in diabetic rats after sucrose challenge. B. The incremental AUC0-120 min in diabetic rats after sucrose administration. Data are expressed as the mean ± S.E, n = 6. *, P < 0.05 vs. control; **, P < 0.01 vs. control; ***, P < 0.001 vs. control.

coadministration of CZ extract. In the control group, blood glucose level increased by an average of 20 mg/dl at 30 minutes after the glucose load. In the group that received CZ extract along with glucose, the 30 minutes post-load glucose level increased by 20.8 mg/dl on an average (Figure 9A), which shows that the glucose absorption is not significantly affected due to CZ extract (Figure 9B).

14.3.10 GLUCOSE LOADING IN DIABETIC RATS

To evaluate the effect of cinnamon on glucose tolerance in diabetic condition and to elucidate whether the observed postprandial glucose suppression is majorly due to α-glucosidase inhibition, postprandial blood glucose variation was measured after loading glucose to the diabetic rats with and without the coadministration of CZ extract. In the control group, blood glucose level increased by an average of 350.1 mg/dl at 30 minutes after the glucose load. In the group that received CZ extract along with glucose, the 30 minutes post-load glucose level increased by 327.8 mg/dl on an average (Figure 10A), which shows that the glucose absorption is not significantly affected due to CZ extract (Figure 10B).

14.4 DISCUSSION

Diabetic individuals are at an increased risk of developing microvascular complications (retinopathy, nephropathy, and neuropathy) and cardiovascular disease (CVD). Abnormalities in insulin and glucagon secretion, hepatic glucose uptake, suppression of hepatic glucose production, and peripheral glucose uptake contribute to higher and more prolonged postprandial glycemic (PPG) excursions than in non diabetic individuals [2]. Elevated PPG even in the absence of fasting hyperglycemia increases the risk of cardiovascular diseases and it is the most common cause of death among the people with diabetes. Acute hyperglycemia induces endothelial dysfunction by generating oxidative stress resulting in impaired vasodilatation [39]. Also, postprandial spikes can result in microvascular damage

A.

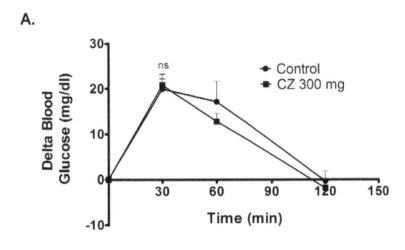

B.

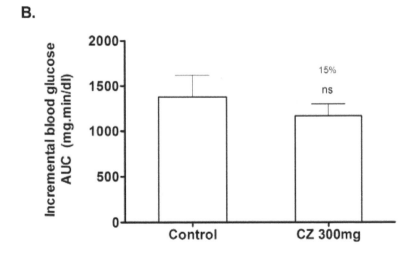

FIGURE 9: Inhibitory effect of CZ on blood glucose after glucose loading in normal rats. The rats fasted for 16 h received glucose (2 g/kg body wt; p.o.) and a dose of CZ (300 mg/kg body wt; p.o.) by gastric intubations. Control animals were given only glucose (2 g/kg body wt; p.o.). Blood glucose was monitored at 0, 30, 60 and 120 min after food administration. There are no significant changes observed in the 30 minutes post-load glucose level between the control group and CZ treated group. Data are expressed as the mean ± S.E, n = 6. ns- not significant.

A.

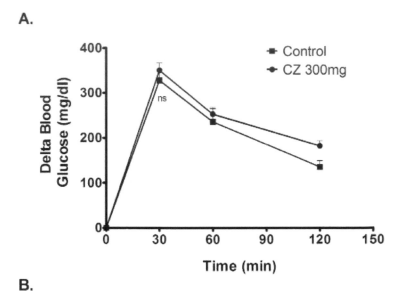

B.

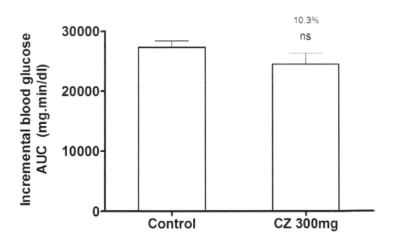

FIGURE 10: Inhibitory effect of CZ on blood glucose after glucose loading in diabetic rats. The diabetic rats fasted for 16 h received glucose (2 g/kg body wt; p.o.) and a dose of CZ (300 mg/kg body wt; p.o.) by gastric intubations. Control animals were given only glucose (2 g/kg body wt; p.o.). Blood glucose was monitored at 0, 30, 60 and 120 min after food administration. There are no significant changes observed in the 30 minutes post-load glucose level between the control group and CZ treated group. Data are expressed as the mean ± S.E, n = 6. ns- not significant.

through oxidation of low density lipoprotein (LDL) and other proatherogenic mechanisms [40].

Diet rich in carbohydrate causes sharp rise in the blood glucose level as the complex carbohydrates in the food is rapidly absorbed in the intestine aided by the α- glucosidase enzyme which breaks disaccharides into absorbable monosaccharides [41]. α-glucosidase inhibitor inhibits the disaccharide digestion and impedes the postprandial glucose excursion to enable overall smooth glucose profile [42].

The antidiabetic action of cinnamon exerted by insulin secretagogue action and insulin resistance amelioration has been previously reported [17,18]. Ahmad Gholamhoseinian [24] screened 200 Iranian medicinal plants in vitro and reported that the cinnamon extract exhibited strong inhibition on yeast α-glucosidase. However, the nature of the enzyme inhibition was not studied. As most of the plant derived inhibitors showing effective inhibition on yeast α-glucosidase do not effectively inhibit the mammalian α-glucosidase, we have prompted to evaluate the effect of cinnamon on mammalian α-glucosidase. In addition, the effect of cinnamon extract on postprandial glucose excursion associated with maltose, sucrose challenge was studied in normal and STZ induced diabetic rats.

The ability of cinnamon bark extract to inhibit the α-glucosidase in vitro has been evaluated using yeast α-glucosidase and mammalian α-glucosidase which are commonly used for investigating α-glucosidase inhibitors from microbes and medicinal plants [43]. In our in vitro studies, cinnamon extract showed remarkable inhibition on both yeast and mammalian α-glucosidase suggesting the presence of potential enzyme inhibiting compound in the extract. To find the mechanism of inhibition, we have formulated double reciprocal plot from the kinetics data and the results indicate the competitive mode of inhibition of CZ extract similar to acarbose which is also a competitive inhibitor.

In our study, we found that the inhibitory action of cinnamon on α-glucosidase to be reversible: the enzyme activity was recovered intact after dialysis as the process of dialysis cleared the inhibitors from the enzyme. The reversible inhibition is the propitious property of α-glucosidase inhibitor because the enzymes remain intact even after the elimination of the inhibitor. In other words, when inhibitor binds irreversibly to the intestinal enzyme, it will lead to hypoglycemia due to chronic carbohydrate malabsorption.

Following the positive in vitro inhibitory results of the cinnamon extract, we have continued to evaluate its effect on postprandial hyperglycemia associated with carbohydrate challenge using rats as our model. The hypothesis is that on administering cinnamon extract to the diabetic rats, postprandial glucose excursion associated maltose or sucrose challenge gets stymied but not during glucose challenge. Because, the α-glucosidase action is crucial for the digestion of maltose and sucrose without which these disaccharides would not be rapidly converted into absorbable glucose. As expected, cinnamon extract blunted acute postprandial hyperglycemic spike in the normal wistar rats loaded with maltose and sucrose but not with glucose. Subsequently, the postprandial hyperglycemia amelioration of cinnamon extract was evaluated in the STZ induced diabetic wistar rats. In general, the postprandial glucose level of STZ induced diabetic rat is poorly controlled due to impaired insulin production [44]. However, in our study, coadministration of maltose or sucrose along with cinnamon extract to the diabetic rats prevented the sharp hike in a dose dependent manner. On the other hand, control animals showed an extremely high level of blood glucose that has been staying high even two hours after the maltose or sucrose load. One of the reasons for observing the suppressed postprandial glucose level in diabetic rats could be due to the damping effect of cinnamon extract on the maltose or sucrose digestion at small intestine. The standard drug, acarbose similarly suppressed the postprandial glucose level. As the observed postprandial glucose suppression could also be possible because of the secretagogue activity and insulin sensitizing property of cinnamon, we have evaluated the effect of cinnamon on glucose loading in the normal and diabetic rats. Cinnamon did not suppress the postprandial hyperglycemia associated with glucose (monosaccharide) loading significantly but on maltose and sucrose (disaccharide) loading, which shows that the major mechanism of action of postprandial glucose suppression is exhibited by inhibition of α-glucosidase. To precisely understand the mechanism of enzyme inhibition, we are on the process of purifying and isolating an active compound(s) and determine its chemical structure for further study.

The phytochemical analysis indicated the presence of flavonoids and glycosides along with other major common secondary metabolites in the extract. Previous reports on α-glucosidase inhibitors isolated from medicinal plants suggest that several potential inhibitors belong to flavonoid glycoside class which has the characteristic structural features to inhibit α-glucosidase enzyme [45] - [46]. Based on the preliminary results obtained from our LC-MS study (data not shown), we speculate that the presence of flavonoid glycosides might have contributed to the α-glucosidase inhibitory effect of the cinnamon extract.

14.5 CONCLUSIONS

Cinnamon bark extract shows competitive, reversible inhibition on α-glucosidase enzyme. It effectively suppresses the maltose and sucrose induced postprandial blood glucose spikes in rats. Cinnamon extract could be used as a potential nutraceutical agent for treating postprandial hyperglycemia. In future, specific inhibitor has to be isolated from the crude extract, characterized and therapeutically exploited.

REFERENCES

1. Parkin CG, Brooks N: Is Postprandial Glucose Control Important?, Is It Practical In Primary Care Settings? Clinical Diabetes 2002, 20:71-76.
2. ADA: Postprandial Blood Glucose. Diabetes Care 2001, 24:775-778.
3. Hanefeld M, Schmechel H, Julius U, Schwanebeck U: Determinants for coronary heart disease in non-insulin-dependent diabetes mellitus: lessons from the diabetes intervention study. Diabetes Res Clin Pract 1996, 30(Suppl):67-70.
4. Hanefeld M, Fischer S, Julius U, Schulze J, Schwanebeck U, Schmechel H, Ziegelasch HJ, Lindner J: Risk factors for myocardial infarction and death in newly detected NIDDM: the Diabetes Intervention Study, 11-year follow-up. Diabetologia 1996, 39:1577-1583.
5. Ortiz-Andrade RR, Garcia-Jimenez S, Castillo-Espana P, Ramirez-Avila G, Villalobos-Molina R, Estrada-Soto S: alpha-Glucosidase inhibitory activity of the methanolic extract from Tournefortia hartwegiana: an anti-hyperglycemic agent. J Ethnopharmacol 2007, 109:48-53.
6. Monnier L, Colette C: Contributions of fasting and postprandial glucose to hemoglobin A1c. Endocr Pract 2006, 12(Suppl 1):42-46.
7. Sies H, Stahl W, Sevanian A: Nutritional, dietary and postprandial oxidative stress. J Nutr 2005, 135:969-972.
8. Nichols BL, Avery SE, Karnsakul W, Jahoor F, Sen P, Swallow DM, Luginbuehl U, Hahn D, Sterchi EE: Congenital maltase-glucoamylase deficiency associated with lactase and sucrase deficiencies. J Pediatr Gastroenterol Nutr 2002, 35:573-579.

9. Hillebrand I, Boehme K, Frank G, Fink H, Berchtold P: The effects of the alpha-glucosidase inhibitor BAY g 5421 (Acarbose) on meal-stimulated elevations of circulating glucose, insulin, and triglyceride levels in man. Res Exp Med (Berl) 1979, 175:81-86.

10. van de Laar FA: Alpha-glucosidase inhibitors in the early treatment of type 2 diabetes. Vasc Health Risk Manag 2008, 4:1189-1195.

11. van de Laar FA, Lucassen PL, Akkermans RP, van de Lisdonk EH, Rutten GE, van Weel C: Alpha-glucosidase inhibitors for type 2 diabetes mellitus. Cochrane Database Syst Rev 2005, CD003639.

12. van de Laar FA, Lucassen PL, Akkermans RP, van de Lisdonk EH, Rutten GE, van Weel C: Alpha-glucosidase inhibitors for patients with type 2 diabetes: results from a Cochrane systematic review and meta-analysis. Diabetes Care 2005, 28:154-163.

13. Suzuki Y, Sano M, Hayashida K, Ohsawa I, Ohta S, Fukuda K: Are the effects of alpha-glucosidase inhibitors on cardiovascular events related to elevated levels of hydrogen gas in the gastrointestinal tract? FEBS Lett 2009, 583:2157-2159.

14. Coniff RF, Shapiro JA, Robbins D, Kleinfield R, Seaton TB, Beisswenger P, McGill JB: Reduction of glycosylated hemoglobin and postprandial hyperglycemia by acarbose in patients with NIDDM. A placebo-controlled dose-comparison study. Diabetes Care 1995, 18:817-824.

15. Toeller M: alpha-Glucosidase inhibitors in diabetes: efficacy in NIDDM subjects. Eur J Clin Invest 1994, 24(Suppl 3):31-35.

16. Matsuda H, Nishida N, Yoshikawa M: Antidiabetic principles of natural medicines. V. Aldose reductase inhibitors from Myrcia multiflora DC. (2): Structures of myrciacitrins III, IV, and V. Chem Pharm Bull (Tokyo) 2002, 50:429-431.

17. Qin B, Nagasaki M, Ren M, Bajotto G, Oshida Y, Sato Y: Cinnamon extract (traditional herb) potentiates in vivo insulin-regulated glucose utilization via enhancing insulin signaling in rats. Diabetes Res Clin Pract 2003, 62:139-148.

18. Couturier K, Batandier C, Awada M, Hininger-Favier I, Canini F, Anderson RA, Leverve X, Roussel AM: Cinnamon improves insulin sensitivity and alters the body composition in an animal model of the metabolic syndrome. Arch Biochem Biophys 2010, 501:158-161.

19. Jayaprakasha GK, Ohnishi-Kameyama M, Ono H, Yoshida M, Jaganmohan Rao L: Phenolic constituents in the fruits of Cinnamomum zeylanicum and their antioxidant activity. J Agric Food Chem 2006, 54:1672-1679.

20. Atta AH, Alkofahi A: Anti-nociceptive and anti-inflammatory effects of some Jordanian medicinal plant extracts. J Ethnopharmacol 1998, 60:117-124.

21. Fichi G, Flamini G, Giovanelli F, Otranto D, Perrucci S: Efficacy of an essential oil of Eugenia caryophyllata against Psoroptes cuniculi. Exp Parasitol 2007, 115:168-172.

22. Ballabh B, Chaurasia OP, Ahmed Z, Singh SB: Traditional medicinal plants of cold desert Ladakh-used against kidney and urinary disorders. J Ethnopharmacol 2008, 118:331-339.

23. Hlebowicz J, Darwiche G, Bjorgell O, Almer LO: Effect of cinnamon on postprandial blood glucose, gastric emptying, and satiety in healthy subjects. Am J Clin Nutr 2007, 85:1552-1556.

24. Gholamhoseinian A, Fallah H, Sharifi-far F, Mirtajaddini M: The Inhibitory Effect of Some Iranian Plants Extracts on the Alpha Glucosidase. Iranian Journal of Basic Medical Sciences 2008, 11:1-9.
25. Trease GE, Evans W: Textbook of Pharmacognosy. 12th edition. London: Bailliere Tindall Ltd; 1989.
26. Harborne J: Phytochemical Methods - A Guide to Modern Techniques of Plant Analysis. London: Chapman and Hall; 1998.
27. Sofowora: Medicinal plants and traditional medicine in Africa. 1982, 68-72.
28. Wall ME, Krider MM, Rothman ES, Eddy CR: Steroidal sapogenins. I. Extraction, isolation and identification. Journal of Biological Chemistry 1952, 198:543-553.
29. Pistia-Brueggeman G, Hollingsworth RI: A preparation and screening strategy for glycosidase inhibitors. Tetrahedron 2001, 57:8773-8778.
30. Shinde J, Taldone T, Barletta M, Kunaparaju N, Hu B, Kumar S, Placido J, Zito SW: Alpha-glucosidase inhibitory activity of Syzygium cumini (Linn.) Skeels seed kernel in vitro and in Goto-Kakizaki (GK) rats. Carbohydr Res 2008, 343:1278-1281.
31. Jo SH, Ka EH, Lee HS, Apostolidis E, Jang HD, Kwon YI: Comparison of Antioxidant Potential and Rat intestinal a-Glucosidases inhibitory Activities of Quercetin, Rutin, and Isoquercetin. International Journal of Applied Research 2009, 2(4):52-60.
32. Shim YJ, Doo HK, Ahn SY, Kim YS, Seong JK, Park IS, Min BH: Inhibitory effect of aqueous extract from the gall of Rhus chinensis on alpha-glucosidase activity and postprandial blood glucose. J Ethnopharmacol 2003, 85:283-287.
33. Kim YM, Jeong YK, Wang MH, Lee WY, Rhee HI: Inhibitory effect of pine extract on alpha-glucosidase activity and postprandial hyperglycemia. Nutrition 2005, 21:756-761.
34. Lee DS, Lee SH: Genistein, a soy isoflavone, is a potent alpha-glucosidase inhibitor. FEBS Lett 2001, 501:84-86.
35. Lee DS: Dibutyl phthalate, an alpha-glucosidase inhibitor from Streptomyces melanosporofaciens. J Biosci Bioeng 2000, 89:271-273.
36. Shah AH, Al-Shareef AH, Ageel AM, Qureshi S: Toxicity studies in mice of common spices, Cinnamomum zeylanicum bark and Piper longum fruits. Plant Foods for Human Nutrition 1998, 52(3):231-239.
37. Sima AA, Chakrabarti S: Long-term suppression of postprandial hyperglycaemia with acarbose retards the development of neuropathies in the BB/W-rat. Diabetologia 1992, 35:325-330.
38. Kamath JV, Rana AC, Chowdhury AR: Pro-healing effect of Cinnamomum zeylanicum bark. Phytother Res 2003, 17:970-972.
39. Monnier L, Mas E, Ginet C, Michel F, Villon L, Cristol JP, Colette C: Activation of oxidative stress by acute glucose fluctuations compared with sustained chronic hyperglycemia in patients with type 2 diabetes. JAMA 2006, 295:1681-1687.
40. Aryangat AV, Gerich JE: Type 2 diabetes: postprandial hyperglycemia and increased cardiovascular risk. Vasc Health Risk Manag 2010, 6:145-155.
41. Dahlqvist A, Borgstrom B: Digestion and absorption of disaccharides in man. Biochem J 1961, 81:411-418.
42. Casirola DM, Ferraris RP: alpha-Glucosidase inhibitors prevent diet-induced increases in intestinal sugar transport in diabetic mice. Metabolism 2006, 55:832-841.

43. Hogan S, Zhang L, Li J, Sun S, Canning C, Zhou K: Antioxidant rich grape pomace extract suppresses postprandial hyperglycemia in diabetic mice by specifically inhibiting alpha-glucosidase. Nutr Metab (Lond) 2010, 7:71.

44. Pospisilik JA, Martin J, Doty T, Ehses JA, Pamir N, Lynn FC, Piteau S, Demuth HU, McIntosh CH, Pederson RA: Dipeptidyl peptidase IV inhibitor treatment stimulates beta-cell survival and islet neogenesis in streptozotocin-induced diabetic rats. Diabetes 2003, 52:741-750.

45. Lee SS, Lin HC, Chen CK: Acylated flavonol monorhamnosides, alpha-glucosidase inhibitors, from Machilus philippinensis. Phytochemistry 2008, 69:2347-2353.

46. Jong-Anurakkun N, Bhandari MR, Kawabata J: a-Glucosidase inhibitors from Devil tree (Alstonia scholaris). Food Chemistry 2007, 103:1319-1323.

This chapter was originally published under the Creative Commons Attribution License. Shihaburdeen, H. M. S., Priscilla, D. H., and Thirumurugan, K. Cinnamon Extract Inhibits α-Glucosidase Activity and Dampens Postprandial Glucose Excursion in Diabetic Rats. Nutrition & Metabolism 2011, 8:46. doi:10.1186/1743-7075-8-46.

CHAPTER 15

GARLIC IMPROVES INSULIN SENSITIVITY AND ASSOCIATED METABOLIC SYNDROMES IN FRUCTOSE FED RATS

RAJU PADIYA, TARAK N. KHATUA, PANKAJ K. BAGUL, MADHUSUDANA KUNCHA, and SANJAY K. BANERJEE

15.1 BACKGROUND

Type 2 diabetes mellitus is the most common form of diabetes comprising 80% of all diabetic population. The World Health Organization has predicted that developing countries would have to bear the major burden of this disease. It has been estimated that there will be a 42% increase from 51 to 72 million individuals affected in the developed countries. In developing countries these figures are much higher and are expected to show 170% increase from 84 to 228 million [1,2]. The long term consequences of type 2 diabetes make it imperative to focus on the development of novel treatment strategies for the management of insulin resistance and the metabolic syndrome.

It is well known that dietary factors play a key role in the prevention of diabetes and other metabolic disorders [3-5]. Among all such agents, garlic has attracted the attention of modern medical science because of its widespread over the counter use. The salutary effects of garlic in type 1 diabetes are well established. Several studies document the efficacy of garlic in reducing blood glucose in various animal models of type 1 diabetes mellitus[6-10]. The hypoglycemic effect of garlic has been attributed to the presence of allicin and sulfur compounds [9]. Intraperitoneal injection of aqueous garlic extract has been reported to increase insulin sensitivity in rats administered low dose fructose orally. However, this parenteral route of administration of garlic does not mimic the effect of dietary garlic on diabetes. Moreover, in this model, only a slight (~8%) increase in blood glucose levels is observed while other metabolic changes associated with the metabolic syndrome have not been investigated [11].

Thus, the present study was designed to investigate the effect of dietary raw garlic homogenate on insulin resistance and the associated metabolic syndrome in an established model of type 2 diabetes mellitus characterized by insulin resistance, metabolic syndrome and oxidative stress [12-14].

15.2 METHODS

15.2.1 PREPARATION OF GARLIC HOMOGENATE

Fresh garlic (*Allium Sativum L.*) was purchased from a fixed shop in a local market in Hyderabad, India. Individual bulbs were put in a grinder to form a juicy paste as described earlier [15]. The garlic homogenate was prepared freshly each day.

15.2.2 ANIMALS AND TREATMENT

All animal experiments were undertaken with the approval of Institutional Animal Ethical Committee of Indian Institute of Chemical Technology, Hyderabad. Male Sprague-Dawley rats (200-250 gms) were purchased from the National Institute of Nutrition (NIN), Hyderabad, India. The animals were housed in BIOSAFE, an animal quarantine facility of the Indian Institute of Chemical Technology, Hyderabad, India. The animal house is maintained at temperature $22 \pm 2°C$ with relative humidity of $50 \pm 15\%$ and 12 hour dark/light cycle. Animals were randomly divided into three groups (n = 7). Control group was fed 65% corn starch diet (Cat no. d11708b, Research diet, USA), whereas Diabetic group was fed 65% fructose diet (Cat no. d11707, Research diet, USA) for the induction of diabetes and associated metabolic disorders, [16] while the third group (Dia+Garl) was fed 65% fructose diet along with raw garlic homogenate (250 mg/kg) for a period of 8 weeks. Raw garlic homogenate was administered orally to rats using an oral gavage.

15.2.3 BIOCHEMICAL ASSAYS

Blood samples from all groups were analysed for various biochemical parameters at different time intervals to confirm the induction of diabetes and metabolic syndrome. Serum glucose and triglyceride levels were determined after 3 and 8 weeks of feeding, while serum insulin, uric acid, total cholesterol, nitric oxide and H_2S were determined after 8 weeks of feeding. Glycated haemoglobin was determined from blood after 8 weeks of feeding. Blood was collected at 3 and 8 weeks from the retro-orbital plexus using small capillary tubes, centrifuged at 4000 rpm for 10 min. at 4°C, and serum was collected for all biochemical assays.

15.2.4 ESTIMATION OF GLUCOSE, URIC ACID, TOTAL CHOLESTEROL AND TRIGLYCERIDE

Serum samples were analysed for estimation of uric acid, total cholesterol and triglyceride levels using an auto blood analyser (Bayer Corp. USA). Triglyceride (Sensitivity: 10 mg/dl), uric acid (Sensitivity: 0.2 mg/dl), and total cholesterol (Sensitivity: 10 mg/dl) kits were obtained from Siemens, India. Blood glucose was measured using glucometer (One Touch Horizon, Singapore).

15.2.5 ESTIMATION OF NITRIC OXIDE (NO)

Nitric oxide was determined by a commercially available kit (Assay design, USA). The sensitivity of the kit is 0.222 μmole/L. Assay is based on reduction of NO_3^- into NO_2^- using nitrate reductase. The azo dye is produced by diazotization of sulfanilic acid (Griss Reagent-1) with NO_2^-

and then subsequent coupling with N-(1-napthyl)-ethylene diamine (Griss Reagent-2). The azo dye was measured calorimetrically at 540 nm. Serum NO level was expressed as µmol/L.

15.2.6 ESTIMATION OF HYDROGEN SULPHIDE (H2S)

Serum H_2S concentration was measured as described by Cai et. al, 2007 [17] after some modifications. Briefly, 0.1 ml serum was added into a test tube containing 0.125 ml 1% zinc acetate and 0.15 ml distilled water. Then 0.067 ml 20 mM N, N-dimethyl-phenylene diamine dihydrochloride in 7.2 M HCL was added. This was followed by addition of 0.067 ml 30 mM FeCl3 in 1.2 M HCL. The absorbance of resulting solution was measured with a spectrophotometer at a wavelength of 670 nm. The H_2S concentration in a solution was calculated according to the calibration curve of sodium hydrogen sulphide (NaHS: 3.12-400µmol) and data were expressed as H_2S concentration in µmol/L.

15.2.7 ESTIMATION OF GLYCATED HAEMOGLOBIN

Glycated haemoglobin was estimated by using ion exchange micro-columns (Biosystem Ltd, Spain). Detection limit of the kit is lower than 4.3%. After preparing the hemolysate, where the labile fraction is eliminated, haemoglobin was retained by a cationic exchange resin. Haemoglobin A1c (HbA1c) was specifically eluted after washing away the haemoglobin A1a and A1b fractions, and was quantified by direct spectrophotometric reading at 415 nm.

15.2.8 ESTIMATION OF INSULIN

Quantitative estimation of serum insulin was done by rat insulin ELISA kits (Mercodia, USA). The sensitivity of the kit is 0.025 µg/l. It is a solid phase two-site enzyme Immunoassay. It is based on the direct sandwich technique in which two monoclonal antibodies are directed against separate antigenic determinants on the insulin molecule. During incubation, insulin in the sample reacts with peroxidase-conjugated anti-insulin anti-bodies and anti-insulin antibodies bound to microtitration well. A simple washing step removes unbound enzyme loaded antibody. The bound conjugate was detected by reaction with 3, 3', 5, 5'-tetramethylbenzidine. The reaction was stopped by adding acid and read using a spectrophotometer at 450 nm.

15.2.9 ESTIMATION OF TBARS, CATALASE AND GSH

At the end of 8 weeks, rats were sacrificed by cervical dislocation, and liver samples were collected and stored at -80°C. Liver samples from each rat were homogenized in freshly prepared phosphate buffer saline with 20 times dilution. Tissue homogenate was used for the estimation of TBARS. Remaining volume of homogenate was centrifugation at 5000 ×g for 15 min at 4°C. The supernatant was collected and used for the estimation of GSH, catalase and protein levels.

The extent of lipid peroxidation (TBARS) in liver was determined by measuring malondialdehyde content based on the reaction with thiobarbituric acid (TBA) [18]. Data were expressed as nmoles per gm liver weight using extinction co-efficient of 1.56×10-5 M-1 cm^{-1}. Catalase activity was determined by measuring the decomposition of hydrogen peroxide at 240 nm [19]. Data was expressed as units per mg of protein. Tissue glutathione (GSH) content in liver homogenate was measured by biochemical assay using a dithionitrobenzoicacid (DTNB) method [20]. Data were expressed as µg per gm liver weight. Protein in supernatant was determined by Bradford method.

15.2.10 INTRAPERITONEAL GLUCOSE TOLERANCE TEST

In a separate experiment, rats from all three groups were injected intraperitoneally with a freshly prepared glucose load of 2 gm/kg of body weight. Blood was collected from the retro-orbital plexus just before injecting the glucose load (0 min) and at 5, 30, 60 and 120 min for the estimation of blood glucose using glucometer (One Touch Horizon, Singapore).

15.2.11 STATISTICAL ANALYSIS

All values are expressed as mean ± SEM. Data were statistically analysed using one way ANOVA for multiple group comparison, followed by Student's unpaired 't' test for group wise comparison. Significance was set at $p \leq 0.05$. Data were computed for statistical analysis by using Graph Pad Prism Software.

15.3 RESULTS

15.3.1 BODY WEIGHT GAIN

There was no significant difference in body weight gain between Control and Diabetic groups after 8 weeks of feeding. However, a significant ($p < 0.05$) decrease in body weight gain was observed in Dia+Garl group when compared to both Control and Diabetic groups (Table 1).

TABLE 1: Body weight gain after 3 and 8 weeks of fructose feeding

	Control	Diabetic	Dia + Garl
Body weight gain after 3 weeks	42.62 ± 14.70	45.54 ± 13.95	18.12 ± 7.51*,†
Body weight gain after 8 weeks.	63.84 ± 14.53	65.78 ± 24.65	35.66 ± 24.79*,†

$p < 0.05$ vs Control group; †$p < 0.05$ vs Diabetic group

15.3.2 GLUCOSE LEVELS

After 3 weeks of feeding, no significant change in blood glucose levels was observed in fructose fed rats (Diabetic group) compared to rats from Control group (Figure 1A). But after 8 weeks of feeding, rats from the Diabetic group showed a significant ($p < 0.05$) increase in blood glucose levels compared to Control rats. However, this increase in serum glucose levels in fructose feeding rats was significantly ($p < 0.05$) decreased after chronic administration of garlic (Dia+Garl group) (Figure 1B).

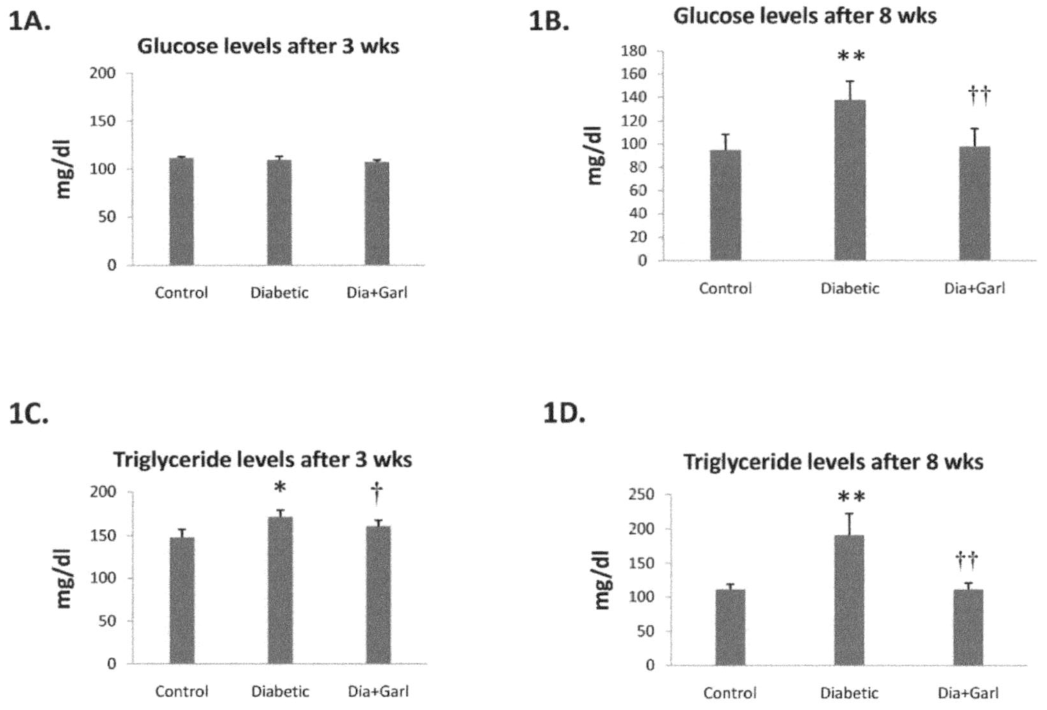

FIGURE 1: Biochemical changes after administration of garlic in fructose fed rats. Effect of garlic on blood glucose levels (A & B) and serum triglyceride levels (C & D) after 3 weeks and 8 weeks of fructose feeding. Data are shown as Mean ± SEM, *p ≤ 0.05, **p ≤ 0.01 vs Control group; †p ≤ 0.05, ††p ≤ 0.01 vs Diabetic group.

15.3.3 TRIGLYCERIDE LEVELS

Serum triglyceride levels were measured at different time intervals during the study. A significant increase in serum triglyceride levels was observed after 3 and 8 weeks of fructose feeding in rats from Diabetic group. However this increased serum triglyceride level in fructose feeding rats was significantly ($p < 0.05$) decreased after chronic administration of garlic (Dia+Garl group) (Figure 1C & 1D).

15.3.4 SERUM INSULIN LEVELS

After 8 weeks, serum insulin levels were significantly ($p < 0.01$) higher in the Diabetic group when compared to the Control group. Chronic administration of garlic (Dia+Garl group) significantly ($p < 0.05$) reduced serum insulin levels when compared to Diabetic group (Figure 2A).

15.3.5 GLYCATED HAEMOGLOBIN

After 8 weeks, no significant increase in blood glycated haemoglobin levels was observed in Diabetic group as compared to Control. However, a significant ($p < 0.05$) decrease in blood glycated haemoglobin levels was observed in Dia+Garl group when compared to Diabetic group (Figure 2B).

15.3.6 TOTAL CHOLESTEROL LEVELS

After 8 weeks, no significant change in serum cholesterol level was observed between Control and Diabetic group. Similarly no change in cholesterol levels was observed after chronic administration of garlic (Figure 2C).

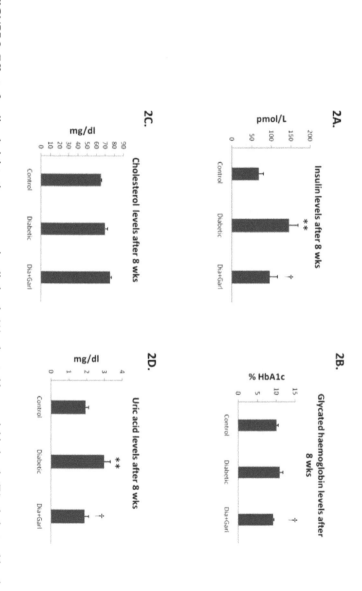

FIGURE 2: Effect of garlic administration on serum insulin levels (A), glycated haemoglobin levels (B), cholesterol levels (C) and uric acid levels (D) after 8 weeks of fructose feeding. Data are shown as Mean ± SEM, **p ≤0.01 vs Control group; †p ≤ 0.05 vs Diabetic group.

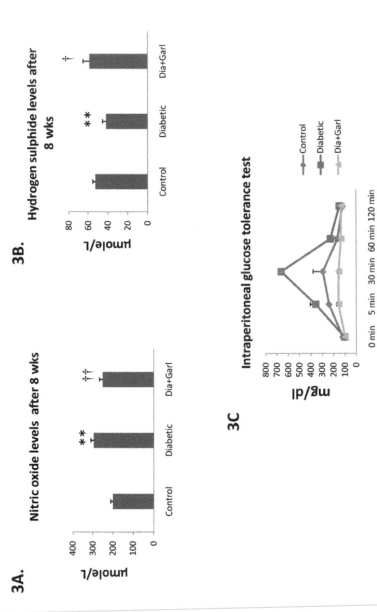

FIGURE 3: Effect of garlic administration on serum nitric oxide levels(A) and serum hydrogen sulphide levels (B) after 8 weeks of fructose feeding. (C) Effect of garlic administration on intraperitoneal glucose tolerance test. Data are shown as Mean ± SEM, **$p \leq 0.01$ vs Control group; †$p \leq 0.05$, ††$p \leq 0.01$ vs Diabetic group.

15.3.7 URIC ACID LEVELS

After 8 weeks, serum uric acid levels were significantly ($p < 0.05$) increased in the Diabetic group as compared to the Control group. Chronic administration of garlic (Dia+Garl group) significantly ($p < 0.05$) reduced serum uric acid levels as compared to Diabetic group (Figure 2D).

15.3.8 NITRIC OXIDE LEVELS

Serum nitric oxide levels were significantly ($p < 0.05$) increased in the Diabetic group after 8 weeks as compared to the Control group. Chronic administration of garlic (Dia+Garl group) significantly ($p < 0.05$) reduced serum nitric oxide levels in fructose fed rats when compared to the Diabetic group (Figure 3A).

15.3.9 HYDROGEN SULPHIDE LEVELS

Serum hydrogen sulphide levels were significantly decreased ($p < 0.05$) in the Diabetic group after 8 weeks as compared to Control group. Chronic administration of garlic (Dia+Garl group) significantly ($p < 0.05$) increased serum hydrogen sulphide levels in fructose fed rats when compared to Diabetic group (Figure 3B).

15.3.10 INTRAPERITONEAL GLUCOSE TOLERANCE TEST

An intraperitoneal glucose load led to a marked increase in blood glucose levels in Diabetic group, at 5 and 30 min, compared to the Control group. Chronic administration of garlic (Dia+Garl group) prevented this rise in serum glucose levels and was observed to be lower than the Control group (Figure 3C).

15.3.11 HEPATIC TBARS, CATALASE AND GSH LEVELS

Hepatic TBARS levels were significantly increased ($p < 0.05$) in the Diabetic group after 8 weeks as compared to Control group. Chronic administration of garlic (Dia+Garl group) significantly ($p < 0.05$) decreased hepatic TBARS levels in fructose fed rats when compared to Diabetic group (Table 2). However, we did not observe any significant change in hepatic catalase activity in any of the groups (Table 2). Hepatic GSH levels were significantly decreased ($p < 0.01$) in the Diabetic group after 8 weeks as compared to Control group. Chronic administration of garlic (Dia+Garl group) significantly ($p < 0.05$) increased hepatic GSH levels in fructose fed rats when compared to Diabetic group (Table 2).

TABLE 2: TBARS, catalase and GSH levels in rat livers after 8 weeks

	Control	Diabetic	Dia + Garl
TBARS (nmol/gm wet tissue)	46.92 ± 5.70	60.51 ± 3.49*	47.22 ± 3.67†
Catalase (U/mg protein)	8.69 ± 1.7	10.88 ± 0.66	11.75 ± 1.25
GSH (µg/gm wet tissue)	176.95 ± 7.45	143.83 ± 3.73**	177.16 ± 16.72†

*$p < 0.05$ and ** $p < 0.01$ vs Control group; † $p < 0.05$ vs Diabetic group*

15.4 DISCUSSION

High fructose corn syrup (HFCS) - a corn-based sweetener that has been on the market since 1970, is a popular food sweetener. Consumption of fructose in the form of HCFS is high in many countries including USA. Between 1970 and 1990, the consumption of HFCS increased over 1,000 percent [21]. High fructose intake over long periods is known to be hazardous for human beings as well as animals [21-23]. In the present study, a fructose rich diet was used for the induction of diabetes, which is characterized by insulin resistance and metabolic syndrome very much similar to human type 2 diabetes mellitus. Previous studies have shown that long-term fructose feeding induces diabetes associated with insulin resistance and metabolic syndrome in experimental animals such as rats and mice [16,24-27].

In the present study, rats were fed with a 65% fructose diet for a period of eight weeks in order to induce diabetes associated with insulin resistance and metabolic syndrome. Although blood triglyceride levels were increased after 3 weeks of high fructose feeding, we observed an increase in blood glucose level only after 8 weeks. Along with triglycerides, there was increase in other biochemical parameters associated with the metabolic syndrome such as uric acid and plasma insulin levels although blood cholesterol and glycated haemoglobin were not significantly affected in this rat model. Most importantly, insulin resistance, an important pathogenic mechanism in human type 2 diabetes and cause of all metabolic complications, was present in this model of diabetes, as evidenced by the altered glucose tolerance test.

Current medical research focuses on correcting insulin resistance, the primary underlying disorder in type 2 diabetes mellitus. Naturally occurring compounds represent a valuable source of such therapeutic agents of which garlic (*Allium sativum*) holds a unique position in history and is well recognized for its therapeutic potential for control of diabetes and its metabolic complications.

Although the antidiabetic effect of raw garlic has been well established in the type 1 experimental diabetic model [6-10] only one experimental study has been conducted so far to evaluate the effect of garlic on insulin resistance in rats [11]. However, in the study by Jalal et al [11] since low dose fructose was administered, only a marginal increase in blood glucose levels (~8%) was observed while the metabolic syndrome was not well characterized. Moreover, in this study, where an aqueous extract of garlic was administered intraperitoneally does not simulate the effect of oral intake of garlic. In the present study we evaluated whether oral administration of raw garlic homogenate improves insulin sensitivity and associated metabolic syndrome in fructose fed rats. The dose of 250 mg/kg was chosen as we have previously shown that garlic homogenate in this dose is effective in an animal model of heart disease and does not have any adverse effects [14,28].

In the present study, oral administration of raw garlic for a period of eight weeks showed salutary effects in an animal model of type 2 diabetes mellitus. There was significant reduction of blood glucose and improvement of insulin sensitivity in garlic treated rats. Other metabolic complications like increased serum triglyceride, insulin and uric acid levels observed in diabetic rats were also normalised after garlic administration. Lowering serum uric acid and triglyceride after garlic administration might be responsible for improving insulin resistance in fructose fed rats as there is evidence that fructose-induced insulin resistance is mediated by fructose-induced hyperuricemia or hypertriglyceridemia [24,25].

Increased weight gain and fat deposition is also responsible for insulin resistance [29]. However, we did not observe any increase of body weight gain and neither did the hepatic histopathological studies reveal any adiposity in fructose fed rats after 8 weeks (data not shown). After 20 weeks of high fructose feeding, Abdullah et al (2009) [30] did observed increased liver adiposity, but without any change in body weight. They also observed lipid deposition in liver section [30]. Hence it is likely that high fructose feeding in rats for 8 weeks is not enough to induce any adiposity and liver fat deposition.

An interesting observation of the present study is that chronic administration of garlic reduced body weight gain significantly compared to both control and diabetic rats. However, the reason for decrease body weight by garlic is not clear. Previously, it has been reported that allicin, one of the components of raw garlic paste, reduced weight gain in fructose fed rats [31,32]. The hypoglycaemic effect of garlic, has also been attributed primarily to the presence of allicin-type compounds [16,33,34]. Our finding on the significantly salutary effect of garlic on intraperitoneal glucose tolerance test is particularly promising and requires further elucidation. The insulin secretagogues activity of garlic and diallyl sulphide (active compound of garlic), may possibly contribute to this effect [35,36]. Reduction of body weight gain could also be responsible for improving insulin sensitivity in fructose fed rats.

NO and H_2S are key players in disease progression [37-40]. Similar to NO, H_2S is considered to be an important vasodilator, inducing endothelium-dependent and K+-ATP channel-dependent vasorelaxation in vivo and in vitro [41]. Increased serum levels of NO [42] and decreased levels of H_2S [38] have been reported in diabetic patients. In the present study, we measured both serum NO and H_2S levels in diabetic rats. Serum NO levels were significantly higher while H_2S levels were significantly lower in diabetic rats compared to the control rats. Importantly, chronic administration of garlic normalised both gaseous molecules in fructose fed rats.

Increasing evidence in both experimental and clinical studies indicates that oxidative stress plays a major role in the pathogenesis of Type 2 diabetes mellitus. Free radicals are generated in diabetes by glucose oxidation. High levels of free radicals and the simultaneous decline of endogenous antioxidants can lead to damage of cellular organelles, and development of insulin resistance [41]. In the present study, high fructose feeding increased oxidative stress as evidenced by elevation of TBARS levels and reduction of GSH levels in diabetic liver in comparison to Control group. However, administration of raw garlic homogenate normalised both the increased TBARS and decreased GSH levels in diabetic liver.

Thus we may conclude that high fructose feeding for 8 weeks induces diabetes along with insulin resistance, metabolic disorder and oxidative stress. Oral administration of raw garlic homogenate increases insulin sensitivity and reduces metabolic complications along with oxidative stress in diabetic rats. Further human studies are essential to establish the role of garlic in controlling type 2 diabetes and its complications.

REFERENCES

1. King H, Aubert RE, Herman WH: Global burden of diabetes, 1995-2025: prevalence, numerical estimates, and projections. Diabetes Care 1998, 21:1414-1431.

2. Johnson RJ, Perez-Pozo SE, Sautin YY, Manitius J, Sanchez-Lozada LG, Feig D I, Shafiu M, Segal M, Glassock RJ, Shimada M, Roncal C, Nakagawa T: Hypothesis: Could excessive fructose intake and uric acid cause type 2 diabetes? Endocr Rev 2009, 30:96-116.

3. Carson JF: Chemistry and biological properties of onion and garlic. Food Res Int 1987, 3:71-103.

4. Maki KC: Dietary factors in the prevention of diabetes mellitus and coronary artery disease associated with the metabolic syndrome. Am J Cardiol 2004, 93:12-17.

5. Riccardi G, Rivellese AA: Dietary treatment of the metabolic syndrome-the optimal diet. Br J Nutr 2000, 83:S143-S148.

6. Banerjee SK, Maulik M, Mancahanda SC, Dinda A, Gupta SK, Maulik SK: Dose dependent induction of endogenous antioxidant by chronic administration of garlic in rat heart. Life Sci 2002, 70:1509-1518.

7. Farva D, Goji LA, Joseph PK, Augusti KT: Effects of garlic oil on streptozotocin-diabetic rats maintained on normal and high fat diets. Ind J Biochem Biophys 1986, 23:24-27.

8. Kumar GR, Reddy KP: Reduced nociceptive responses in mice with alloxan induced hyperglycemia after garlic (Allium sastivum) treatment. Ind J Exp Biol 1999, 37:662-666.

9. Mathew PT, Augusti KT: Studies on the effect of allicin (diallyldisulphide-oxide) on alloxan diabetes I. Hypoglycemic action and enhancement of serum insulin effect and glycogen synthesis. Ind J Biochem Biophys 1973, 10:209-212.

10. Ohaeri OC: Effect of garlic oil on the levels of various enzymes in the serum and tissue of streptozotocin diabetic rats. Bio sci Rep 2001, 2:19-24.

11. Jalal R, Bagheri SM, Moghimi A, Rasuli MB: Hypoglycemic effect of aqueous shallot and garlic extracts in rats with fructose-induced insulin resistance. Clin Biochem Nutr 2007, 41:218-223.

12. Elliott SS, Keim NL, Stern JS, Teff K, Havel PJ: Fructose, weight gain, and the insulin resistance syndrome. Am J Clin Nutr 2002, 76:911-922.

13. Martinez FJ, Rizza RA, Romero JC: High-fructose feeding elicits insulin resistance, hyperinsulinism, and hypertension in normal mongrel dogs. Hypertension 1994, 23:456-463.

14. Thorburn AW, Storlien LH, Jenkins AB, Khouri S, Kraegen EW: Fructose-induced in vivo insulin resistance and elevated plasma triglyceride levels in rats. Am J Clin Nutr 1989, 49:1155-1163.

15. Banerjee SK, Maulik M, Manchanda SC, Dinda AK, Das TK, Maulik SK: Garlic induced alteration in rat liver and kidney morphology and associated changes in endogenous antioxidant status. Food chem toxicol 2001, 39:793-797.

16. Veerapur VP, Prabhakar KR, Thippeswamy BS, Bansal P, Srinivasan KK, Unnikrishnan MK: Antidiabetic effect of Dodonaea viscosa (L). Lacq. aerial parts in high fructose-fed insulin resistant rats: a mechanism based study. Ind J Exp Biol 2010, 48:800-810.

17. Cai WJ, Wang MJ, Moore PK, Jin HM, Yao T, Zhu YC: The novel proangiogenic effect of hydrogen sulfide is dependent on akt phosphorylation. Cardiovas Res 2007, 76:29-40.

18. Okhawa H, Oohishi N, Yagi K: Assay for lipid peroxides in animal tissues by thiobarbituric acid reaction. Anal Biochem 1979, 95:351-358.

19. Aebi H: Catalase. In Methods of Enzymatic Analysis. Edited by HU Bergmeyer. Chemic Academic Press Inc., Verlag; 1974:673-85.

20. Ellman GL: Tissue sulphydryl groups. Arch Biochem Biophy 1959, 82:70-77.

21. Bray GA, Nielsen SJ, Popkin BM: Consumption of high-fructose corn syrup in beverages may play a role in the epidemic of obesity. Am J Clin Nutr 2004, 79:537-543.

22. George AB: How bad is fructose? Am J Clin Nutr 2007, 86:895-896.

23. Johnson RJ, Segal MS, Sautin Y: Potential role of sugar (fructose) in the epidemic of hypertension, obesity and the metabolic syndrome, diabetes, kidney disease, and cardiovascular disease. Am J Clin Nutr 2007, 86:899-906.

24. Bruckdorfer KR, Kang SS, Yudkin J: Plasma concentrations of insulin, corticosterone, lipids and sugars in rats fed on meals with glucose and fructose. Proc Nutr Soc 1973, 32:12-13.

25. Nakagawa T, Hu H, Zharikov S, Tuttle KR, Short RA, Glushakova O, Ouyang X, Feig DI, Block ER, Acosta JH, Patel JM, Johnson RJ: A causal role for uric acid in fructose-induced metabolic syndrome. Am J Physiol 2006, 290:625-631.

26. Reungjui S, Roncal CA, Mu W, Srinivas TR, Sirivongs D, Johnson RJ, Nakagawa T: Thiazide diuretics exacerbate fructose-induced metabolic syndrome. J Am Soc Nephrol 2007, 18:2724-2731.

27. Asdaq SMB, Inamdar MN: The potential benefits of a garlic and hydrochlorothia-
    zide combination as antihypertensive and cardio protective in rats. J Nat Med 2010,
    65:81-88.
28. Banerjee SK, Maulik SK: Effect of garlic on cardiovascular disorders: a review. Nutr
    J 2002, 19:1-4.
29. Everson SA, Goldberg DE, Helmrich SP, Lakka TA, Lynch JW, Kaplan GA, Salonen
    JT: Weight gain and the risk of developing insulin resistance syndrome. Diabetes
    Care 1998, 21:1637-1643.
30. Abdullah MM, Riediger NN, Chen Q, Zhao Z, Azordegan N, Xu Z, Fischer G, Oth-
    man RA, Pierce GN, Tappia PS, Zou J, Moghadasian MH: Effects of long-term
    consumption of a high-fructose diet on conventional cardiovascular risk factors in
    Sprague-Dawley rats. Mol Cell Biochem 2009, 327:247-256.
31. Elkayam A, Mirelman D, Peleg E, Wilchek M, Miron T, Rabinkov A, Oron-Herman
    M, Rosenthal T: The Effects of Allicin on Weight in Fructose-Induced Hyperinsulin-
    emic, Hyperlipidemic, Hypertensive Rats. Am J Hypertension 2003, 16:1053-1056.
32. Rosenthal T, Elkayam A, Wilchek M, Miron T, Peleg E, Rabinkov A, Mirelman
    D: Allicin prevents weight gain in fructose-induced hypertensive, hyperinsulinemic
    rats. Am J Hypertens 2001, 14:A219-A220.
33. Eidi A, Eidi M, Esmaeli E: Antidiabetic effect of garlic (Allium sativum L.) in nor-
    mal and streptozotocin-induced diabetic rats. Phytomedicine 2006, 13:624-629.
34. Madkor HR, Mansour SW, Ramadan G: Modulatory effects of garlic, ginger, tur-
    meric and their mixture on hyperglycaemia, dyslipidaemia and oxidative stress in
    streptozotocin-nicotinamide diabetic rats. Br J Nutr 2011, 105:1210-1217.
35. Jain RC, Vyas CR: Garlic in alloxan-induced diabetic rabbits. Am J Clin Nutr 1975,
    28:684-685.
36. Liu C, Hse H, Li C, Chen P, Sheen L: Effects of garlic oil and diallyl trisulfide on
    glycemic control in diabetic rats. Eur J Pharmacol 2005, 516:165-173.
37. Hanan M, Abd EG, Maha M, Sawahli E: Nitric oxide and oxidative stress in brain
    and heart of normal rats treated with doxorubicin: role of aminoguanidine. J Bio-
    chem Mol Toxicol 2004, 18:69-77.
38. Jain SK, Bull R, Rains JL, Bass PF, Levine SN, Reddy S, McVie R, Bocchini JA:
    Low levels of hydrogen sulfide in the blood of diabetes patients and streptozotocin-
    treated rats causes vascular inflammation? Antioxid Redox signal 2010, 12:1333-
    1337.
39. Mancardi D, Pla AF, Moccia F, Tanzi F, Munaron L: Old and New Gasotransmit-
    ters in the Cardiovascular System: Focus on the Role of Nitric Oxide and Hydrogen
    Sulfide in Endothelial Cells and Cardiomyocytes. Curr Pharm Biotechnol, in press.
40. Mattapally S, Banerjee SK: Nitric oxide: Redox balance, protein modification and
    therapeutic potential in cardiovascular system. IIOABJ 2011, 2:29-38.
41. Whiteman M, Gooding KM, Whatmore JL, Ball CI, Mawson D, Skinner K, Tooke
    JE, Shore AC: Adiposity is a major determinant of plasma levels of the novel vaso-
    dilator hydrogen sulphide. Diabetologia 2010, 53:1722-1726.
42. Chiarelli F, Cipollone F, Romano F, Tumini S, Costantini F, Ricco diL, Pomilio M,
    Pierdomenico SD, Marini M, Cuccurullo F, Mezzetti A: Increased circulating nitric

oxide in young patients with type 1 diabetes and persistent microalbuminuria: relation to glomerular hyperfiltration. Diabetes 2000, 4:1258-1263.
43. Maritim AC, Sanders RA, Watkins JB: Diabetes, oxidative stress, and antioxidants: a review. J Biochem Mol Toxicol 2003, 17:24-38.

This chapter was originally published under the Creative Commons Attribution License. Padiya, R., Khatua, T. N., Bagul, P. K., Kuncha, M., and Banerjee, S. K. Garlic Improves Insulin Sensitivity and Associated Metabolic Syndromes in Fructose Fed Rats. Nutrition & Metabolism 2011, 8:53. doi:10.1186/1743-7075-8-53.

CHAPTER 16

A BILBERRY DRINK WITH FERMENTED OATMEAL DECREASES POSTPRANDIAL INSULIN DEMAND IN YOUNG HEALTHY ADULTS

YVONNE E. GRANFELDT and INGER M. E. BJÖRCK

16.1 BACKGROUND

One important nutritional characteristic of carbohydrate foods concerns their impact on glycaemic regulation and insulin demand. Whereas the glycaemic response to starchy foods are influenced mainly by the rate of starch digestion and absorption, the gastric emptying rate and/or the motility in the small intestine [1], that of fruits may also be influenced by other characteristics. Consequently, the carbohydrate composition; starch, glucose, fructose and sucrose [2,3], the degree of ripeness, affecting the distribution of starch to low molecular weight carbohydrates, and the food structure [4] play a role. Additionally, the type and amount of organic acids present in berries might affect glycaemic regulation, in accordance with the benefits seen with organic acids produced upon sour-dough fermentation [5,6]. The glucose and insulin responses to carbohydrate foods have been extensively tested most of them being rich in starch rather than sugars [7]. The glycaemic and insulin responses to sugars are particularly relevant in juices rather than in intact vegetable or fruits, as drinks and juices may allow consumption of higher amounts of carbohydrates, thus having a greater impact on glycaemia. A major challenge of nutrition science is the combat of diet related disorders, in particular, diseases connected to the insulin resistance syndrome. Quality parameters of importance in this connection are the postprandial glucose and insulin responses, where food characterised by a low glycaemic index (GI) or glycaemic load (GL) have been found to induce benefits on several risk makers for this syndrome as judged from interventions in healthy and type 2 diabetic-subjects [8]. In fact, oscillatory hyperglycaemic episodes are considered to trigger production of inflammatory markers and oxidative stress, events that are increasingly being associated with endothelial damage, and risk of cardiovascular disease [9]

Several members of the Vaccinium genus, including *Vaccinium myrtillus*, bilberry (European blueberry), closely related to blueberries, *Vaccinium angustifolium*, are considered to possess anti-diabetic activity, and are used in traditional medicine for the treatments of diabetic symptoms [10]. However, the majority of human and animal studies on blueberries and bilberries have focused on the anti-oxidative properties [11-14] as evaluated based on serum antioxidant status, and not on the potential effects on glycaemic control. Some in vitro results are available, though, showing potential anti-diabetic capacity of blueberries caused by the presence of specific bioactive components displaying insulin-like properties [15]. Further, recent studies in diabetic mice have shown decreased blood glucose with bilberry extract [16] and with fermented blueberry juice [17].

Although some studies have investigated the glycaemic response after mixed berries [18] and certain fruits [2,7,19,20], human data on glycaemic and insulinemic response to blueberries, bilberries or products made from these berries, are to our knowledge not available. The present study was performed to determine the glycaemic and insulinemic responses in healthy humans after single meal intakes of fermented oat meal drinks containing different amounts of bilberries (0, 10 or 47%) or rosehip (10%).

16.2 METHODS

16.2.1 EXPERIMENTAL DESIGN

The study was divided in two series, series 1 with two fermented oatmeal drinks added with bilberry and rosehip, respectively, and series 2 with a fermented oatmeal reference drink without fruit, and with 2 oat meal drinks with bilberry added in different amounts. The effect of carbohydrate equivalent servings of these drinks on blood glucose and insulin responses was studied at breakfast in healthy young subjects. White wheat bread was used as a reference in both series allowing for calculation of glycaemic and insulinemic indices.

16.2.2 SERIES 1

The two test products were; 1) a bilberry drink based on bilberry (10%), and, oatmeal (5%), fermented with Lactobacillus plantarum 299v (BFOMD), and 2) a rose hip drink based on rose-hip (10%), and oatmeal (5%), fermented with Lactobacillus plantarum 299v (RFOMD). The drinks were provided by (Skånemejerier, Malmö, Sweden) (ProViva®). As reference, a white wheat bread was baked under standardised conditions [21]. The test meals were standardised to contain 30 g available carbohydrates corresponding to bilberry drink (302 g), rosehip drink (300 g) and reference bread (70,3 g) (Table 1). The content of fluid in the two drinks was compensated for by providing 300 g of water with the reference bread meal.

TABLE 1: Composition of available carbohydrate in test meals and reference meal (series 1 and 2) (g wet weight)

Product	Glucose[1]	Fructose[1]	Sucrose[1]	Available starch[2]	Total available carbohydrates
Series 1					
Reference bread 70.2 g				30.0	30.0
BFOMD 302 g	2.7	2.9	20.5	3.7	29.8
RFOMD 300 g	0.9	0.8	26.9	1.35	29.9
Series 2					
Reference bread 70.0 g				30.0	30.0
FOMD 270.3 g	1.9	2.2	21.1	1.1	26.3
BFOMD 270.3 g	1.9	2.2	21.1	4.9	30.1
BBFOMD 307,7 g[3]	5.7	6.9	14.1	3.2	29.9

1Analysis with HPAEC (high pressure anion exchange chromatography)
2Available starch analysed according to [43]
3BFOMD contributed to two third of the available carbohydrates (180.2 g) and homogenised bilberries with one third (127.5 g)

16.2.1.1 SUBJECTS

Nine healthy, non-smoking volunteers, 7 women and 2 men, took part in the study. Their average age was 32.7 ± 9.9 years (mean $\pm$ SD) and their mean body mass index 23.0 ± 2.4 kg/m^2 (mean $\pm$ SD). The night before every test breakfast, the subjects were requested to eat a standardised late evening meal, based on 2-3 slices of white wheat bread. After 10 pm, the subjects were allowed to drink only water. The reference- and test breakfast meals were served randomised after an overnight fasting. The tests were performed approximately one week apart and commenced at the same time in the morning. All meals were consumed steadily and completed within 12-14 min. Tea, coffee or water (150 ml) was served after each meal. The test subjects were allowed to choose between water, tea or coffee, and maintained the same drink through-out the study.

16.2.2 SERIES 2

The test products in series 2 were 1) a bilberry drink based on bilberry (10%), and, oatmeal (5%), fermented with Lactobacillus plantarum 299v (BFOMD) (ProViva®) 2) a fermented oatmeal drink (5%) (FOMD). The fermented oatmeal drink was supplemented with glucose (1,9 g/serving), fructose (2.2 g/serving) and sucrose (21.1 g/serving) to mimic the sugar composition in the bilberry drink, and 3) a bilberry drink BBFOMD, BFOMD added with frozen, thawed and homogenised bilberries. In the BBFOMD, the BFOMD contributed with two-thirds of the available carbohydrates, and homogenised bilberries with one-third. The test meals were standardized to contain 30 g available carbohydrates. Thus, the volunteers were served; 1) BFOMD (270.3 g), 2) FOMD (270.3 g), 3) BBFOMD (307,7 g), homogenised bilberries (127.5 g) added to BFOMD (180.2 g) and 4) 70.0 g reference bread (Table 1). The content of fluid in the two drinks was compensated for with 300 g of water being served with the reference bread. Tea, coffee or water (150 ml) was served after each meal. The test subjects were allowed to choose between these drinks and retained the same drink throughout the study.

16.2.2.1 SUBJECTS

Eleven healthy, non-smoking volunteers, 7 women and 4 men, took part in the study. Their average age was 26.2 ± 4.6 years (mean ± SD) and their mean body mass index 23.5 ± 2.9 kg/m2 (mean ± SD). The night before each test breakfast, the subjects were requested to eat a standardised late evening meal, based on 2-3 slices of white wheat bread. After 10 pm, the subjects were allowed to drink only water. The tests were performed approximately one week apart and commenced at the same time in the morning. All meals were consumed steadily and completed within 12-14 min.

16.2.3 SAMPLING AND ANALYSIS

A fasting blood sample was taken before the meal was served. After the breakfast, blood samples were taken at 15, 30, 45, 70, 95 and 120 min for analysis of glucose, and at 15, 30, 45, 95 and 120 min for analysis of insulin. Capillary blood was used.

Blood glucose concentrations were determined with a glucose oxidase peroxidase reagent [21] and serum insulin concentrations with an enzyme immunoassay kit (Mercodia Insulin Elisa; Mercodia AB, Uppsala, Sweden).

The Ethics Committee of the Faculty of Medicine at Lund University approved the study.

16.2.4 STATISTICAL ANALYSIS

The incremental areas under the curves were determined for blood glucose and serum insulin (GraphPad Prism version 4.03; GraphPad Software, San Diego, CA, USA). GI and II were calculated from the area under the glucose/insulin response (0-120 min) after consumption of 50 g of carbohydrates from a test food divided by the area under curve after consumption of 50 g of carbohydrates from white wheat bread (reference) and with each subject being their own reference. All areas below the baseline were excluded from the calculations. The relationship between the insulin and glucose response (II/GI) was used to predict the insulin demand for the test product. Values are presented as mean ± SEM. All statistical calculations were performed in MINITAB Statistical Software (release 13 for Windows; Minitab Inc., State College, PA). Significances were evaluated with the general linear model (analysis of variance) followed by Tukey's multiple comparisons test. Values of P < 0.05 were considered significant.

TABLE 2: Series 1.Glycaemic and insulinaemic data following breakfast meals with BFOMD, RFOMD and white wheat bread reference.

	Reference bread	BFOMD	RFOMD
Blood glucose:			
Fasting value (mmol/L)	4.5 ± 0.1[a]	4.4 ± 0.1[a]	4.5 ± 0.1[a]
Incremental area under curve (0-	88.1 ± 11.6[a]	78.6 ± 9.4[a]	73.4 ± 8.8[a]
120 min) (mmol min/L) GI (0-120 min) (%)	100[a]	95 ± 10[a]	87 ± 8[a]
Serum insulin:			
Fasting value (pmol/L)	65 ± 7[a]	81 ± 10[a]	80 ± 10[a]
Incremental area under curve (0-	14.7 ± 2.4[a]	9.4 ± 2.1[b]	10,0 ± 1.6[ab]*
120 min) (nmol min/L) II (0-120 min) (%)	100[a]	65 ± 6[b]	79 ± 16[ab]*
II/GI	1	0.68	0.9

Mean values ± SEM, n = 9. Mean values with different letters in each row are significantly different (ANOVA followed by Tukey's test), P < 0.05,
* comparing reference bread with RFOMD P = 0.0673

16.3 RESULTS

16.3.1 SERIES 1

The blood glucose responses after the drinks BFOMD and RFOMD, and the white reference bread are shown in Figure 1. At 15 min, the glucose response after the BFOMD was significantly higher than after the white wheat bread (P < 0.05). A similar tendency was observed at 30 min, but did not reach significance (P = 0.0595). The glucose response after both fruit

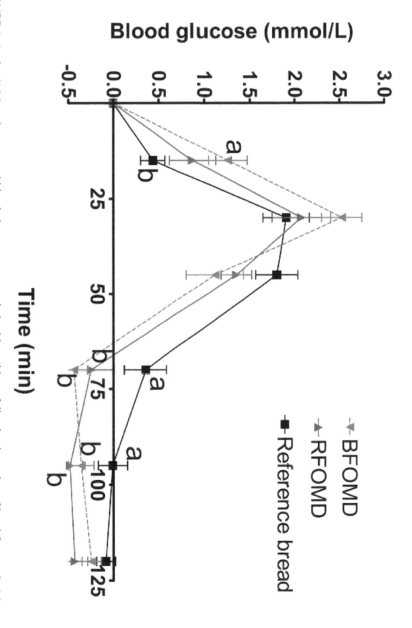

FIGURE 1: (series 1) Mean incremental blood glucose responses in healthy subjects following ingestion of breakfast meals. Mean values ± SEM, n = 11. Mean values with different letters at each time are significantly different (ANOVA followed by Tukey's test), P < 0.05.

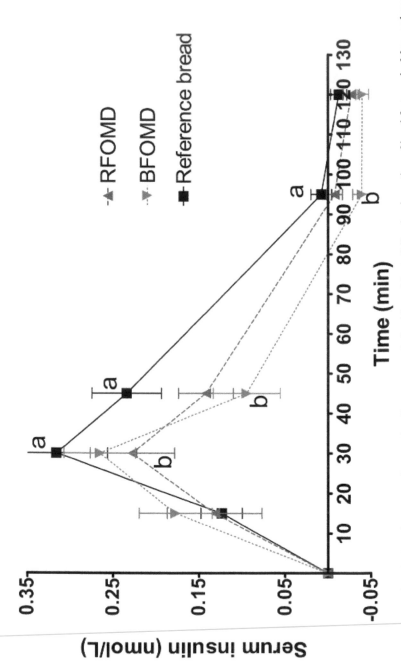

FIGURE 2: (series 1) Mean incremental serum insulin responses in healthy subjects following ingestion of breakfast meals. Mean values ± SEM, n = 11. Mean values with different letters at each time are significantly different (ANOVA followed by Tukey's test), P < 0.05.

drinks decreased more rapidly than after the reference bread. At 45 min there was a tendency to lower blood glucose response after the BFOMD compared with the reference bread (P = 0.0526), and at 70 and 95 min the blood glucose responses after both fruit drinks were lower than with the reference bread (P < 0.05).

The incremental areas under the postprandial glucose curves (0-120 min) and GI for the two fruit drinks were not significantly different from that with the reference bread (Table 2).

The insulin responses are shown in Figure 2. At 30 min, the postprandial insulin response after the RFOMD was significantly lower than after the reference bread (P < 0.05), and at 45 and 95 min the insulin responses after the bilberry drink (BFOMD) were significantly lower than after the reference bread (P < 0.05). The incremental area under the postprandial insulin curve (0-120 min) after the BFOMD drink was significantly smaller than the corresponding area after the reference bread (P < 0.05). The corresponding area after the RFOMD drink also tended to be smaller than after the reference bread, but the difference did not reach significance (P = 0.0673) (Table 2).

16.3.2 SERIES 2

The post prandial blood glucose responses are shown in Figure 3. At 30 min the glucose response after the FOMD was significantly higher than that after the BBFOMD (P < 0.05). Further, the FOMD gave a higher response than the reference bread at 15 min and a lower at 70 min (P < 0.05). The blood glucose responses after the two drinks with bilberry (BFOMD and BBFOMD) were not at any time point significantly different from each other, nor from the fermented oat meal base (FOMD) (except for at 30 min mentioned above) or from the reference bread. However, the incremental glucose area in the early postprandial phase (0-45 min), was significantly smaller after the BBFOMD (34.7 mmol min/L) compared with the FOMD (56.8 mmol min/L) (Table 3). Comparing the glycaemic areas (0-120 min), there was a tendency to a smaller area after BBFOMD than that after the reference bread but the difference did not reach significance (P = 0.0684).

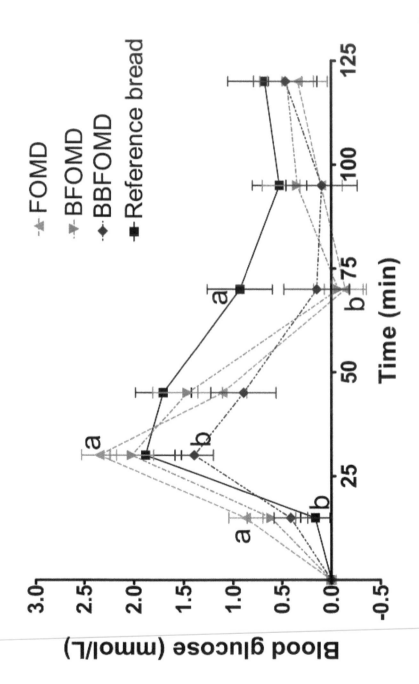

FIGURE 3: (series 2) Mean incremental blood glucose responses in healthy subjects following ingestion of breakfast meals. Mean values ± SEM, n = 11. Mean values with different letters at each time are significantly different (ANOVA followed by Tukey's test), P < 0.05.

TABLE 3: Series 2: Glycaemic and insulinaemic data following breakfast meals with FOMD, BFOMD, BBFOMD and white wheat bread

Variables	White wheat reference bread	FOMD	BFOMD	BBFOMD
Blood glucose:				
Fasting value (mmol/L)	4.8 ± 0.1[a]	4.8 ± 0.2[a]	4.9 ± 0.1[a]	5.1 ± 0.2[a]
Peak value at 30 min (mmol/L)	1.9 ± 0.4[ab]	2.4 ± 0.2[a]	2.2 ± 0.2[ab]	1.4 ± 0.2[b]
Incremental area under curve (0-45 min) (mmol min/L)	44.5 ± 6.9[ab]	56.8 ± 3.9[a]	51.5 ± 5.5[ab]	34.7 ± 3.4[b]
Incremental area under curve (0-120 min) (mmol min/L)	114.9 ± 15.1[a]	87.3 ± 8.5[a]	96.5 ± 14.3[a]	75.6 ± 13.7[a]*
GI (0-120 min) (%)	100[a]	95 ± 19[a]	94 ± 16[a]	79 ± 17[a]*
Serum insulin:				
Fasting value (pmol/L)	58.2 ± 5.5[a]	61.3 ± 7.9[a]	62.4 ± 8.9[a]	66.7 ± 9.3[a]
Incremental area under curve (0-120 min) (nmol min/L)	13.8 ± 2.5[a]	9.5 ± 1.5[b]	7.7 ± 1.2[b]	5.9 ± 1.0[b]
II (0-120 min) (%)	100[a]	76 ± 7[b]	63 ± 8[b]	49 ± 6[b]
II/GI	1	0.80	0.67	0.62

Mean values ± SEM, n = 11. Mean values with different letters in each row are significantly different (ANOVA followed by Tukey's test), $P < 0.05$
* P = 0.0684 comparing reference bread with BBFOMD.

Serum insulin responses are shown in Figure 4. The postprandial insulin responses after 15 min were significantly higher after the FOMD than after the two drinks with bilberries (BFOMD and BBFOMD) ($P < 0.05$). At 30 min the serum insulin response after the reference bread and the FOMD were significantly higher than that after the BBFOMD. At 45 min, all drinks, including the FOMD, BFOMD and BBFOMD, gave lower insulin responses than the reference bread. Also, the area under the insulin curves and the II-values were significantly smaller after FOMD and the bilberry drinks (BFOMD and BBFOMD) compared to the reference bread ($P < 0.05$, Table 3).

16.4 DISCUSSION

The two fruit drinks in series 1, BFOMD and RFOMD, gave a postprandial blood glucose response similar to that after an equivalent amount of

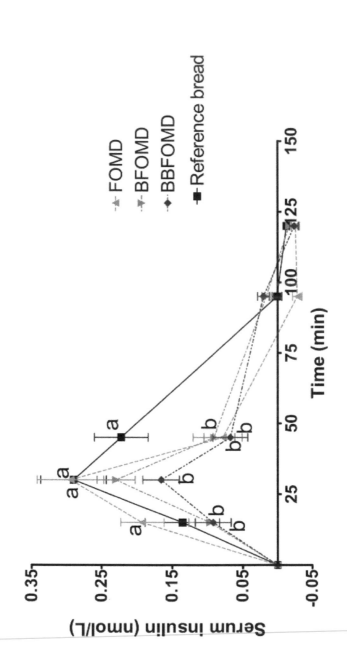

FIGURE 4: (series 2) Mean incremental serum insulin responses in healthy subjects following ingestion of breakfast meals. Mean values ± SEM, n = 11. Mean values with different letters at each time are significantly different (ANOVA followed by Tukey's test), P < 0.05.

carbohydrate from white bread. The GI was thus 97 and 89 for BFOMD and RFOMD, respectively. However, the high blood glucose responses were not accompanied by corresponding high insulin responses. Consequently, a tendency to a lower insulin response was present after both fruit drinks compared to that after white bread, even though only the area under the insulin curve after the BFOMD (0-120 min), was significantly smaller than after white reference bread. The II was determined to 65 (P < 0.05) and 79 (P = 0.0673) for BFOMD and RFOMD, respectively.

When calculating GI's, according to the content of digestible carbohydrates and their GI values (table 4), the BFOMD received a GI of 95 and the RFOMD a GI of 97. Accordingly, the determined GI values in series 1 (97 and 89 for BFOMD and RFOMD respectively) are in good agreement with those calculated. Previously, also Gannon et al [3] found that glucose response to fruits (oranges and apples) can be predicted from the constituent carbohydrates present, whereas the insulin response cannot. However, in contrast to our finding with low insulin responses Gannon et al found higher observed insulin responses than predicted from glycaemia in the case of orange- and apple juice.

In series 2, the GI and II for a BFOMD, matching that in series 1, were determined to 92 and 64, respectively compared with GI = 97 and II = 65 in series 1. Thus, the favourable effect of a fermented oat meal drink with bilberry on insulin demand in series 1 could be repeated. The FOMD gave a high GI (GI = 95) whereas that of the BBFOMD was lower (GI = 79). Insulin indices (II) for the FOMD and the BBFOMD were; II = 76 and 49, respectively. The II for the BBFOMD was remarkably low, consequently when using white bread as a reference, beverages like soft drinks (II = 97-118) [19], and other fruit/berry based drinks like orange-(II = 78) [19] or apple juice (II = 54, (estimated from insulin areas) [3] BBFOMD has a low II. Apple juice and BBFOMD resembling, that after e.g. pasta products (II = 35-53) [22].

The content of carbohydrates in the different products in the present study was similar, the amount of sugars being high (approximately 90%) and that of starch low. When calculating GI values according to the content of digestible carbohydrates and their GI values (table 4), the three test drinks in series 2, FOMD, BFOMD and BBFOMD received; 95, 95 and 90 respectively, to be compared with the determined GI values; 95, 94 and 79.

TABLE 4: Available carbohydrate composition, and calculated GI (Series 1 and 2)

Product	Glucose	Fructose	Sucrose	Available starch	Calculated GI[1]
Series 1					
BFOMD					
Proportion of total amount of digestible carbohydrates (%)	9.1	9.9	68.6	12.5	
Contribution to GI	12.8	2.7	66.5	12.5	95
RFOMD					
Proportion of total amount of digestible carbohydrates (%)	3.0	2.7	89.8	4.5	
Contribution to GI	4.2	0.8	87.1	4.5	97
Series 2					
FOMD					
Proportion of total amount of digestible carbohydrates (%)	7.2	8.4	80.2	4.2	
Contribution to GI	10.2	2.3	77.8	4.2	95
BFOMD					
Proportion of total amount of digestible carbohydrates (%)	6.3	7.3	70.1	16.3	
Contribution to GI	8.9	2.0	68.0	16.3	95
BBFOMD					
Proportion of total amount of digestible carbohydrates (%)	19.1	23.1	47.2	10.7	
Contribution to GI	26.9	6.2	45.8	10.7	90

[1]*Glucose GI = 141, Fructose GI = 27, Sucrose GI = 97, Starch GI = 100 [7]*

The observed GI value for the BBFOMD was thus lower than expected; but for the other drinks the calculated and determined GI values were in good agreement. The decreased acute glycaemic response with a larger amount of bilberries (145.5 g) is noteworthy, and a hypothesis may be that the bilberries cause an increased uptake of glucose into the peripheral cells. Thus, Martineau et al [15] found insulin-like properties of ethanol extracts from Canadian low bush blueberries (*Vaccinium angustifolium*), another member of the Vaccinium genus. The insulin-like properties were evident from enhancement of glucose uptake in differentiated muscle cells and adipocytes using an in vitro assay. Bilberry extract was also shown to reduce blood glucose level and enhance insulin sensitivity in type 2 diabetic mice via activation of AMPK (AMP-activated protein kinase), an enzyme central in the regulation of fuel preference, in adipose tissue, muscle- and liver cells [16]. These results support our finding of a decreased acute glycaemic response with bilberry and show that bilberry (*Vaccinium myritillus*) may contain active molecules with potential anti-diabetic properties. Such an effect, if present, is also coherent with the lowered insulin demand with bilberry drinks seen in both series 1 and 2. The existence of a dose-response relation between intake of bilberries and the corresponding glucose/insulin responses is currently in progress.

In this study we show an inconsistency between glycaemic and insulinaemic responses, especially in the case of the products containing bilberries. Consequently a high glucose response was accompanied by a comparatively low insulin response. The insulin demand, if expressed as a relationship between insulinaemia and glycaemia (II/GI), was low, 0.62-0.68. To our knowledge, this apparent discrepancy with a low insulin response in parallel to a high glucose response has only been reported previously for fermented whole-grain oat [23], certain rye products [24] and for cinnamon added to a rise pudding [25]. Earlier studies generally have shown a good correlation between glucose and insulin responses. Thus, studies with cereals [26], and certain fruits like mango, melon, pineapple, kiwi, apple and black grapes [19,27] indicate good agreement between GI and II. In contrast studies with oranges and apples [28] as well as juice from these fruits [3,29] were reported to display unexpectedly high insulin responses. Similarly, a discrepancy between GI and II, with unexpectedly high insulin responses has been shown, for milk and milk products [30-32]. Consequently despite extremely low GI (GI = 15-30) for regular and fermented milk, the II values were high (II = 90-98) [31], probably due to an insulinotrophic effect of whey protein [33]. One cause for the beneficial metabolic effects of a low glycaemic diet is probably a lower insulin response [34,35], and increased insulin sensitivity [36]. The present findings of low insulin demand following bilberry drinks might thus indicate advantageous metabolic properties.

The fermented oat meal base of the drinks, with sugars added to mimic the BFOMD was included as a reference drink (FOMD). A comparison between the FOMD and the BFOMD, gave no significant differences in glucose or insulin responses. Also, the insulin response to the FOMD (II = 76) was significantly lower than for white bread (II = 100). In a previous study with oats (oat porridge and oat flakes) no differences in glucose- or insulin responses were seen compared with a white bread [37]. This indicates that the fermentation process per se may decrease insulin response to oats. It is also supported by a study with fermented whole grain oat showing a lower insulin response than would be expected from the glucose values [23]. However, the magnitude of insulin decrease in the post prandial phase was more pronounced when more bilberries were included in the meal. Of interest in this respect are results from an in vitro study evaluating the effect of fermented blueberry juice (intrinsic micro flora of blueberries) on glucose uptake and transport into muscle cells and adipocytes. Treatment of cells with fermented juice potentiated glucose uptake by 48% in C2C12 (mouse myoblast cell line) myotubes, and by 142% in 3T3-L1 adipocytes, whereas non fermented juice had no effect on glucose transport [38]. Treatment of cells with fermented blueberry juice was shown to activate AMPK. The authors thus suggest an insulin-independent pathway to be the mechanism for an increased glucose uptake [38]. A follow-up study in obese and diabetic mice showed that fermented blueberry juice decreased hyperglycaemia, in part due to increased adiponectin levels. However, no positive effects were seen on insulin levels [17]. In the presently reported study, we saw a decreased early glucose response after BBFOMD. However, insulin responses were significantly lower, or close to being significantly lower for the drink with rose hip (RFOMD) (P = 0.0673), and for all fermented test drinks whether containing bilberries or not. Whether the low insulin demand (II/GI) shown in the present study could be an effect of fermentation of bilberries is currently under investigation.

All tested drinks had a low pH, or about 3 for the bilberry containing drinks, and about 4 for the fermented oat meal drink (FOMD). A low pH may lower post prandial glycaemia and hormonal responses due to e.g. a lowering of the rate of gastric emptying [39]. However, such a mechanism should preferably affect both blood glucose and insulinaemia to a similar extent.

Berries like bilberries and blueberries are known to be a rich source of bioactive molecules like phenolic and antocyanin contents [40] and phenolic acids [41]. Besides that they both are powerful antioxidants they may also exert effects on other important biological systems as glucose-and insulin response. Bilberries mixed with blackcurrants, cranberries and strawberries (150 g), other berries rich in antocyanins, have recently been shown to decrease the peak glucose increment of 35 g sucrose in healthy subjects [18]. Also in type-2 diabetic mice, anthocyanins in bilberry have been suggested to reduce blood glucose levels and enhance insulin sensitivity [16]. Water soluble polyphenols isolated from cinnamon has been shown to have strong insulin-enhancing activity on cultured fat cells in vitro, [42]. Also, when tested in healthy subjects, cinnamon (3 g) added to a rice pudding (300 g) was shown to reduce post prandial serum insulin, but not glucose, levels compared to a rice pudding without cinnamon [25]. It is impossible to draw any conclusions regarding the effects of the antocyanins and/or polyphenols present in the oat meal based fruit drinks (RFOMD, BFOMD, BBFOMD) in the present study. However, it cannot be excluded that such components might have contributed to the low insulin demand seen after the fermented oat meal drink added with bilberries.

To our knowledge no meal studies have been published showing impact of bilberries on glycaemic and insulinaemic responses in humans.

16.5 CONCLUSIONS

In the present study in healthy volunteers, we found that fermented oatmeal drinks added with bilberries reduced insulin demand to a considerable extent, with the fermented oat meal blueberry drink enriched with bilberries also being capable of reducing glycaemia. The mechanism remains obscure, and provides an interesting area for further investigations.

REFERENCES

1. Björck I, Elmståhl HL: The glycaemic index:importance of dietary fibre and other food properties. Proc Nutr Soc 2003, 62(1):201-206.
2. Lunetta M, Mauro D, Crimi S, Mughini L: No important differences in glycemic responses to common fruits in type 2 diabetic patients. Diabetic medicine 1995, 12:674-678.

3. Gannon MC, Nuttall FQ, Krezowski PA, Billington CJ, Parker S: The serum insulin and plasma glucose responses to milk and fruit products in type 2 (non-insulin-dependent) diabetic patients. Diabetologia 1986, 29:784-791.

4. Haber GB, Heaton KW, Murphy D, Burroughs LF: Depletion and disruption of dietary fibre, effects on satiety, plasma-glucose, and serum-insulin. Lancet 1977, 679-682.

5. Östman EM, Nilsson M, Elmståhl HGL, Molin G, Björck IM: On the effect of lactic acid on blood glucose and insulin responses to cereal products: mechanistic studies in healthy subjects and in vitro. J Cereal Sci 2002, 36(3):339-346.

6. Östman EM, Elmståhl HGL, Björck IME: Barley bread containing lactic acid improves glucose tolerance at a subsequent meal in healthy men and women. J Nutr 2002, 132:1173-1175.

7. Foster-Powell K, Holt SH, Brand-Miller JC: International table of glycemic index and glycemic load values: 2002. Am J Clin Nutr 2002, 76(1):5-56.

8. Livesey G, Taylor R, Hulshof T, Howlett J: Glycemic response to health-a systematic review and meta-analysis: relations between dietary glycemic properties and health outcomes. Am J Clin Nutr 2008, 87(suppl):258S-268S.

9. Ceriello A, Testa R: Antioxidant anti-inflammatory treatment in typ 2 diabetes. Diabetes Care 2009, 32(S2):S232-236.

10. Cicero A, Derosa G, Gaddi A: What do herbalists suggest to diabetic patients in order to improve glycemic control? Evaluation of scientific evidence and potential risks. Acta Diabetol 2004, 41:91-98.

11. Kay C, Holub B: The effect of wild blueberry (Vaccinium angustifolium) consumption on postprandial serum antioxidant status in human subjects. Br J Nutr 2002, 88:389-397.

12. Erlund I, Marniemi J, Hakala P, Alfthan G, Meririnne E, Aro A: Consumption of black currant, lingonberries and bilberries increases serum quercetin concentrations. Eur J Clin Nutr 2003, 57:37-42.

13. Wu X, Kang J, Xie C, Burries R, Ferguson ME, Badger TM, Nagarajan S: Dietary blueberries attenuate atherosclerosis in apolipoprotein E-deficient mice by upregulating antioxidant enzyme expression. J Nutr 2010, 140:1628-1632.

14. Basu A, Du M, Leyva MJ, Sanchez K, Betts NM, Wu M, Aston CE, Lyons TJ: Blueberries decrease cardiovascular risk factors in obese men and women with metabolic syndrome. J Nutr 2010, 140:1582-1587.

15. Martineau LC, Couture A, Spoor D, Benhaddou-Andaloussi A, Harris C, Meddah B, Leduc C, Burt A, Voung T, Mai Le P, et al.: Anti-diabetic properties of the Canadian lowbush blueberry Vaccinium angustifolium Ait. Phytomedicine 2006, 13:612-623.

16. Takikawa M, Inoune S, Horio F, Tsuda T: Dietary anthocyanin-rich bilberry extract ameliorates hyperglycemia and insulin sensitivity via activation of AMP-activated protein kinase in diabetic mice. J Nutr 2010, 140:527-533.

17. Voung Tri, Benhaddou-Andaloussi A, Brault A, Harbilas D, Martineau LC, Vallerand D, Ramassamy C, Matar C, S HP: Antiobesity and antidiabetic effects of biotransformed blueberry juice in KKAy mice. Int J Obes 2009. online publication, 18 augusti 2009

18. Törrönen R, Sarkkinen E, Tapola N, Hautaniemi E, Kilpi K, Niskanen L: Berries modify the postprandial plasma glucose response to sucrose in healthy subjects. Br J Nutr 2010, 103:1094-1097.

19. Brand-Miller JC, Pang E, Broomhead L: The glycaemic index of foods containing sugars: comparison of foods with naturally-occurring v. added sugars. Br J Nutr 1995, 73:613-623.

20. Wolever TMS, Vuksan V, Relle LK, Jenkins AL, Josse RG, Wong GS, Jenkins DJA: Glycaemic index of fruits and fruit products in patients with diabetes. International Journal of Food Sciences and Nutrition 1993, 205-212.

21. Granfeldt YE, Björck IME: Glycemic response to starch in pasta: a study of mechanism of limited enzyme availability. J Cereal Sci 1991, 14:47-61.

22. Granfeldt YE, Björck IME, Hagander B: On the importance of processing conditions, product thickness and egg addition for the glycaemia and hormonal responses to pasta: a comparison with bread made from 'pasta ingredients'. Eur J Clin Nutr 1991, 45:489-499.

23. Alminger M, Eklund-Jonsson C: Whole-grain cereal products based on a high-fibre barley or oat genotype lower post-prandial glucose and insulin responses in healthy humans. European Journal of Nutrition 2008, 47:294-300.

24. Leinonen K, Liukkonen K, Poutanen K, Uusitupa M, Mykkänen H: Rye bread decreases postprandial insulin response but does not alter glucose response in healthy Finnish subjects. Eur J Clin Nutr 1999, 53:262-267.

25. Joanna Hlebowicz, Anna Hlebowicz, Sandra Lindstedt, Ola Björgell, Peter Höglund, Holst Jens J, Gassan Darwiche, Lars-Olof A: Effects of 1 and 3 g cinnamon on gastric emptying, satiety, and postprandial blood glucose, insulin, glucose-dependent insulinotropic polypeptide, glucagon-like peptide 1, and ghrelin concentrations in healthy subjects. Am J Clin Nutr 2009, 89:815-821.

26. Björck IME, Liljeberg HGM, Östman EM: Low glycaemic-index foods. Br J Nutr 2000, 83(Suppl 1):S149-S155.

27. Hlebowicz J, Hlebowicz A, Lindstedt S, Björgell O, Höglund P, Holst JJ, Darwiche G, Almér L-O: Effects of 1 and 3 g cinnamon on gastric emptying, satiety, and postprandial blood glucose, insulin, glucose-dependent insulinotropic polypeptide, glucagon-like peptide 1, and ghrelin concentrations in healthy subjects. Am J Clin Nutr 2009, 89:815-821.

28. Gregersen S, Rasmussen O, Larsen S, Hermansen K: Glycemic and insulinaemic responses to orange and apple compared with white bread in non-insulin-dependent diabetic subjects. Eur J Clin Nutr 1992, 46:301-303.

29. Johnston KL, Clifford MN, Morgan LM: Possible role for apple juice phenolic compounds in the acute modification of glucose tolerance and gastrointestinal hormone secretion in humans. J Sci Food Agric 2002, 82:1800-1805.

30. Schrezenmeir J, Tato F, Tato S, Küstner E, Krause U, Hommel G, Asp N-G, Kasper H, Beyer J: Comparison of glycemic response and insulin requirements after mixed meals of equal carbohydrate content in healthy, type-1, and type-2 diabetic man. Klin Wochenschr 1989, 67:985-994.

31. Östman EM, Liljeberg Elmståhl HGM, Björck IME: Inconsistency between glycemic and insulinemic responses to regular and fermented milk products. Am J Clin Nutr 2001, 74:96-100.

32. Liljeberg Elmståhl HGM, Björck IME: Milk as a supplement to mixed meals may elevate postprandial insulinaemia. Eur J Clin Nutr 2001, 55:994-999.

33. Nilsson M, Stenberg M, Frid AH, Holst JJ, Björck IM: Glycemia and insulinemia in healthy subjects after lactose equivalent meals of milk- and other food proteins: on the role of plasma amino acids and incretins. Am J Clin Nutr 2004, 80:1246-1253.

34. Järvi AE, Karlström BE, Granfeldt YE, Björck IME, Asp N-G, Vessby BOH: Improved glycemic control and lipid profile and nomalized fibrinolytic activity on a low-glycemic index diet in type 2 diabetic patients. Diabetes Care 1999, 22(1):10-18.

35. Slabber M, Barnard HC, Kuyl JM, Dannhauser A, Schall R: Effects of a low-insulin-response, energy-restricted diet on weight loss and plasma insulin concentrations in hyperinsulinemic obese females. Am J Clin Nutr 1994, 60(1):48-53.

36. Livesey G, Taylor R, Hulshof T, Howlett J: Glycemic response and health - a systematic review and meta-analysis: the database, study characteristics, and macronutrient intakes. Am J Clin Nutr 2008, 87(suppl 1):223S-236S.

37. Granfeldt YE, Hagander B, Björck IM: Metabolic responses to starch in oat and wheat products. On the importance of food structure, incomplete gelatinization or presence of viscous dietary fibre. Eur J Clin Nutr 1995, 49:189-199.

38. Voung Tri, Martineau Louis C, Charles Ramassamy, Chantal Matar, Haddad PS: Fermented Canadian lowbush blueberry juice stimulates glucose uptake and AMP-activated protein kinase in insulin-sensitive cultured muscle cells and adipocytes. Canadian Journal of Physiology and Pharmacology 2007, 85:956-965.

39. Liljeberg H, Bjorck I: Delayed gastric emptying rate may explain improved glycaemia in healthy subjects to a starchy meal with added vinegar. Eur J Clin Nutr 1998, 52(5):368-371.

40. Prior RL, Cao G, Martin A, Sofic E, McEwen J, O'Brien C, Lischner N, Ehlenfeldt M, Kalt W, Krewer G, et al.: Antioxidant Capacity As Influenced by Total Phenolic and Anthocyanin Content, Maturity, and Variety of Vaccinium Species. J Agric Food Chem 1998, 46:2686-2693.

41. Mattila P, Hellström J, Törrönen R: Phenolic acids in berries, fruits, and beverages. J Agric Food Chem 2006, 54:7193-7199.

42. Andersson Richard A, Broadhurst Leigh C, Polansky Marilyn M, Schmidt Walter F, Alam Khan, Flanagan Vincent P, Schoene Norberta W, Graves Donald J: Isolation

and Characterizatio of Polyphenol Type-A Polymers from cinnamon with insulin-like Biological Activity. J Agric Food Chem 2004, 52:65-70.

43. Holm J, Björck IME, Drews A, Asp N-G: A rapid method for the analysis of starch. Starch/Stärke 1986, 38:224-226..

This chapter was originally published under the Creative Commons Attribution License. Granfeldt, Y., and Björck, I. M. E. A Bilberry Drink with Fermented Oatmeal Decreases Postprandial Insulin Demand in Young Healthy Adults. Nutrition Journal 2011, 10:57. doi:10.1186/1475-2891-10-57. http://www.nutritionj.com/content/10/1/57

CHAPTER 17

BRAZIL NUTS INTAKE IMPROVES LIPID PROFILE, OXIDATIVE STRESS, AND MICROVASCULAR FUNCTION IN OBESE ADOLESCENTS: A RANDOMIZED CONTROLLED TRIAL

PRISCILA A. MARANHΓO, LUIZ G. KRAEMER-AGUIAR,
CECILIA L. DE OLIVEIRA, MARIA C. C. KUSCHNIR,
YASMINE R. VIEIRA, MARIA G. C. SOUZA, JOSELY C. KOURY,
and ELIETE BOUSKELA

17.1 BACKGROUND

Worldwide prevalence of obesity in adolescence is actually high and increasing [1]. In Brazil, about 29% of adolescents tested between 2008-2009 have excessive weight [2]. Obesity, especially abdominal, even in young subjects leads to metabolic alterations, such as insulin resistance and dyslipidaemia, increasing risk factors for cardiovascular disease (CVD) in adulthood [3].

Excessive abdominal adiposity is characterized by accumulation of adipose tissue and is associated to a low-grade inflammatory process and oxidative stress, both well-established pathogenetic factors for cardiovascular diseases [4]. Morpho-functional microvascular alterations related to metabolic disorders have already been described and findings on skin were related to microvascular dysfunction on coronary bed [5], occurring even in the absence of disglycaemic states [6] pointing to obesity per se as independent risk factor for microangiopathy. Possibly, longer duration of excessive adiposity is involved on it as well. Recently, we have unraveled that overweight/obese young women have microvascular dysfunction linked to adiposity levels and glucose homeostasis [7]. Certainly, obesity in adolescence and its long-term damage to target organs in adulthood deserves special focus and strategies to reduce future cardiovascular risks.

Excessive visceral fat induces the release of cytokines, such as tumor necrosis factor-alfa (TNFα) and interleukin-6 (IL-6) leading to increased production of reactive oxygen species (ROS) and subsequent induction of tissue oxidative stress [8]. Some authors consider this pathophysiological process as a major mechanism underlying cardiovascular obesity-related comorbidities [9]. The antioxidant system is directly linked to environmental factors and nutrient intake. Some minerals such as selenium are also involved in decreasing levels of hydrogen peroxide and reducing lesions to cellular membranes [10]. Altered metabolic pathway of very low density lipoprotein cholesterol (VLDL) secondary to excessive visceral adiposity results on higher levels of low density lipoprotein (LDL) particles, which are more aggressive to endothelium. Additionally, the oxidative stress increases oxidation of these LDL particles[11], a process identified as a risk factor for atherosclerosis [12].

Bioactive substances existing in nuts have already been identified [13] and their beneficial effects on inflammation [14] and on endothelial function [15] demonstrated. The Brazil nut (*Bertholletia excelsa*) comes from the Amazon region and has a complex matrix, composed of bioactive substances, such as selenium, α- e γ- tocopherol, phenolic compounds, folate, magnesium, potassium, calcium, proteins and mono (MUFA) and polyunsaturated (PUFA) fatty acids [14]. Its composition is different from other nuts and data to corroborate its beneficial effects, especially in obese adolescents with special focus on microcirculatory function are lacking.

The aims of this study were to investigate the influence of Brazil nuts consumption on nutritive skin microcirculation, serum antioxidant capacity, lipid and metabolic/cardiovascular risk profiles in obese female adolescents.

17.2 SUBJECTS AND METHODS

17.2.1 STUDY POPULATION

Seventeen female adolescents (15.4 ± 2.0 years) were selected, after spontaneous interest to participate (male adolescents showed very little interest), from outpatient care clinics for Prevention and Assistance on Cardiovascular and Metabolic Disease in Adolescence (NESA, State University of Rio de Janeiro, Brazil). Main inclusion criterion was being above 95th percentile for BMI according to age [16]. Main exclusion criteria were: use of any nutritional intervention and/or drugs; presence of chronic diseases (diabetes mellitus and/or hypertension) or lactose intolerance; a verbally informed weight reduction six months before entering the study; being beyond stage IV for Tanner pubertal development [17] and dietary habits of an excessive consumption of any kind of nut. All volunteers gave their written informed consent and this study was approved by the Ethics Committee for Clinical Research of Pedro Ernesto University Hospital (COEP 1950/2007).

17.2.2 EXPERIMENTAL DESIGN

The study was a 16-week non-blinded pilot trial with two randomly selected groups of obese female adolescents: Brazil nut (BNG, n = 08) ingested 15-25 g/day (equivalent to 3 to 5 units/day) of Brazil nuts and placebo (PG, n = 09) one capsule/day containing lactose. Lactose was chosen as placebo due to its lack of therapeutic effects to improve adolescents' compliance to the study. Both groups were informed that they would receive a supplemental dietary intake composed of Brazil nuts but only one group would receive it on its natural form. Nuts were consumed as snacks or with meals in salads. Before the beginning and at the end of the study, the usual food intake of each participant was assessed by a dietary inquiry and, during the study, adolescents were advised not to change their dietary habits.

Anthropometry, blood, urine and microvascular parameters were analyzed at baseline (T0) and after 16 weeks (T1).

17.2.3 BRAZIL NUTS DIET

Brazil nuts (*Bertholletia excelsa*) consumption was calculated to achieve 10% of the energy from MUFAs in the diet. Total energy intake was calculated by energy expenditure to overweight children and adolescents of 3 to 18 years according to dietary reference intake (DRI). Adolescents received

15-25 g/day (equivalent to 3 to 5 units/day) in bags. Serum selenium levels and returned empty bags were used as markers of compliance for BNG.

Brazil nut composition was determined in 100 g by Adolph Lutz Institute-Brazil assays (1985). Lipid, carbohydrates and protein content (mean ± SD) was 50.6 ± 0.08 g [coefficient of variation (CV) 1.6%], 25.9 ± 0.6 g (CV 2.4%) and 16.8 ± 0.2 g (CV 1.4%). Selenium was measured by flame atomic absorption spectrometry. Saturated (SFA - 15.3 g), monounsaturated (MUFA - 27.4 g) and polyunsaturated (PUFA) fatty acids (21 g) contents were calculated according to Brazilian Table of Food Composition [18]. Therefore, nut intake (15-25 g/day) was about: 124 ± 31 kcal with 5.2 ± 1.3 g of carbohydrates and 10.1 ± 2.5 g of lipids. The latter was composed of 4.2 ± 1.0 g of PUFA, 5.5 ± 1.4 g of MUFA and 3.0 ± 0.7 g of SFA and 108.5 ± 27 µg of selenium.

17.2.4 ANTHROPOMETRY

The same trained examiner collected anthropometric measurements in duplicate, waist circumference, height, weight as previously reported [19]. BMI was defined as the ratio between weight in kg and squared height in meters.

17.2.5 LABORATORY ANALYSIS

All laboratory measurements were performed in duplicate after 10-12 hours fast using an automated method (Modular Analytics E 170 and P, Roche, Basel, Switzerland). Fasting plasma glucose (FPG), total cholesterol (TC), triglycerides (TG) and high-density lipoprotein (HDL) cholesterol were measured respectively, by enzyme-colorimetric GOD-PAP [inter-assay coefficient of variation (IACV) = 1.09%], enzymatic GPO-PAP (IACV = 2.93%), enzymatic GPO-PAP (IACV = 1.29%) and enzyme-colorimetric without pre-treatment (IACV = 3.23%). Plasma LDL-cholesterol was calculated according to Friedwald equation [20]. C-reactive protein and serum insulin were respectively measured by imunoturbidimetry (IACV =

8%) and eletrochemiluminescence (IACV = 10.6%). Homeostasis model assessment (HOMA-IR) was calculated (fasting serum insulin (μUI/ml) X FPG (mmol/l)/22.5). Serum antioxidant capacity was determined by glutathione peroxidase (GPx) through Elisa [GPX3 (human) Elisa Kit, Axxora, LLC, USA, diluted in 1:200 (IACV = 3.9%; sensitivity = 0.1 ng/ml]. Oxidized-LDL (ox-LDL) levels were also analyzed through Elisa (Kit Mercodia, Sweden), diluted in 1:6561 (IACV = 6.13%; sensitivity = 0.05 ng/ml). Oxidative stress was measured by a competitive enzyme-linked immunosorbant assay in duplicate determining levels of isoprostane nominated as 8-epi-prostaglandin F2α (8-epi-PGF$_{2\alpha}$) in urine samples (BIOXYTECH urinary 8-epi-PGF$_{2\alpha}$ - OxisResearch, Portland, OR, USA). Urinary samples were acidified during collection with HCl 6 mol/l (final pH = 2.0), diluted to 1:2 before measurements and results were corrected for creatinine levels in each sample assessed by enzymatic colorimetric Modular Evo (Roche) with an intra-assay CV of 5% (2-14%) [19] Serum selenium was determined through atomic absorption spectrometry using a SpectraAA - 640Z - VARIAN (IACV = 11%).

17.2.6 MICROVASCULAR FUNCTION ASSESSMENT

Nailfold videocapillaroscopy (NVC) was carried out and analyzed according to a standardized, well-validated methodology on the 4th finger of the left hand [7]. The exam, always made by the same observer who was not aware of any patient data, recorded continuously microvascular parameters for later measurements using the Cap Image software [21]. Functional capillary density (FCD), number of capillaries/mm^2 with flowing red blood cells, was evaluated using x250 magnification and an area of 3 mm of the distal row of capillaries into three different areas [Intra-assay coefficient of variation - IACV = 5.5 $\pm$ 2.5%]. Capillary diameters [afferent (AF), apical (AP) and efferent (EF)], red blood cell velocity (RBCV) at rest, after 1 min arterial occlusion (RBCVmax) and time taken to reach it (TRBCVmax) were measured, with a final magnification of x680, before and during the post-occlusive reactive hyperemia (PORH) response after 1 min ischemia. Before RBCV assessment on an individual capillary loop, a pressure cuff (1 cm wide) was placed around the proximal phalanx and connected to a mercury manometer. Conceptually, AF, AP and EF are considered morphological and FCD, RBCV, $RBCV_{max}$ and $TRBCV_{max}$ functional microcirculatory parameters. Capillary diameters and basal RBCV were measured three times each and IACV for all measurements ranged from 16.9 to 17.1%. At PORH, each variable was tested once. NVC was repeated on nine subjects in different days and the IACV ranged from 12.3% to 17.3% and from 2.0% to 9.0% between morphological and functional parameters, respectively.

17.2.7 STATISTICAL ANALYSIS

Data are expressed as median [1st-3rd] and analyzed by Graphpad Prism 4.0, 2003 and intra- or inter-group analysis compares significant results in different time points within the same group or between BNG and PG, respectively. Comparisons between groups at T0 and T1 and intra-group differences were determined using Mann-Whitney U test and Wilcoxon matched pair tests, respectively. GPower 3.1.10 software was used for power analysis and sample size estimation. The statistical power for comparisons between two dependent groups was above 0.9 with an a error probability of 0.01 for RBCV, estimating a total sample size of 7 patients/group. Significant differences were assumed to be present at $p < 0.05$.

17.3 RESULTS

Adolescents included in the study had 15.4 ± 2.0 years and BMI of 35.6 ± 3.3 kg/m^2. At T0, there were no significant differences between groups (inter-group) on anthropometrical-laboratorial-microvascular variables (tables 1, 2 and 3). On the counterpart, at T_1, we observed lower values for total ($p = 0.003$) and LDL-cholesterol ($p = 0.03$), and also for TG ($p = 0.05$) for BNG compared to PG.

TABLE 1: Anthropometric measurements and metabolic profile of obese female adolescents at baseline and after 16 weeks of Brazil nuts (BNG) or placebo (PG) intake

	BNG		PG	
	T0	T1	T0	T1
Body mass (kg)	86.3 [82.2-94.3]	86.3 [80.3-96.9]	91.7 [82.1-110.8]	93.2 [83.7-108.7]
Height (m)	1.55 [1.52-1.66]	1.55 [1.53-1.67]	1.64 [1.56-1.69]	1.64 [1.56-1.69]
BMI (kg/m²)	35.3 [33.9-36.0]	35.3 [33.3-36.2]	34.0 [33.0-39.1]	35.6 [32.8-38.9]
Waist circumference (cm)	105.0 [93.0-117.8]	112.0 [99.5-116.0]	111.0 [104.0-117.8]	115.0 [107.5-120.0]
Insulin (mcU/ml)	15.9 [14.5-23.6]	15.5 [12.2-22.2]	18.0 [15.2-20.1]	18.0 [10.9-22.8]
Fasting glucose (mg/dl)	89.5 [81.5-94.0]	86.0 [82.5-93.0]	87.0 [82.0-90.5]	90.0 [82.0-94.0]
HOMA	3.6 [2.9-5.2]	3.5 [2.6-6.5]	3.9 [3.1-4.4]	3.7 [2.2-5.2]
CRP (mg/dl)	0.31 [0.15-0.63]	0.37 [0.20-0.68]	0.38 [0.14-0.78]	0.52 [0.29-0.79]
Cholesterol (mg/dl)	152.0 [140.5-159.0]	136.0 [129.5-141.5]*†	167.0 [133.5-184.5]	170.0 [148.5-177.0]
HDL-c (mg/dl)	44.0 [41.5-48.5]	43.5 [40.0-50.5]	45.0 [36.0-48.0]	45.0 [41.0-47.5]
LDL-c (mg/dl)	86.5 [80.5-98.0]	72.5 [71.0-83.0]*†	106.0 [77.0-123.0]	103.0 [87.0-109.0]
TG (mg/dl)	84.5 [68.0-98.5]	69.0 [58.5-83.5]†	113.5 [87.0-132.5]	106.0 [77.0-153.0]

* *significant results intra-group (p < 0.05)*
† *significant results inter-group (p < 0.05)*

Body mass, BMI, waist circumference and metabolic profile were all kept unchanged during the follow-up on both groups. Additionally, we have noticed that Brazil nuts consumption did not influence glucose homeostasis (FPG, insulin, HOMA-IR) or CRP levels, although its positive influence was observed on LDL- (p = 0.01) and total cholesterol (p = 0.01) levels (table 1).

TABLE 2: Biomarkers with antioxidant capacity in obese female adolescents at baseline and after 16 weeks of Brazil nuts (BNG) or placebo (PG) intake

	BNG		PG	
	T0	T1	T0	T1
8-epi-PGF2α	143.2	100.0	34.0	88.4
(pg/μmol/g of creatinine)	[102.9-198.8]	[80.0-131.4]	[61.5-126.6]	[65.0-149.5]
	622.4	514.9	648.8	646.9
LDL-ox (ng/ml)	[457.2-665.0]	[440.3-624.6]†	[515.9-737.9]	[595.7-883.5]
	15.5	16.7	17.2	17.3
GPX-3 (ng/ml)	[12.4-19.8]	[12.8-17.3]	[13.7-19.7]	[13.9-20.1]
	110.5	133.0	118.0	126.0
Selenium (μg/L)	[87.5-131.5]	[104.5-178.0] *	[107.5-148.5]	[106.5-146.5]

Signicant intra-group results (p < 0.05)
† Signficant inter-group results (p < 0.05)

TABLE 3: Microcirculatory parameters on obese female adolescents at baseline and after 16 weeks of Brazil nuts (BNG) or placebo (PG) intake

	BNG		PG	
	T0	T1	T0	T1
Functional capillary density	10.2	8.8	9.9	10.2[
(n/mm²)	[7.8-11.2]	[8.2-11.8]	[8.0-17.4]	8.4-13.4]
	15.7	15.2	15.1	14.0
Afferent diameter (μm)	[13.9-18.6]	[11.8-19.5]	[12.9-18.6]	[11.6-16.1]
	19.7	22.1	21.9	20.0
Apical diameter (μm)	[18.1-22.3]	[18.2-24.1]	[18.5-27.5]	[18.4-25.6]
	20.4	21.6	18.4	21.5
Efferent diameter (μm)	[17.5-22.9]	[19.2-25.3]	[15.9-21.5]	[17.8-21.8]
	1.43	1.63	1.47	1.54
RBCV (mm/s)	[1.38-1.48]	[1.6-1.7] *†	[1.41-1.52]	[1.40-1.56]
	1.68	1.84	1.71	1.79
RBCVmax (mm/s)	[1.64-1.73]	[1.81-1.94] *	[1.56-1.79]	[1.76-1.84]
		4.0	7.5	5.5
TRBCVmax (s)	7.5[6.0-8.0]	[4.0-7.0]	[5.5-8.5]	[5.0-7.0]

Significant results intra-group (p < 0.05)
† Significant results inter-group (P < 0.05)

At T0, antioxidant capacity biomarkers were similar between groups. Gpx, LDL-ox and 8-epi-PGF$_{2α}$ did not change during the follow-up in either group (Table 2). It should be highlighted that at T1, serum levels of

LDL-ox were lower on BNG when compared to PG ($p = 0.02$), while on PG, serum selenium levels, a marker of compliance for the proposed intervention, was kept unchanged [118 (107.5-148.5 µg/l) vs. 126 (106.5-146.5 µg/l), $p = 0.74$] but on BNG we have observed a significant increase [110.5 (87.5-131.5 µg/l) vs. 133 (104.5-178 µg/l), $p = 0.02$] (Figure 1).

Morphological microvascular parameters (capillary diameters - Table 3) did not change during the follow-up in both groups. Obese female adolescent in PG had all functional microvascular parameters unaltered during the follow-up but those supplemented with Brazil nuts (BNG) increased RBCV and RBCVmax during PORH 13.9% and 9.52%, respectively (intra-group results) (Figure 2). Additionally, RBCV at T1 was higher on BNG when compared to PG at the same time point [1.63(1.6-1.7) vs. 1.54 (1.4-1.56) mm/s; $p = 0.02$].

17.4 DISCUSSION

Obesity is associated to metabolic disturbances, including insulin resistance, dyslipidemia and low-grade inflammation, all of them putative factors for endothelial and microvascular dysfunction (MD). Frequent nut intake is associated with many health benefits in adults [22-24], although no studies on adolescents could be found. To the best of our knowledge, this study is the first one to investigate effects of Brazil nuts intake on metabolic-lipid profiles, antioxidant status and microvascular parameters in obese female of this specific age period.

Nuts are rich in lipids, mainly in MUFAs and PUFAs and have high energy density. Comparing fat composition of diets rich on SFAs, high content of MUFAs and PUFAs in foodstuffs is potentially beneficial to health [25]. In spite of the caloric composition, some authors have demonstrated that consumption of nuts for short periods (less than 4 weeks) did not increase body mass[26]. In the present study, both groups kept body mass, waist circumference and BMI unchanged suggesting that although Brazil nuts have high energy density, possibly the satiety feeling due to nuts composition, medium-chain-triglycerides[27], fiber and protein [26] reduced energy intake from other sources. Positive influences on lipid profile by other types of nuts have been already demonstrated [28]. It has been

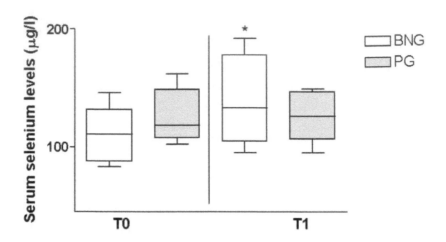

FIGURE 1: Differences in serum selenium levels observed on female obese adolescents during follow-up in both groups. *p = 0.02 (intra-group) BNG = Brazil Nuts Group; PG = Placebo Group.

reported that consumption of Brazil nuts during 15 days did not improve LDL- and total cholesterol and its benefits were noticed on transfer of cholesterol into HDL pool. In our study, supplementation of Brazil nuts during 16 weeks to obese female adolescents positively influenced the lipid profile such as, total cholesterol (TC), LDL-C and TG, but the present study did not assess cholesteryl esters.

MD has already been described in obesity [29], metabolic syndrome at normoglycemia [19] and type 2 diabetes mellitus using NVC. This technique is a non-invasive method to assess microvascular morphology and function and data obtained using this diagnostic tool has been already associated to cardiovascular risk [30]. In our investigation, adolescents were seen every 4 weeks what is roughly during the same phase of the menstrual cycle and no especial precaution were taken in this direction because it has already been shown that microvascular function is not dependent on menstrual cycle in ovulatory women [31]. Even tested for short-term period in obese female adolescents, we could observe an improvement of

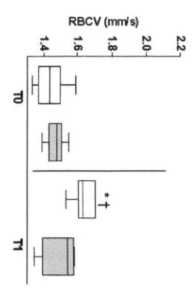

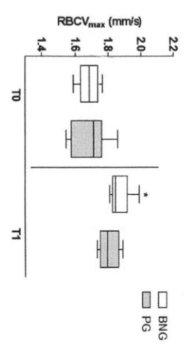

FIGURE 2: Resting Red Blood Cell Velocity (RBCV) and RBCVmax during post-occlusive reactive hyperemia response (PORH) on obese female adolescents at baseline and after 16 weeks of Brazil nuts (BNG) or placebo (PG) intake. *p = 0.03 (intra-group) and †p = 0.03 (inter-group).

microvascular reactivity by supplementing their diet with Brazil nuts and to our knowledge this is the first clinical study that showed its positive influence on nutritive skin microvascular function, although the effects of PUFAs and MUFAs contained in nuts have been well established on macro and microvascular functions [15].

Capillary morphology was not influenced by Brazil nuts consumption. The analysis of the hemodynamic behavior of the microcirculation was pursued by measuring RBCV before and after 1 min arterial occlusion as well as the time taken to reach RBCVmax. We have chosen 1 min arterial occlusion because our purpose was to measure the effect of an increased shear stress during the reactive hyperemia response and to decrease the discomfort for our patients (1 min ischemia is more tolerable than 4-6 min) [32]. Longer duration of the occlusion is commonly used to evaluate capillary recruitment [33]. In the BNG, a significant improvement on RBCV at baseline and during PORH could be detected in conjunction with a trend towards a positive influence on time for reperfusion (TRBCVmax), which needs further elucidation. PORH is thought to be determined at the level of small arterioles [34] and to be independent of the autonomic nervous system [35]. After the onset of reperfusion there is a sharp rise in blood flow followed by a gradual return to its baseline level, influenced by accumulation of vasodilator metabolites (including nitric oxide) and formation of ROS, normally washed out or destroyed by circulating blood and smooth muscle cell reactivity. During reperfusion, the myogenic response, due to rapid stretch of microvascular smooth muscle cells, is responsible, at least partially, for the return of blood flow to its baseline values. Our sample size did not allow us to correlate variables to find out associations between them, but we suppose that higher intake of MUFAs and PUFAs by obese adolescents on BNG resulting in an improved lipid profile and lower levels of LDL-ox, positively influenced microvascular reactivity. To strengthen this hypothesis we have previously noticed a multiple association between total and LDL-cholesterol levels and capillary microflow on obese metabolic syndrome subjects [19], suggesting that interventions with beneficial effects on the lipid profile could positively influence RBCV and RBCVmax. Consumption of nutrients rich in PUFAs (especially linolenic acids) such as fish oil and also Brazil nuts was associated to reduction in fibrinogen levels and in blood viscosity [36].

Nuts as sources of MUFAs have also been inserted in diets aiming to improve not only lipid profile but also insulin sensitivity, but in healthy, type 2 diabetes mellitus patients [37], and our own results in obese adolescents, nuts-enriched diets did not influence glucose homeostasis and CRP, an acute-phase protein, associated to atherosclerosis, metabolic disorders and CV disease [38].

Brazil nuts consumption during 15 days increased selenium levels, showing that shorter periods with higher amounts of nuts per day were followed by increments of 370% on selenium levels [24]. Physiological values for selenium ranged from 53 to 161.1 µg/l [39]. In the present study, selenium levels increased on BNG and were considered as a marker of good compliance although a significant difference could not be found inter-group (perhaps the intervention time was too short). Oxidative imbalance has been pointed as a causal factor for CVD. Nutrients and bioactive substances with antioxidant action are able to avoid arachidonic acid oxidation [33] and improvement of lipid peroxidation could reduce plasmatic and urinary levels of 8-epi-PGF$_{2\alpha}$, an oxidative product of arachidonic acid related to cell membranes. Urinary 8-epi-PGF$_{2\alpha}$ levels, also called isoprostanes, have shown large variation already at baseline inter- (two adolescents at PG had reduced urinary 8-epi-PGF$_{2\alpha}$ levels) and also intra-group in a reduced sample size and no significant changes could be detected on either group. However, we have noticed that its levels in BNG showed a decremental direction while in PG results went in the opposite way. It should be recalled that isoprostanes, derived from the arachidonic acid, are only one form of metabolite related to oxidative stress thus not able to reflect the whole oxidative status on cell membranes. Although our data do not explain improvements in the nutritive microvascular reactivity by an amelioration of urinary isoprostanes, we suggest that this pathophysiological focus deserves further elucidation with the use of other biomarkers of oxidative stress.

While investigated markers of oxidative stress were not altered by Brazil nuts consumption (Table 2), we could notice that LDL-ox levels were significantly reduced in BNG, probably due to reduction on oxidative stress status. The exact mechanism involved on it was not our aim, but the knowledge that LDL-ox is associated to endothelial dysfunction and atherosclerosis rise possible questions about the beneficial long-term effects of Brazil nuts consumption on obesity.

Limitations of the present study warrant mention. Although a power analysis was done to strengthen our results, the reduced number of patients investigated limits our conclusion to the population studied. Although the present study was randomized, it was not blinded and we have to consider that some changes in diet that we were not aware of, might have happened. Long-term beneficial cardiovascular outcomes could be inferred by these data.

Finally, treatment strategies for obese adolescent should focus on lifestyle interventions, aiming weight reduction, even if some supplements with established beneficial effects are added to diets.

17.5 CONCLUSION

Our results show that short-term intake of Brazil nuts added to diet of an obese female adolescent group did not change body mass or waist circumference, but, as a nutrient rich in bioactive substances, it positively influenced lipid profile and nutritive microvascular reactivity.

REFERENCES

1.   Daniels SR, Arnett DK, Eckel RH, Gidding SS, Hayman LL, Kumanyika S, et al.: Overweight in children and adolescents: pathophysiology, consequences, prevention, and treatment. Circulation 2005, 111:1999-2012.
2.   IBGE: Pesquisa de Orçamentos Familiares 2008-09 - Antropometria e Estado Nutricional de crianças e adolescentes do Brasil. 2010.
3.   World Health Organization: Obesity: Prevention and managing: The global epidemic. Report of a WHO Consultation on Obesity 2007. Report of a WHO Consultation on Obesity.
4.   Yusuf S, Hawken S, Ounpuu S, Dans T, Avezum A, Lanas F, McQueen M, Budaj A, Pais P, Varigos J, Lisheng L: Effect of potentially modifiable risk factors associated with myocardial infarction in 52 countries (the INTERHEART study): case-control study. Lancet 2004, 364:937-952.
5.   Pasqui AL, Puccetti L, Di RM, Bruni F, Camarri A, Palazzuoli A, Biagi F, Servi M, Bischeri D, Auteri A, Pastorelli M: Structural and functional abnormality of systemic microvessels in cardiac syndrome X. Nutr Metab Cardiovasc Dis 2005, 15:56-64.

6. Kraemer-Aguiar LG, Maranhao PA, Cyrino FZ, Bouskela E: Waist circumference leads to prolonged microvascular reactive hyperemia response in young overweight/obese women. Microvasc Res 2010, 80(3):427-432.

7. Kraemer-Aguiar LG, Maranhao PA, Sicuro FL, Bouskela E: Microvascular dysfunction: a direct link among BMI, waist circumference and glucose homeostasis in young overweight/obese normoglycemic women? Int J Obes (Lond) 2010, 34:111-117.

8. Stapleton PA, James ME, Goodwill AG, Frisbee JC: Obesity and vascular dysfunction. Pathophysiology 2008, 15:79-89.

9. Ceriello A: Hypothesis: the "metabolic memory", the new challenge of diabetes. Diabetes Res Clin Pract 2009, 86(Suppl 1):S2-S6.

10. Roberts CK, Sindhu KK: Oxidative stress and metabolic syndrome. Life Sci 2009, 84:705-712.

11. Dandona P, Mohanty P, Ghanim H, Aljada A, Browne R, Hamouda W, Prabhala A, Afzal A, Garg R: The suppressive effect of dietary restriction and weight loss in the obese on the generation of reactive oxygen species by leukocytes, lipid peroxidation, and protein carbonylation. J Clin Endocrinol Metab 2001, 86:355-362.

12. Berliner JA, Heinecke JW: The role of oxidized lipoproteins in atherogenesis. Free Radic Biol Med 1996, 20:707-727.

13. Kocygit A, Koylu AA, Keles H: Effects of pistachio nuts consumption on plasma lipid profile and oxidative status in healthy volunteers. Nutr Metab Cardiovasc Dis 2006, 16(3):202-209.

14. Ros E: Nuts and novel biomarkers of cardiovascular disease. Am J Clin Nutr 2009, 89:1649S-1656S.

15. Ros E, Nunez I, Perez-Heras A, Serra M, Gilabert R, Casals E, Deulofeu R: A walnut diet improves endothelial function in hypercholesterolemic subjects: a randomized crossover trial. Circulation 2004, 109:1609-1614.

16. CDC: Center Of Disease Control and Prevention National Center For Health Statistics. CDC Growth Charts 2000; 2007.

17. Tanner J: Growth at adolescence. Oxford: Blackwell; 1962.

18. Nepa Unicamp: Tabela Brasileira de Composição de Alimentos (TACO). 2006.

19. Kraemer-Aguiar LG, Laflor CM, Bouskela E: Skin microcirculatory dysfunction is already present in normoglycemic subjects with metabolic syndrome. Metabolism 2008, 57:1740-1746.

20. Friedwald W, Levy AL, Frederickson DS: Estimation of concentrations of low density cholesterol in plasma, without use of the preparative ultracentrifuge. Clinical Chemistry 1972, 18:499-502.

21. Klyscz T, Junger M, Jung F, Zeintl H: Cap image--a new kind of computer-assisted video image analysis system for dynamic capillary microscopy. Biomed Tech (Berl) 1997, 42:168-175.

22. Ip C, Lisk DJ: Bioactivity of selenium from Brazil nut for cancer prevention and selenoenzyme maintenance. Nutr Cancer 1994, 21:203-212.

23. Thomson CD, Chisholm A, McLachlan SK, Campbell JM: Brazil nuts: an effective way to improve selenium status. Am J Clin Nutr 2008, 87:379-384.

24. Strunz CC, Oliveira TV, Vinagre JC, Lima A, Cozzolino S, Maranhao RC: Brazil nut ingestion increased plasma selenium but had minimal effects on lipids, apolipoproteins, and high-density lipoprotein function in human subjects. Nutr Res 2008, 28:151-155.

25. Ros E, Mataix J: Fatty acid composition of nuts--implications for cardiovascular health. Br J Nutr 2006, 96(Suppl 2):S29-S35.

26. Brennan AM, Sweeney LL, Liu X, Mantzoros CS: Walnut consumption increases satiation but has no effect on insulin resistance or the metabolic profile over a 4-day period. Obesity (Silver Spring) 2010, 18:1176-1182.

27. St-Onge MP: Dietary fats, teas, dairy, and nuts: potential functional foods for weight control? Am J Clin Nutr 2005, 81:7-15.

28. Rajaram S, Haddad EH, Mejia A, Sabate J: Walnuts and fatty fish influence different serum lipid fractions in normal to mildly hyperlipidemic individuals: a randomized controlled study. Am J Clin Nutr 2009, 89:1657S-1663S.

29. de Jongh RT, Serne EH, Ijzerman RG, Jorstad HT, Stehouwer CD: Impaired local microvascular vasodilatory effects of insulin and reduced skin microvascular vasomotion in obese women. Microvasc Res 2008, 75:256-262.

30. Ijzerman RG, de Jongh RT, Beijk MA, van Weissenbruch MM, Delemarre-van de Waal HA, Serne EH, Stehouwer A: Individuals at increased coronary heart disease risk are characterized by an impaired microvascular function in skin. Eur J Clin Invest 2003, 33:536-542.

31. Ketel IJ, Stehouwer CD, Serne EH, Poel DM, Groot L, Kager C, Hompes PG, Homburg R, Twisk JW, Smulders YM, Lambalk CB: Microvascular function has no menstrual-cycle-dependent variation in healthy ovulatory women. Microcirculation 2009, 16:714-724.

32. Fagrell B, Intaglietta M: The dynamics of skin microcirculation as a tool for the study of systemic diseases. Bibl Anat 1977, 231-234.

33. de Jongh RT, Serne EH, Ijzerman RG, de VG, Stehouwer CD: Impaired microvascular function in obesity: implications for obesity-associated microangiopathy, hypertension, and insulin resistance. Circulation 2004, 109:2529-2535.

34. Meininger GA: Responses of sequentially branching macro- and microvessels during reactive hyperemia in skeletal muscle. Microvasc Res 1987, 34:29-45.

35. Walmsley D, Wiles PG: Reactive hyperaemia in skin of the human foot measured by laser Doppler flowmetry: effects of duration of ischaemia and local heating. Int J Microcirc Clin Exp 1990, 9:345-355.

36. Hostmarck A, Bjerkedal A, Kierulf P, Flaten H, Ulshagen K: Fish oil and plasma fibrinogen. British Medical Journal 1988, 297:180-181.

37. Lovejoy JC, Most MM, Lefevre M, Greenway FL, Rood JC: Effect of diets enriched in almonds on insulin action and serum lipids in adults with normal glucose tolerance or type 2 diabetes. Am J Clin Nutr 2002, 76:1000-1006.

38. Gabay C, Kushner I: Acute-phase proteins and other systemic responses to inflammation. N Engl J Med 1999, 340:448-454.

39. Versiek J, Cornelis R: Trace elements in plasma or serum. CRC press; 1989.

This chapter was originally published under the Creative Commons Attribution License. Maranhão, P. A., Kraemer-Aguiar, L. G., de Oliveira, C. L., Kuschnir, M. C. C., Vieira, Y., R., Souza, M. G. C., Koury, J. C., and Bouskela, E. Brazil Nuts Intake Improves Lipid Profile, Oxidative Stress and Microvascular Function in Obese Adolescents: A Randomized Controlled Trial. Nutrition & Metabolism 2011, 8:32. doi:10.1186/1743-7075-8-32.

AUTHOR NOTES

CHAPTER 1

Acknowledgments

This work was supported in part by the Irish Health Research Board (reference HRC/2007/13) and the European Commission, Framework Programme 6 (LIPGENE), contract number FOOD-CT-2003-505944.

Conflict of Interest
The author declares no conflict of interest.

CHAPTER 2

Competing interests
Corneel Vandelanotte is supported by a National Health and Medical Research Council of Australia (#519778) and National Heart Foundation of Australia (#PH 07B 3303) post-doctoral research fellowship.

All Authors declare that they have no competing interests that are directly relevant to the content of this review.

Authors' contributions
PJT contributed to the literature review, prepared initial draft of the manuscript, rated and tabulated primary articles and prepared final draft for publication. GSK contributed to the conceptual phase and editing of the manuscript and drafts. CV contributed in the conceptual phase of the paper as well as editing drafts of manuscript. CMC contributed to conceptual phase and editing of the manuscript and drafts. KWM contributed to the conceptual phase and editing of the manuscript and drafts. ESG contributed to the literature review and in the editing of the manuscript. MK contributed to the conceptual phase and editing of the manuscript and drafts.

MJN contributed to the project development, preparation and editing of the manuscript and ranking and tabulation of primary articles. All authors read and approved the final manuscript.

Acknowledgments

Queensland Health provided funding to conduct this project.

CHAPTER 3

Competing interests

The authors declare that they have no competing interests.

Authors' contribution

AAA and BJS each contributed equally to the formulation and writing of the manuscript. Both authors read and approved the final manuscript.

CHAPTER 5

Competing interests

The authors are employed full time by Archer Daniels Midland Company (ADM). ADM is a major oilseed and grain commodity processor and produces, among other products, fructose-containing sweeteners.

Authors' contributions

The two authors, SZS and MWE, have made similar contributions to the review. Both authors have read and approved the final manuscript.

Acknowledgments

The authors kindly thank Drs. Walter Glinsmann, Sheldon Hendler and Brent Flickinger for helpful discussions on the manuscript.

CHAPTER 6

Acknowledgments

Author gratefully acknowledges Dr.S.G.Bhat, Head, Department of Biochemistry and Nutrition and Dr.V.Prakash, Director, CFTRI, Mysore for their encouragement in preparing this review. The author also acknowledges

Dr. Santo V.Nicosia and Dr. D.Coppola, Moffitt Cancer Research Center, University of South Florida, Tampa, FL, USA for supporting the work on ascorbyl stearate in his laboratory.

CHAPTER 7

Competing interests
The authors declare that they have no competing interests.

Author contributions
The authors contributions are as follows: DLK served as the Principal Investigator and is responsible for oversight of all study related activities, data analysis and manuscript preparation. VYN was responsible for the protocol development, data analysis, interpretation, manuscript preparation, and critical review of the paper. ZF was responsible for study management, ultrasound reading, data collection, and manuscript preparation. SD contributed to manuscript preparation. AGS contributed to manuscript preparation. All Authors have read and approved the final manuscript.

Acknowledgments
We wish to thank the study participants for taking part in the study. Additionally the technical assistance of Dr. Yuka Yazaki and Mrs. Michelle Pinto-Evans is greatly appreciated.

Funding sources
Funding for this study was provided by the Egg Nutrition Center and the Centers for Disease Control & Prevention (Grant#U48-CCU115802).

CHAPTER 8

Competing interests
The authors declare that they have no competing interests.

Authors' contributions
LT participated in the conduct of the study, the analysis of the data. SC participated in the conduct of the study. EY participated in the conduct of the study. CC participated in the study design and statistical analysis.

GT participated in the study coordination. ZL conceived of the study, participated in its design and coordination, and drafting the manuscript. All authors read and approved the final manuscript.

Acknowledgments

Funding was provided by Herbalife, International, Los Angeles, California. LT was supported by NIH Training Grant No. DK0718033. SC, and EY were supported by NIH Training Grant No. T32 DK 07688.

CHAPTER 9

Acknowledgments

The authors gratefully acknowledge Nicolaas Busscher at University of Kassel and Friedrich-Karl Lücke at University of Applied Sciences Fulda for valuable comments on the earlier stage of the manuscript.

CHAPTER 11

Competing interests

The authors declare that they have no competing interests.

Authors' contributions

KE designed and conducted research, analyzed data and drafted the manuscript. SW helped to design research, conducted research, helped to draft the manuscript. IR conducted research, helped to draft the manuscript. UB helped to design research, helped to draft the manuscript. All authors read and approved the final manuscript.

Expert and Affiliation

Lindsay H. Allen - Western Human Nutrition Research Center, USA; Narendra K. Arora - International Clinical Epidemiology Network, IN-CLEN, India; Zulfiqar A. Bhutta - Aga Khan University, Pakistan; Rodolfo F. Florentino - Nutrition Foundation of the Philippines; Guillermo Meléndez - Mexican Health Foundation, Mexico; Noel W. Solomons - Program Director for Central America, International Nutrition Foundation, Guatemala; Edgar Vasquez-Garibay - University of Guadalajara, Mexico.

Acknowledgments

Thanks also to Richard Hurrell (Swiss Federal Institute of Technology, Switzerland) and Monika Potter (Dietitian) who provided valuable advice. Thanks to Paul Kelly for English language editing.

The study was supported by the Nestlé Nutrition Institute. The supporting source had no influence on study design; in the collection, analysis, and interpretation of the data; in the writing of the manuscript; and in the decision to submit the manuscript for publication.

CHAPTER 12

Competing interests

The authors declare that they have no competing interests.

Authors' contributions

DSS and IG contributed equally to the study design, collection of data, development of the animals and vegetables sampling, analyses, interpretation of results and preparation of the paper. Both authors read and approved the final manuscript.

Acknowledgments

This work was supported by CNCSIS - UEFISCSU, project number 1116/2009 PNII - IDEI, code 896/2008.

CHAPTER 13

Competing interests

The authors declare that they have no competing interests.

Authors' contributions

WC was responsible for the conception and the design of the model programme, interpretation of data, field works, drafting and approval of the manuscript. PS participated in design the programme on the aspect of iodine enrichment for animal feeds, field works, data analysis and drafting the manuscript. PT participated in the analysis of urine samples by ICPMS. JW participated in field works and specimens collection. All authors read and approved the final manuscript.

Acknowledgments

The authors would like to thank Dr. Saksom Attamangkune for the technical assistance on the aspect of iodine enriched eggs and the supports from the staff of Department of Medical Sciences, Kasetsart University and Napu Sub-district Municipality are also appreciated

CHAPTER 14

Competing interests

The authors declare that they have no competing interests.

Authors' contributions

MSS carried out the in vitro and in vivo experiments, participated in its design, analysed and interpreted the data and drafted the manuscript. KT conceived of the study, designed, coordinated, involved in drafting the manuscript and revised it critically. HPD participated in the in vivo studies and helped drafting the manuscript.All authors read and approved the final manuscript.

Acknowledgments

The VIT University has provided the lab facility and funded this study.

CHAPTER 15

Competing interests

The authors declare that they have no competing interests.

Author details

RP (M.Sc) is Junior Research Fellow (ICMR), TNK (M. Pharm) is Senior Research Fellow (CSIR), PKB is doing MS in Pharmacology (NIPER, Hyderabad), MK (PhD) is Senior Technical Assistant and SKB (PhD) is Principle Investigator and Scientist in the Division of Pharmacology and Chemical Biology, Indian Institute of Chemical Technology (IICT), Hyderabad-500607, India.

Authors' contributions

RP, TNK, PKB and MK carried out animal experimentation, biochemical estimation and statistical analysis of results. SKB conceived the study, and participated in its design, coordination and drafted the manuscript. The authors read and approved the manuscript.

Acknowledgments

Financial support was provided by Ramalingaswami Fellowship fund (SKB) from Department of Biotechnology (DBT), Junior Research Fellowship (RP) from Indian Council of Medical Research (ICMR), and Senior Research Fellowship (TNK) from Council of Scientific and Industrial Research (CSIR), Govt. of India and IICT institute fund. We wish to thank Dr J S Yadav, Director, IICT, Hyderabad for providing all kind of support for this work and gratefully acknowledge Dr Mohua Maulik for her suggestions and critical review of the manuscript.

CHAPTER 16

Competing interests
The authors declare that they have no competing interests.

Authors' contributions
YEG participated in the design, conducted research, analyzed data and statistical analyses, wrote the paper I.MEB participated in the design and wrote the paper. Both authors read and approved the final manuscript.

Acknowledgments and founding
We thank Anna Berggren Probi AB and Lena Nyberg, Skånemejerier for providing of Proviva products. The study was sponsored by Dr Persfoods foundation, Probi, Lund Sweden and Skånemejerier, Malmö, Sweden

CHAPTER 17

Competing interests
The authors declare that they have no competing interests.

Authors' contributions
PM - Performed microvascular analyses, statistical analyses and draft the manuscript; LGK - Participated to draft the manuscript and performed statistical analyses; CL - Participated in study design; MCK- Performed patients selection; YR - Carried out immunoassays; MG - Carried

out immunoassays; JK- Performed statistical analyses and participated in study design; EB - Participated in study design and to draft the manuscript. All authors read and approved the final manuscript.

Acknowledgments

Authors would like to thank Drs. Tatiane Bertoni de Toledo and Fabiana Barreto Lima for their help in recruiting patients and EMBRAPA for selenium measurements and Funding: National Research Council (CNPq), Foundation to Support Research in the State of Rio de Janeiro (FAPERJ) and Agency to Finance Studies and Projects (FINEP). During this study Ms. Priscila Maranhão received a fellowship from the Coordination to Improve Graduate Personnel (CAPES).

INDEX

α-glucosidase xxiii–xxiv, 289–290, 292, 295–300, 309–311

A

adiponectin (ADIPOQ) 11, 18–19, 31–33, 348
adoption study 26
allicin 315, 328, 330–331
American Heart Association (AHA) 24, 183
amino acid 68, 70, 72, 78–79, 84–87, 95–97, 104, 218, 220–222, 228–229
    taurine 93, 96–97, 108
    arginine 181, 185, 218, 220–221, 225, 227, 229
    glutamine 220, 225, 229, 232
Analysis of Variance (ANOVA) 178, 180, 190, 283, 285–286, 338–340, 342–344
ascorbic acid 107, 145–167, 251
Asia 235, 238
atherosclerosis 24, 26, 29, 31–32, 153–154, 158, 164–165, 170, 181, 184–186, 350, 356, 367
autism xiv, 91, 107

B

bilberry xxiii, 333–353
bioaccumulation xix, 259, 261, 263, 265–267, 269, 271
biomagnification 259

bipolar disorder xv–xvi, 91, 93, 96–98, 103, 106–108
blood acetate 133–134
blood pressure 43, 51, 171, 174–175, 177–178, 180, 199
blueberry xxiii, 334, 346, 348–350, 352
body mass index (BMI) xiv, xix, xxvi, 2, 18, 21, 26, 28–29, 32, 38, 42, 44–46, 51–52, 55, 60, 171, 177–178, 180, 191, 196, 199–200, 361, 363, 369, 381
bone density xvi–xvii, 187, 190, 195, 198, 200
brachial artery 173–175, 177–178, 180, 184–185
Brazil ix, xxv–xxvi, 355–359, 361–367, 369, 371
brazil nut 356, 358, 369
breastfeeding xviii, 233–234
breath hydrogen 116

C

caffeine 98, 107, 199, 218, 225, 227
calcium xix, 67, 95, 184, 194, 196, 199, 251, 273, 356
Calorie King™ 59
cancer xvii, 55, 63, 150, 154, 156–157, 159, 165–168, 369
capsaicin 219
carbohydrate xv, xxiii, 19–20, 24–25, 32, 45, 68–73, 75–76, 80, 82–88, 115, 127, 129, 133, 139–142, 174, 185, 188–189, 195, 197, 199–200,